Selective Organic Transformations

VOLUME 2

Selective Organic Transformations

EDITOR:

B. S. THYAGARAJAN

UNIVERSITY OF IDAHO
MOSCOW, IDAHO

VOLUME 2

WILEY–INTERSCIENCE

A DIVISION OF JOHN WILEY & SONS, INC.

NEW YORK · LONDON · SYDNEY · TORONTO

Library of Congress Catalog Card Number: 72–79147

ISBN 0–471–86688–1

Printed in the United States of America.

10 9 8 7 6 5 4 3 2 1

This volume is dedicated to the many able organic chemists who have contributed to the growth of mechanistic concepts relating to selectivity and specificity in organic reactions.

Editor

"I have found you an argument; but I am not obliged to find you an understanding."

Samuel Johnson

"Or if you choose, a machine might be imagined where the assumptions were put in at one end, while the theorems came out at the other, like the legendary Chicago machine where the pigs go in alive and come out transformed into hams and sausages."

Henri Poincaré

"If now we imagine an observer who discovers that the future course of a certain phenomenon can be predicted by Mathieu's equation but who is unable for some reason to perceive the system which generated the phenomenon, then evidently he would be unable to tell whether the system in question is an elliptic membrane or a variety artist."

Sir Edmund T. Whittaker

"Reasoning draws a conclusion and makes us grant the conclusion, but does not make the conclusion certain, nor does it remove doubt so that the mind may rest on the intuition of truth, unless the mind discovers it by the path of experience."

Roger Bacon

PREFACE

The domain of organic chemistry abounds in asymmetric functionality. Many different chemical operations contribute to the construction or destruction of such dissymmetric assemblages of functional units in a unique and specific manner.

Varied though the reactive sites may be, dependent on the orientational factors, the configurational stability, or the population of proximate reactive centers, the strategy and skill of the organic chemist has found a tool for each specific task in any molecular environment. The art of organic synthesis has unveiled many complex transformations that involve specificity and selectivity in the skillful assemblage of structural units of staggering proportions. A masterly exposition of the challenges of this field was given in a scintillating commentary entitled, "General Methods for the Construction of Complex Molecules" by Professor E. J. Corey in his plenary lecture before the Fourth International Symposium on the Chemistry of Natural Products at Stockholm in 1966.

Concomitant with the challenges of immediate synthetic goals, selective organic transformations pose the parallel objective of determining answers to intriguing questions like, "Specificity—why and how? Selectivity—why and how?" The latter quest is often the slower and more arduous effort. The mechanisms of many such selective organic transformations have been unraveled by numerous investigators. The result is a body of knowledge that has grown rapidly in recent years.

Perceptive analyses of the stereochemical and electronic driving forces responsible for these unique transformations hold continuing fascination for investigators in this area. An interplay of mechanistic ideas between varied molecular environments often leads to complimentary and overlapping growth in mechanistic postulates.

As in every other area of organic chemistry, mechanistic studies in the field of selective organic transformations have proliferated to such proportions that specialized surveys are required to assist the active investigator, the more advanced student, and the teacher.

The present series is therefore an attempt to offer periodic critical evaluations of selective organic transformations, largely from the mechanistic point of view. The original investigations surveyed in these volumes may well seed the growth of others that will challenge, sustain, or enlarge some of these postulates. It is hoped future volumes will bear ample testimony to this possibility.

I heartily acknowledge the generous cooperation of the contributing authors in making possible the appearance of this volume. Their ready willingness to provide these stimulating surveys is a service rendered by them to the many who are interested in investigations in the respective topics of the survey. I hope that this will indicate the need for a continuing series in *Selective Organic Transformations* which many will profitably turn to as a valuable secondary information source.

I shall welcome suggestions for appropriate surveys for inclusion in volumes to follow.

B. S. THYAGARAJAN

Moscow, Idaho
January 1970

CONTENTS

Selective Organic Transformations

VOLUME 2

Stereoselective Epoxide Cleavages

J. G. BUCHANAN*

University of Newcastle upon Tyne, England

HENRY Z. SABLE

Case Western Reserve University, Cleveland, Ohio

* Present address: Heriot-Watt University, Edinburgh, Scotland.

I. INTRODUCTION

A great deal is known about the ring-opening reactions of oxiranes, or vicinal epoxides (1–4). In this chapter we shall not attempt to cover all the known reactions (4), but will treat a smaller number of cases from the point of view of reaction mechanism and the various factors determining the nature of the products. The term "epoxide" will be used in the limited sense to describe compounds containing the oxirane ring. We shall use the terms "regiospecific" and "regioselective" as defined by Hassner (5) to describe the preference for the formation of one or other positional isomer by nucleo-philic ring-opening, while noting the existence of an alternative scheme of nomenclature (6). With regard to the stereochemical outcome of a reaction, "stereospecific" and "stereoselective" are used in the sense defined by Zimmerman et al. (7) and discussed by Eliel (8).

II. THE MECHANISM OF REACTION OF NUCLEOPHILIC REAGENTS WITH EPOXIDES UNDER BASIC, NEUTRAL, AND ACIDIC CONDITIONS

Ethylene oxide undergoes ring-opening with nucleophiles under a num-ber of different reaction conditions (1–4). Formally, reaction with HX is represented by $1 \rightarrow 2$. The ease with which reaction takes place is in contrast to the behavior of acyclic ethers which are very resistant to nucleophilic attack, especially under nonacidic conditions. The most probable explanation is that

$$CH_2 \overset{O}{\underset{}{-}} CH_2 + HX \longrightarrow HOCH_2 \overset{}{-} CH_2X$$
$$\quad \textbf{1} \qquad\qquad\qquad\qquad \textbf{2}$$

the strain associated with the three-membered ring is relieved during the reaction. Two types of reactions can be distinguished.

In the first, which occurs under basic or neutral conditions, the ethylene oxide molecule undergoes attack by an anion or neutral molecule in a bimolecular reaction (3). There is a strong effect due to solvation of the epoxide oxygen,

$$1 + NH_3 \longrightarrow HOCH_2CH_2NH_2$$
$$1 + Cl^{\ominus} + H_2O \longrightarrow HOCH_2CH_2Cl + OH^{\ominus}$$

particularly by protic solvents. Ethylene oxide (1) reacts with diethylamine to give 3 in the presence of methanol, but not in its absence (9). Parker and

$$1 + HN(C_2H_5)_2 \longrightarrow HOCH_2CH_2N(C_2H_5)_2$$
$$\mathbf{3}$$

Rockett (10) have shown that protic solvents strongly increase the rate of reaction of styrene oxide with benzylamine. In the reaction of ethylene oxide with alcohols and the corresponding alkoxides in an inert solvent the rate shows a complex dependence on the alcohol concentration (11), consistent with electrophilic assistance, as in 4.

4

It appears, therefore, that under neutral and basic conditions epoxide ring-opening is S_N2 in character and this is supported by the stereochemical consequences in suitable examples (Section IV). However, the electrophilic assistance by solvent and the effect of polar substituents on the ring-opening (Section III) imply that bond-breaking has progressed to a larger extent than usual in the transition state. Such a reaction is termed borderline S_N2 (3).

Second, epoxide ring-opening is markedly catalyzed by acids, and proceeds by way of the protonated oxide 5, as in the hydrolysis of ethylene oxide. Two mechanisms have been proposed for this reaction (12). Kinetic evidence on this and other acid-catalyzed reactions is inconclusive in deciding between the A1 and A2 mechanisms (3), but the stereochemical evidence (Section IV) is in favor of a modified A2 mechanism, in which the transition state has a greal deal of carbonium ion character. In certain cases a free carbonium ion may be involved.

In an unsymmetrically substituted epoxide, such as **6**, unequal amounts of the two possible products are formed and we consider, in Sections III–VI, the factors responsible for this regioselectivity.

$$
\underset{\textbf{6}}{R-\overset{O}{\overset{\diagup\diagdown}{CH-CH}}-R'} \xrightarrow{HX} \underset{\textbf{7}}{R-\underset{\underset{OH}{|}}{CH}-\underset{\underset{X}{|}}{CH}-R'} + \underset{\textbf{8}}{R-\underset{\underset{X}{|}}{CH}-\underset{\underset{OH}{|}}{CH}-R'}
$$

III. STERIC, POLAR, AND CONJUGATIVE EFFECTS ON REGIOSELECTIVITY

It is very difficult to separate the different factors leading to regio-selectivity in epoxide ring-opening. Substituent effects have been considered to consist of a combination of inductive, field, resonance, and polarizability effects (13). In discussions of the influence of substituents on epoxide scission, the inductive and resonance effects are generally given most weight (see Section VII.A) and in this section we also take into account steric effects. Most epoxide-opening reactions studied are of the modified S_N2 or even of the S_N1 type, with a range of transition states (3). In all cases bond-breaking is relatively more important in the transition state than in a normal bimolecular substitution reaction, due to the strain in the three-membered oxirane ring. The carbon atom at which the displacement is taking place will bear a δ^+ charge in the transition state.

Inductive or conjugative effects due to vicinal substituents may either stabilize or destabilize this partial charge, and as a result they may accelerate or retard the rate of reaction at the adjacent end of the oxirane ring. Under conditions of acid catalysis bond-breaking becomes even more important, with resulting changes in regioselectivity. In the extreme, when a phenyl or vinyl substituent is present, a resonance-stabilized carbonium ion may be formed under acid conditions, leading to S_N1 type ring-opening.

As the first example let us consider the factors determining the regio-selectivity of the reaction **9** → **10** + **11** in cases where R is a weakly electron-donating group such as alkyl.

$$
\underset{\textbf{9}}{R-\overset{O}{\overset{\diagup\diagdown}{CH-CH_2}}} \xrightarrow{HX} \underset{\textbf{10}}{R-\underset{\underset{OH}{|}}{CH}\cdot CH_2X} + \underset{\textbf{11}}{R\cdot\underset{\underset{X}{|}}{CH}\cdot CH_2OH}
$$

In reactions with basic reagents, where protonation of the ring oxygen does not precede attack by the nucleophile, the products are mainly of type **10**, shown by the behavior of propylene oxide (3,14). This is probably a primary steric effect, due to the greater steric accessibility of the terminal carbon atom, in a transition state represented by **12**. It is assumed that under

basic conditions electrophilic assistance by solvation of the epoxide oxygen, or by reaction with metal cations, does not affect the regioselectivity.

Under acidic conditions, when protonation precedes nucleophilic attack, there is a considerable proportion of the "abnormal" isomer **11** formed (3,14). The transition states are now derived from the conjugate acid of the epoxide and are represented by **13** and **14**, in which bond-breaking has proceeded to a greater extent than in **12**. An alkyl group is able to stabilize a developing carbonium ion at a vicinal carbon atom, as in **13**, but will have little effect on the nonvicinal carbon atom, as in **14**. This will tend to counteract the steric preference for nucleophilic attack at the primary carbon atom. Bond-making by X^- will be more important in **14** than in **13**, and this is shown by the behavior of propylene oxide towards aqueous hydrochloric and hydrobromic acids (15). Hydrobromic acid, whose anion is the more nucleophilic (16), gives a higher proportion of normal product **10**.

Styrene oxide (**15**) has been extensively studied. Under alkaline conditions there is preferential attack at the primary carbon, but the regioselectivity is not so great as in propylene oxide (3). It appears that the steric effect predominates despite the possibility of conjugative stabilization of the δ^+ charge in the transition state.

Similarly, butadiene monoxide (**16**) yields mainly the ether **17** with sodium methoxide (4). Under acidic conditions both **15** and **16** undergo abnormal opening almost exclusively (3); epoxide **16** gives the ether **19** with dilute methanolic sulfuric acid (17). In these cases the transition states must have considerable carbonium ion character, stabilized by conjugation as in **18** and **20**.

Whereas sodium azide in aqueous dioxane reacts with propylene oxide at the primary carbon, the abnormal product results from **15** and **16** (18). In the latter two cases it has been suggested that the polarizability of azide ion can lead to diffusion of charge in the transition state of abnormal opening (3,18,19). A similar effect may be important also in the reaction of sodiomalonic ester with **15**. Contrary to earlier reports (20), comparable amounts of the normal and abnormal products are formed (21). In the case of 1,2-epoxytetrahydronaphthalene (**21**) there is no steric difference between the two epoxide carbon atoms, and attack at the benzylic carbon predominates (22a); in the same way 3,4-epoxycycloalkenes are attacked by lithium aluminum hydride at the allylic carbon (22b). There are, however, some discordant results in this series (3) and further work, using modern analytical methods, is clearly desirable.

21

Conjugative stabilization by oxygen is probably responsible for the regioselectivity observed in the reaction of a range of epoxyethers with nucleophiles (3). The epoxide **23**, for example, reacts with lithium aluminum hydride to give **22** (23). It is surprising that reaction of **23** with *n*-butylamine shows the opposite regioselectivity (24), yielding the amino ketone **24**, and it would be interesting to know if the use of a solvent, particularly a protic one, had any effect on the regioselectivity (9,10).

So far, we have dealt in detail with the effects of electron-donating groups on regioselectivity, and now turn to electron-withdrawing groups. In the transition states **25** and **26** proposed by VanderWerf and his students (19,25), Y is a nucleophile, E an electrophile, and bond-breaking is predominant. The regioselectivity represented by **25** is preferred if R is electron-donating or if R′ is weakly electron-withdrawing. In these circumstances, the transition

$$\text{R}-\overset{\delta+}{\underset{}{\text{CH}}}-\text{CH}\rightarrow\text{R}'$$

25

$$\text{R}-\text{CH}-\overset{\delta+}{\text{CH}}\rightarrow\text{R}'$$

26

state corresponding to the opposite regioselectivity, **26**, is destabilized by electron-withdrawal by R'. The VanderWerf model is most satisfactory in cases where R and R' are groups incapable of conjugative interaction with the epoxide ring carbons, in spite of the fact that many of VanderWerf's own examples are drawn from styrene and stilbene oxides.

There are many examples of regioselectivity due to electron-withdrawing groups (3). The 3-alkoxycyclohexene oxides have been very carefully studied (Section VII.A), and several other examples are given in Section VII. Kinetic investigations (3) show that an electron-withdrawing group normally accelerates the overall rate of ring-opening. Parker and Isaacs (3) consider the epoxide **27** in which R is electron-withdrawing. Considering attack at the normal position (β) first, we see that the group R will have little effect on bond-breaking since both the CH_2 group and the oxygen atom will be in-

$$\overset{\delta-}{\text{R}}\leftarrow\overset{\delta+}{\underset{\alpha}{\text{CH}}}\diagdown\overset{\overset{\delta\delta+}{\text{O}}}{\underset{\underset{\delta\delta+}{\text{CH}_2\,\beta}}{|}}$$

27

fluenced inductively. However, bond-making will be assisted by this effect, even though bond-breaking is still important in the transition state. In attack at the abnormal position (α) the electron-withdrawal by R greatly inhibits bond-breaking. In agreement with this Addy and Parker (26a) have shown that in water at pH 7, **28** reacts with chloride ion to give **29** (84%) and **30** (16%), whereas **31** gives **32** (100%) at about five times the rate for **28**. Similar behavior has been observed for azide ion and the corresponding propane derivatives (26b).

$$\underset{\textbf{28}}{CH_3\cdot CH_2\cdot \overset{O}{\overset{\diagup\diagdown}{CH\!-\!CH_2}}} \longrightarrow$$
$$\underset{\textbf{29}}{CH_3\cdot CH_2\cdot CH(OH)\cdot CH_2Cl} + \underset{\textbf{30}}{CH_3\cdot CH_2\cdot CHCl\cdot CH_2OH}$$

$$\underset{\textbf{31}}{CH_3\cdot CHCl\cdot \overset{O}{\overset{\diagup\diagdown}{CH\!-\!CH_2}}} \longrightarrow \underset{\textbf{32}}{CH_3\cdot CHCl\cdot CH(OH)\cdot CH_2Cl}$$

A very striking case is that of *p*-nitrostyrene oxide (**33**) in which, contrary to these ideas, the product of reaction with lithium borohydride in ether is mainly (62%) the alcohol **34** (25). It is postulated that in this reaction bond-breaking is so inhibited by electron-withdrawal by the nitro group that bond-making becomes predominant in the transition state, and attack occurs mainly at the most electron-deficient carbon atom.

Many examples have been cited (3) which show that an adjacent carbonyl, carboxamide, or ester causes the epoxide to be attacked at the nonadjacent end. This is ascribed to the δ^+ charge on the carbonyl carbon atom in these species, which opposes the formation of the transition state δ^+ on the adjacent oxirane carbon (see Section VII.G.2 for other examples). On the other hand, an adjacent carboxylate (anionic) group has the opposite effect. Because of the diffusion of negative charge over the whole carboxylate group, the vicinal δ^+ will be stabilized (13,27) and opening will be favored in this position. A clear example of this effect is seen in the conversion of 2,3-epoxy-carboxylic acids **35** into α-amino-β-hydroxyacids **36** (28). With ammonia or benzylamine as the nucleophile, attack occurred only at the α-position.

R" = H or PhCh$_2$

IV. THE STEREOCHEMISTRY OF THE RING-OPENING REACTION

In most of the examples of ring-opening of an epoxide in which the stereochemistry of the product is known, the entering nucleophile has been found to cause inversion of configuration (1,3,4). This result would be expected under alkaline or neutral conditions, where bond-making is relatively important. The fact that inversion normally occurs also under acidic conditions is a strong argument against the formation of a free carbonium ion during opening of those oxides. Although bond-breaking has occurred to a large extent in the transition state of ring-opening under acidic conditions, the entering nucleophile must approach within bonding distance in the transition state. A particularly significant example is that of the *trans*-decalin derivative **37** which forms the *trans*-diol **38** as the sole product of reaction with aqueous sulfuric acid in acetone (29). In this case two factors, the presence of the methyl groups and the weak nucleophilicity of dilute sulfuric acid, might have been expected to promote S_N1 opening, but no *cis*-diol **39** was observed, nor was any product of pinacol rearrangement [cf. (30a)].

However, if a developing carbonium ion can be stabilized by conjugation, e.g., as a benzylic cation, a number of different stereochemical consequences can be observed (30b). Berti and his colleagues have examined the reaction of (+)-(R)-styrene oxide (**41**) with hydrogen chloride in different solvents (31). Regiospecific ring opening was observed in all cases to give a mixture of the 2-chloro-2-phenylethanols **43** and **45**, in agreement with an S_N1 or borderline S_N2 mechanism (3). The stereochemistry of the product depended markedly on the solvent. In chloroform and benzene the product was mainly **43** (i.e.,

inverted) arising by a borderline S_N2 intermediate **40**. In dry ethers, retention was mainly observed, yielding **45**, but addition of water caused formation of **43**; a solvated ion-pair intermediate **42**, which is water-sensitive, was proposed. Racemization may also proceed by way of any symmetrical carbonium ion stage, e.g., **44**.

Stereoselective formation of 1-phenylethyl halides (e.g., **48**) with preferred retention of configuration, which has been observed for the reaction of

1-phenylethanol (32) and the phenyl ether 46 (33) with hydrogen halides, by way of an ion-pair intermediate 47, is analogous to the proposed sequence

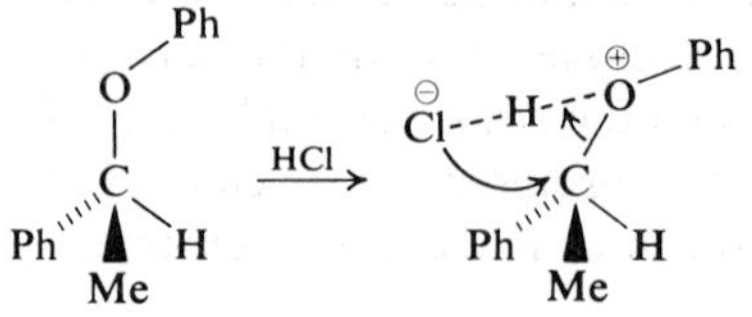

41 → 42 → 45. Ion-pair intermediates involving a benzylic cation have also been proposed to account for regiospecific and predominantly *cis* addition of deuterium bromide to *cis*- and *trans*-1-phenylpropene (34). In more complex epoxides, such as the stilbene oxides (49) (31,35,36), dypnone oxides (50) (37), and *N,N*-dialkylphenylglycidamides (e.g., 51), (38,39), neighboring

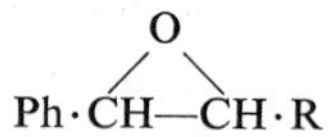

$$\begin{array}{c} O \\ \diagup \quad \diagdown \\ Ph \cdot CH \!-\! CH \cdot R \end{array}$$

49; R = Ar
50; R = CO·Ph
51; R = CONEt$_2$

group participation, with formation of phenonium (3,31) or oxonium intermediates (3,37–39), has been postulated to account for the stereochemical results. This seems unnecessary at the present stage (3,31,35,37). The role of solvent in these acid-catalyzed ring-openings is by no means clear. Some further examples of these complexities, taking conformational factors into account, are given in Section VII.H.

Under neutral or alkaline conditions styrene oxides undergo attack mainly at the primary carbon (3). It would be expected that when attack did occur at the secondary carbon atom a borderline S_N2 mechanism, requiring inversion of configuration, would be involved.

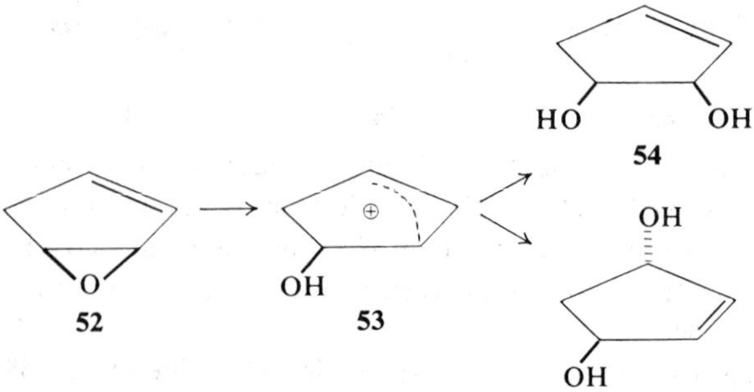

Stabilization of a carbonium ion transition state as an allylic cation also occurs. The highly reactive 3,4-epoxycyclopentene (**52**) yields a mixture of *cis*-1,2- and *trans*-1,3-diols, **54** and **55**, with water, presumably via the cation **53** (40,41). It is interesting that the same two products are obtained from the treatment of cyclopentadiene with lead tetraacetate (41,42), a process not thought to involve an epoxidic intermediate.

Similarly, the lumisterol epoxide **56**, on treatment with benzoic acid in benzene, yields the monobenzoate **57** of the *cis*-diol (43).

56   **57**

V. CONFORMATIONAL ASPECTS OF THE RING-OPENING REACTION. FUSED AND FLEXIBLE RING SYSTEMS

In by far the majority of cases of epoxide ring-opening an inversion by the entering nucleophile occurs at the point of attack. Thus, when cyclohexene oxide (**58**) reacts with HX, the product **59** has a *trans*-relationship of X and OH. Since the product is able to adopt its most stable conformation, X and OH will finally have a *trans*-diequatorial relationship, irrespective of the initial conformation of the adduct.

58   **59**

In 1951, Fürst and Plattner (44) noted that the ring-opening was regioselective in the steroid epoxides which had been studied at that time, the major product having a diaxial arrangement of groups (45). The most clear-cut example of this is the behavior of the 5α-cholestane 2,3-epoxides **60** and **62** towards acids, HX, leading invariably to the 2β,3α-isomer **61** and **63**, respectively, (45,46). Similarly, lithium aluminum hydride in ether yields the 3-α-ol **64** and 2-β-ol **65**, respectively, (47). Because of the rigidity of the steroid ring system the Fürst-Plattner rule, as it has come to be called, has mechanistic implications which were discussed by Fürst and Plattner (44) and have been very clearly stated by Angyal (48).

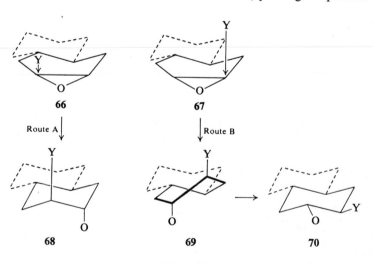

Three aspects must be considered: (*1*) the predominance of the diaxial isomer in the reactions of fused cyclohexene oxide systems; (*2*) the formation of some of the other, diequatorial, isomer in such systems; and (*3*) the behavior of flexible systems.

Figure 1 shows a cyclohexene oxide, rigidly held in its stable half-chair form by fusion to another ring, undergoing reaction with nucleophile Y. For the present purpose, the charges have been ignored. When Y attacks the molecule, the most favorable transition state will be *anti*-parallel, involving the approach of Y from an axial direction (49), as in **66** or **67**, a result in keeping with a simplified quantum mechanical picture of the reaction (50). Route A, in which the transition state is chairlike, yielding the product **68**,

Figure 1

will be favored over Route B, in which the transition state is boat- (or skew-) like. The boat **69**, the initial product in Route B, will readily change to the chair **70**. Route A represents "diaxial" opening of the epoxide ring and is energetically preferred over Route B, the so-called "equatorial" opening (48,51). According to Angyal's view, Routes A and B both involve diaxial opening, in contrast to views expressed by Cookson (51) and by Parker and Isaacs (3). When steric or polar factors are also present in the system the energy difference between the two pathways may be reduced or even reversed, and "exceptions" to the Fürst-Plattner rule may be found. Mechanistically, the Fürst-Plattner rule is always obeyed, in the sense that nucleophilic attack always comes from an axial direction, even in examples in which the regioselectivity of the reaction is due entirely to steric or electronic effects. The relative importance of the various factors is discussed in Section VII.

It may be noted, in passing, that epoxide formation from halohydrins by treatment with base takes place from both diaxial and diequatorial isomers. The slower rate of reaction of the latter, in fused ring systems, is probably due to the higher energy of the boatlike transition state necessary to achieve coplanarity of reacting centers (52).

In the case of flexible cyclohexene oxides, the situation is more complex, as shown in Figure 2. In Route A (cf. Figure 1) the initial diaxial product **72**

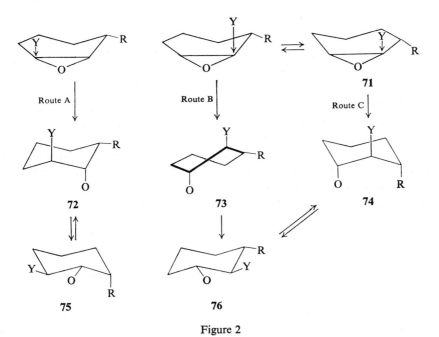

Figure 2

can undergo conformational change to the diequatorial form **75**. Furthermore, nucleophile Y can attack the other carbon atom not only by Route B, as before, but also by Route C, in which the transition state is related to the other half-chair form **71**. The initial diaxial product **74** undergoes ring inversion to the diequatorial form **76**.

The strong conformational influence in epoxide ring-opening in flexible systems is shown by the reactions of the 4-*t*-butylcyclohexene oxides (53–55), in particular with lithium aluminum hydride (55). Attack is at C_1 in the case of *trans*-4-*t*-butylcyclohexene oxide (**77**) with the formation of the axial alcohol **78**, while in the case of the *cis*-compound **79** the axial alcohol **80** results.* In both of these cases reaction through a transition state derived from the alternative half-chair form (Route C) is prevented because of the instability of an axial *t*-butyl group in the transition state. These reactions resemble the behavior of the fused cyclohexene oxide ring in the 5α-cholestene-2,3-epoxides, in that only Route A is operative. When the 4-butyl group is replaced by phenyl (56), or even methyl (57), a strong conformational preference remains.

Conformational control, leading to regioselectivity, in the reaction of a flexible cyclohexene oxide, is exemplified by the inositol epoxides (58) in which steric and polar effects are essentially the same at the two epoxide carbon atoms. Thus, the hydrolysis of 1,2-anhydro-*myo*-inositol (**81**) by aqueous sulfuric acid yields mainly (±)-inositol (**82**), the product expected on conformational grounds (48,59), and the same regioselectivity is observed in the reaction of **81** with hydrobromic acid (60) and ammonia (61). Some

* The existence of an important side reaction in these reductions is discussed in Sections VII.E.1 and VII.E.2.

examples of anomalous behavior in the inositol series are given in Section VI.

The question arises, in flexible systems, whether Route B or Route C operates when ring-opening by Route A is retarded by steric or electronic effects (55,62–66). Since the energy barrier to conformational change is relatively low, the Curtin-Hammett principle (67) predicts that the pathway of reaction is determined only by the relative energy levels of the transition states leading to **72**, **73**, and **74** (Figure 2). We are here concerned with the relative energies of the transition states leading to **73** (Route B) and **74** (Route C). In a flexible cyclohexene oxide in which carbon atoms 3 to 6 bear no bulky substituents that have a strong preference for an equatorial orientation, the transition state leading to **74** should be more favorable than that leading to **73**, i.e., Route C is preferred. Such a mechanism has been proposed for the reaction of *trans*-3-methoxycyclohexene oxide (**83**) with hydrogen chloride (63), which undergoes attack at C_1 due to polar effects (see Section VII.A).

In the reduction of 1-methyl-*trans*-4-*t*-butylcyclohexene oxide (**85**) with lithium aluminum hydride (64), attack at C_1 is sterically inhibited by the methyl group, in comparison with epoxide **77**. Nevertheless, Route A is still preferred, and the secondary alcohol **84** is the major product (60%). The tertiary alcohol **87**, formed in 32% yield, probably arises by way of a transition state related to **86**, i.e., by Route B. Rickborn (64b) has recently discussed this problem in detail. In the case of limonene epoxide **88**, reaction with lithium aluminum hydride should be possible through a transition state related to **89**, corresponding to Route C of Figure 2, and it is perhaps surprising that only 12% of the alcohol **90** is formed (66).

It has been emphasized that it is the relative transition state energies which determine the regioselectivity of epoxide ring-opening and not the relative proportions of conformers in the ground state. Nevertheless, many of the nonbonded interactions that determine the proportions of ground-state conformers are also present in the corresponding transition states (68). It might be expected, therefore, that in systems lacking steric or polar effects

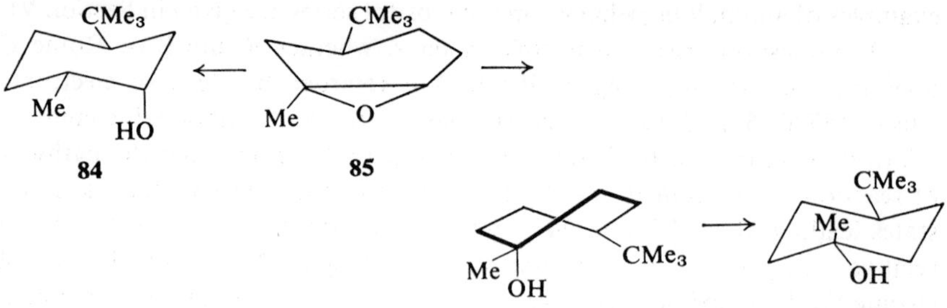

84   85

86   87

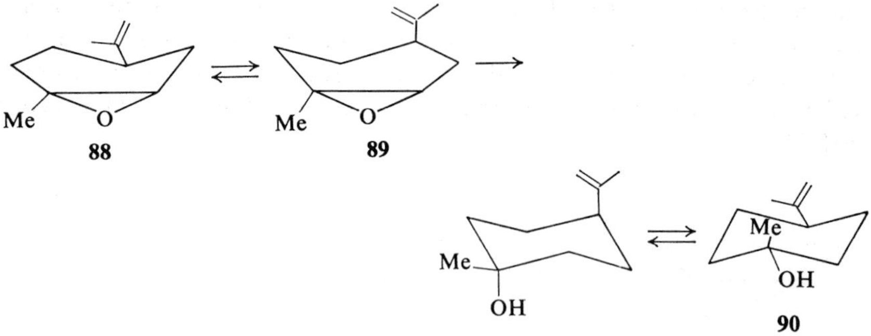

88   89

90

the regioselectivity can be predicted from a knowledge of the proportion of conformers in the ground state.

The foregoing discussion has centered on the conformational behavior of cyclohexene oxides. Although the situation with regard to other cyclo-alkene oxides is less well defined, the condition of *anti*-parallel attack is probably always required when epoxide ring-opening takes place with inversion of configuration. Transition states that may be involved in the case of the 3-methoxycyclopentene oxides have been discussed by Moir, Bannard, and their co-workers (69) (see also Section VII.B).

Aryl-substituted epoxides do not necessarily undergo inversion on ring-opening, especially under acidic conditions (Section IV). It is interesting, therefore, to study conformational effects on the stereochemistry of the ring-opening. Whereas styrene oxide (**41**) reacts with hydrogen chloride in chloroform mainly with inversion of configuration (31), the *cis*-chlorohydrin **93** is the main product from **91** (70). The carbonium ion **92** may be long-lived, due

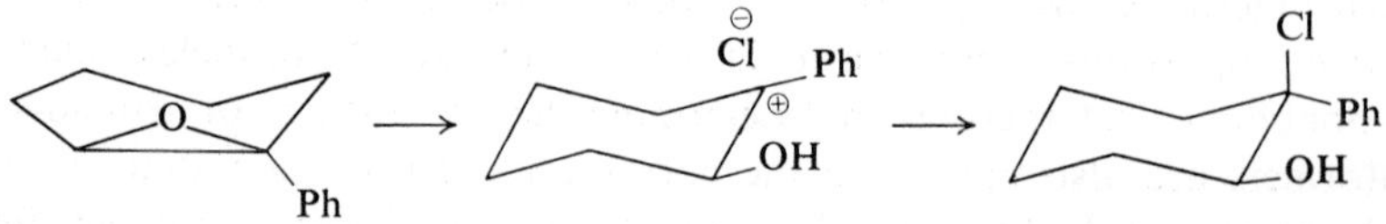

91   92   93

to stabilization by resonance involving the π-electrons of the benzene ring; as a result the hydroxyl group can become equatorial before the next step in the reaction occurs, and axial attack of the chloride ion then yields **93**. Some further examples are discussed in Section VII.H.

VI. NEIGHBORING GROUP PARTICIPATION IN THE RING-OPENING REACTION

Epoxide scission is a particular type of displacement reaction, and as with such reactions in general, the possibility of intramolecular processes, such as group migration, molecular rearrangements, and neighboring group participation, must be considered.

Participation of neighboring acyloxy groups in displacement of halogen was first observed in connection with the acetohalogen sugars (71) and a mechanism for the effect was proposed by Isbell (72). The studies of Winstein and his school (73) provided firm evidence for the intervention of cyclic acyloxonium ions in such reactions (**94** → **95**). The subsequent fate of the intermediate oxonium species depends on the reaction conditions. For ex-

94   **95**

ample, the "wet" and "dry" Prévost reactions (74) (see Section IX) give derivatives of *cis* and *trans* glycols, respectively. Because of diffusion of the positive charge in the acyloxonium ion **95**, subsequent attack can occur on (*1*) C-1, (*2*) C-2, or (*3*) the carbonyl carbon, giving, respectively, (*1*) over-all inversion, (*2*) over-all retention of configuration, and (*3*) inversion only at the atom to which substituent X was attached; as long as any trace of water is present, Reaction (*3*) occurs quantitatively. Related to this phenomenon are the configurational inversions of carbohydrate esters in liquid hydrogen fluoride (75) and of cyclitols in boiling acetic acid containing sulfuric acid (76). The participating neighboring group may also be an acetamido group, a carbamate (77), or a thioureido (78) group, and the cyclic intermediate, instead of being transitory, may actually be found in the products. In these cases aziridines, oxazolines, or thiazolines may be formed. Treatment of *trans*-1,2-acylamido alcohols with $SOCl_2$ also yields oxazolines (79) by a similar mechanism.

Winstein extended the neighboring group concept to cover addition reactions (80), most readily exemplified in the formation of the oxazoline **97**, from 3-benzamidocyclohexene (**96**) and *N*-bromosuccinimide in acetic acid [(80a,80d); see Section IX]. Neighboring group participation in the scission of epoxides bearing a vicinal *trans* acetoxy group was suggested by Buchanan

(81) to account for the formation of the guloside **100** from the treatment of **98** with hydrogen chloride in acetone. The reaction is rapid, stereoselective, and

no chlorohydrin is produced, in contrast with the behavior of the deacetylated epoxide (81a). The acetoxonium ion intermediate **99** was therefore proposed, and several other examples have since been described (82,83).

In the conformationally mobile cyclopentanoid system, the regioselectivity of epoxide hydrolysis has been shown (84,85) to be governed by steric hindrance and polar effects (see Section VII.B). However, the presence of a vicinal *trans* acetoxyl or acetamido group causes a complete reversal of selectivity (86). Hydrolysis of the epoxydiol **101a** gave tetrols **102** and **103** in which **103** predominated by approximately 20:1, whereas the diacetate **101b** gave 98% of **102**. Steric hindrance by the vicinal *trans* substituent, and the electronegativity of the same group are apparently overcome in this reversal of selectivity. In the case of **105**, the electronegativity of the substituents adjacent to both ends of the oxirane ring is the same, and steric hindrance

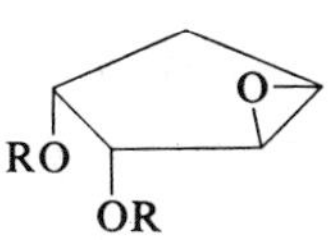

101a,b 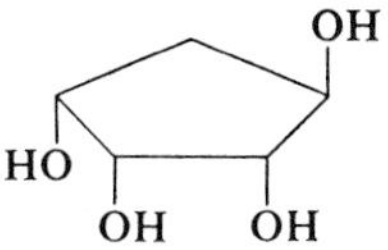   102 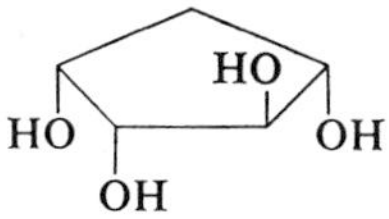   103 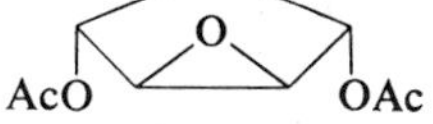   104

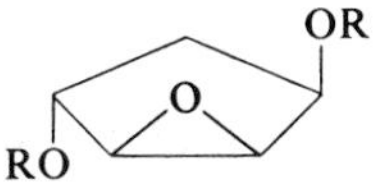

105a,b 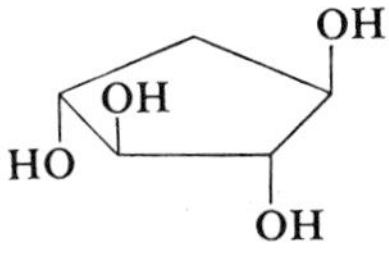   106 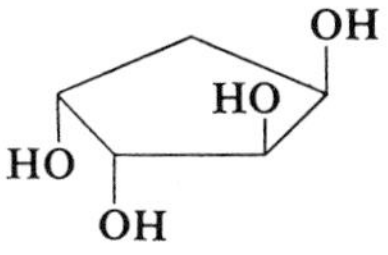   107 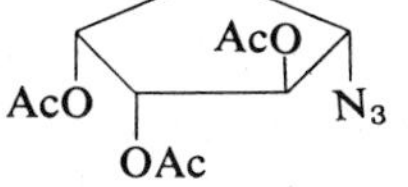   108

a; R = H
b; R = Ac

alone is the basis of the preferential (95%) formation of **106** from the hydrolysis of **105a**. Hydrolysis of the acetylated compound **105b**, however, gives
predominantly **107** and less than 2% of **106**. The reversal of selectivity by the
participating group may be due partly to the low nucleophilicity of water.
Treatment of the diacetate **101b** with NaN_3 gave 73% yield of the azidotriol
triacetate **108**. However, when the diacetate **104** was treated similarly, only the
tetrol **103** was produced, i.e., the epoxide was hydrolyzed. It is unlikely that
the powerful nucleophile N_3^- was excluded while the weak nucleophile H_2O
was able to attack. In this case the steric hindrance is so severe that only the
intramolecular reaction could occur, leading to a net hydrolysis.

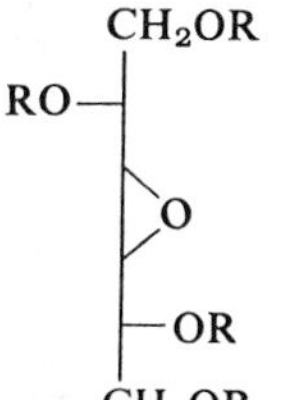

109a,b 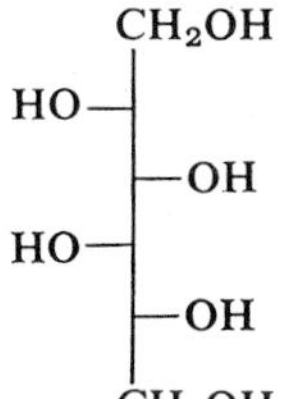   110 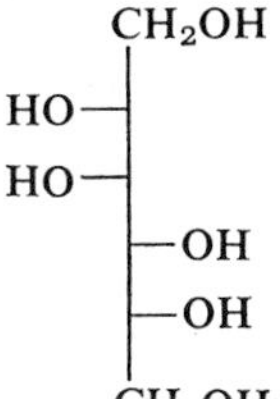   111 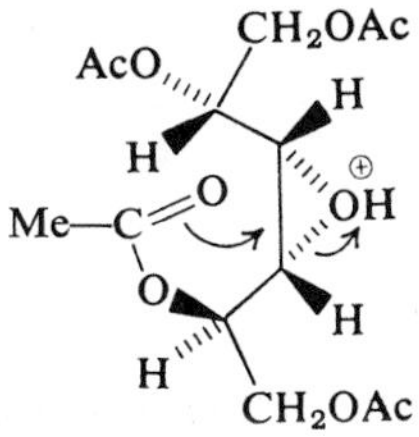   112

a; R = H
b; R = Ac

The anhydroaltritol **109b** undergoes hydrolysis with aqueous acetic acid
under conditions which hardly affect **109a**. The products, after deacetylation,
are **110** (80%) and **111** (15%) (87). As shown in **112**, the C-5 acetoxy group,
being in an *erythro* relationship to the epoxide ring, participates more readily
than the C-2 acetoxy group.

Participation involving a six-membered ring has been shown for the
anhydroaltroside **113**. The product of hydrolysis with acetic acid, and removal
of protecting groups, is **115**, by way of the ion **114** (82b).

113   114   115

Proof of the formation of an acetoxonium ion has been obtained in two cases by conversion into an ethylidene compound. 3α-Acetoxy-4β,5-epoxy-5β-cholestane (116) was treated with boron trifluoride etherate in benzene, followed by lithium borohydride in tetrahydrofuran (88). The product 118 arose by reduction of the acetoxonium ion 117. A similar sequence converted the anhydrolyxoside 119 into 120 (89).

116   117   118

119   120

Other examples of this reaction occur among steroid epoxides (88,90), showing the participation of other ester functions, e.g., carbonate (90b). Although the participating group must clearly become axial in the transition state, the pseudoequatorial acetoxy group in 3β-acetoxy-4α,5-epoxy-5α-

121   122   123

cholestane (121) participates without difficulty (88,90a). When the acetoxonium ion salt (122) is hydrolyzed so that the resulting monoacetate does not undergo acyl migration the product is 123, the expected axial acetate (91).

A particularly clear-cut example of participation is the conversion of the anhydrogalactoside 124 into the oxazolidone 125 in 90% yield by aqueous acetic acid (92).

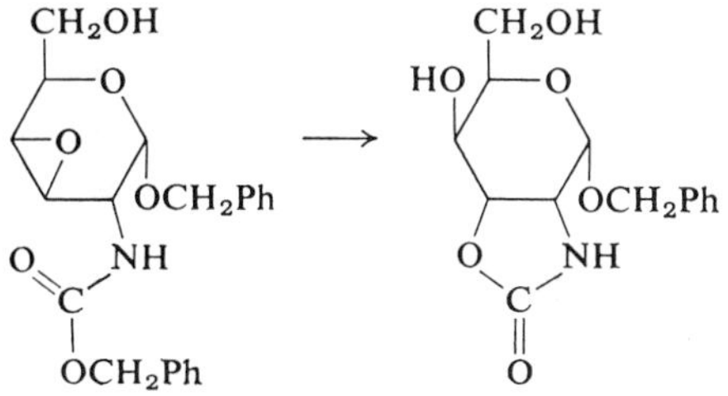

124   125

In the inositol epoxides, conformational effects are predominant in determining the regioselectivity of ring scission (Section V), but the anomalous behavior of some epoxides with sodium borohydride in methanol has been reported (93). 1,2-Anhydro-*myo*-inositol (126), for example, readily yields the all-equatorial methyl ether 129 of *scyllo*-inositol, in complete contrast to the regioselectivity with other reagents (Section V). The effective reagent is probably NaB(OMe)$_4$ (93,94), and Angyal (95) has proposed the elegant scheme 127 → 129 to account for the results. It should be noted that both nucleophilic and electrophilic assistance is provided in 127 → 128. Similar

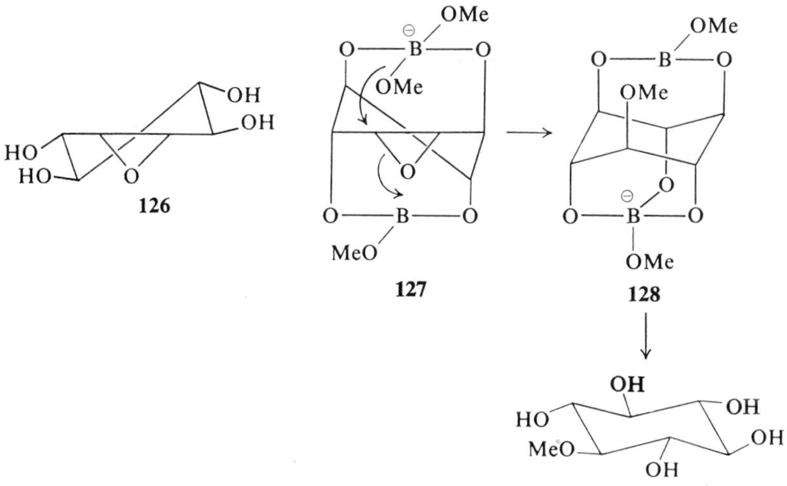

126   127   128

129

schemes have been proposed (95) to account for the regioselectivity and relative rates of the other examples (93).

Angyal and Stewart (96) found that whereas hydrolysis of **130** in aqueous acid yielded comparable amounts of the two isomers **131** and **132**, treatment with sodium benzoate in dimethylformamide gave **131** and **132** in the ratio 22:78. Only the methyl group renders **130** unsymmetrical and it appears to be too far removed to exert a steric influence. The solvent solvates anions poorly (97a)

| 130 | 131 | 132 | 133 |

and it has been proposed (96) that in the formation of **132** the incipient anion is solvated by hydrogen bonding with the *syn*-axial hydroxyl group, as in **133**; no such solvation is possible in the case of **131**. This is an example of intra-molecular electrophilic assistance.

Many of the points mentioned briefly in this section are treated *in extenso* in a recent review on Neighboring Group Participation in Carbohydrates (97b).

VII. EXAMPLES OF THE RELATIVE IMPORTANCE OF VARIOUS FACTORS IN THE RING-OPENING REACTION

A. The Microcosm of 3-Methoxycyclohexene Oxide and Some Closely Related Species

Many of the influences bearing on the regiospecificity of epoxide scission and on the closely related addition of hypohalous acid to alkenes are exemplified in the extensive studies of Bannard, Moir, and their associates (63,69, 98–102) on 3-alkoxycyclohexene **134** and the 3-alkoxycyclohexene oxides **135** and **136**. Both of the methoxycyclohexene oxides **135a** and **136a** open prefer-

| 134a,b | 135a,b | 136a,b |

a; R = Me
b; R = Et

entially at Position 1 under nucleophilic attack by OH^- or CH_3O^- or by $LiAlH_4$, and this observation was ascribed (99) to the polar effect of the alkoxy substituent. The electronegative ethereal oxygen, by withdrawing electrons from C-3, imposes a δ^+ charge on the carbon atom. The entering nucleophile attacks one of the bridgehead atoms of the oxirane ring, and in the transition state a δ^+ develops on that carbon atom (103). Repulsive interaction between these two positive centers will lead to a substantial increase in the energy of activation. Since the repulsion between the two positive centers diminishes as some function of the distance between them, the energy of activation for the path involving **137a** will be less than that for **137b**, so the former will predominate (states of protonation are omitted in these figures).

137a   137b

138   139   140   141

Although conformational influences were neglected at first (99), it was pointed out later (63,84) that the principles outlined in Section V must apply here, and that in fact, not only does the inductive effect oppose the formation of **139** from the *syn*-epoxide **135**, but that the transition state related to **139** is of higher conformational energy than that related to **138**. This point is elaborated below. It is possible but not certain that steric hindrance may also be involved in the regioselectivity of scission of the *anti*-epoxide **136**.

The destabilization of the transition state by the inductive effect of the electronegative substituent X cannot be subjected to a quantitative theoretical treatment at present, for at least two reasons. First, it is not certain whether the propagation of the effect to carbon atoms separated by more than one bond from the induced δ^+ occurs through space or through the intervening bonds (13,104–109). Even if one could decide between these alternatives, calculation of the interaction would require knowledge of the dielectric constant of the medium; that the latter is discontinuous adds to the problem. Second, it is not certain whether the δ^+ should be considered simply as coulombic point charges, or as the positive ends of dipoles. Different mathematical treatments apply to these two possibilities.

From careful reexamination of the earlier studies (63) the authors obtained useful quantitative data, which permit evaluation of the various influences, as well as an assessment of any variation between acidic and basic

TABLE 1
Nucleophilic Scission of the *Syn*- and *Anti*-3-Methoxycyclohexene Oxides[a]

| | | Products from | | | |
| | | syn-Oxide **135a** | | anti-Oxide **136a** | |
Nucleophile Y		**138**	**139**	**140**	**141**
Acidic catalysis					
a	Cl⁻	100	0	91.5	8.5
b	Br⁻	100	0	89.5	10.5
c	OAc⁻	100	0	87.6	12.4
d	OMe⁻	100	0	89.6	10.4
Basic catalysis					
e	NH₃	100	0	90.0	10.0
d	OMe⁻	100	0	90.7	9.3

[a] Modified from the data of Bannard et al. (63). The values are given as per cent of the recovered yield of products resulting from nucleophilic attack at C-1 and C-2.

catalysis (Table 1). *A priori*, one might have expected that of the two epoxides, the *anti* compound **136a** would be less likely than **135a** to give products arising from nucleophilic attack at the end of the oxirane ring adjacent to the methoxyl group, because of steric hindrance ascribable to the vicinal *trans* substituents, in addition to the polar effect of that group. The data, however, show just the opposite. If one assumes that the inductive destabilization of the transition states corresponding to **137b** is identical for the species derived from **135** and **136**, the results illustrate the extent to which an unfavorable conformation may be much more serious than steric hindrance in determining the regioselectivity. In related studies (101,102) on the addition of HOBr and HOCl to **134**, the *syn* and *anti* epihalonium ions corresponding to **135** and **136** were formed in roughly equal amounts, again indicating that an equatorially oriented methoxyl group exerts little if any steric hindrance. In explaining the results of Table 1, the authors propose that the transition state is more productlike than oxidelike, and they point out that in transition state **142** leading to **138** there are no unfavorable interactions for the entering nucleophile or the developing hydroxyl group, whereas in **143**, which they propose as the transition state leading to **139**, there is 1,3-diaxial interaction. Similarly they propose that in **144** leading to **140** there is serious 1,3-diaxial interaction, which is absent in **145** leading to **141**. The effect of the conformation of the transition state is therefore synergistic with the inductive effect, in the scission of **135**, opposing the formation of **139**. However, in the scission of **136** these effects are in opposition. Conformational factors favor the route

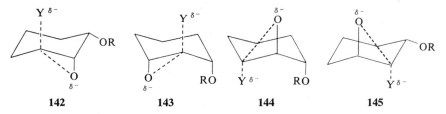

142   143   144   145

$136 \rightarrow 145 \rightarrow 141$, whereas the inductive effect favors $136 \rightarrow 144 \rightarrow 140$; consequently finite amounts of each product are formed. The roughly 10:1 preponderance of 140 is a measure of the importance of the inductive effect. In the study (101,102) of addition of hypohalous acids to 134 the Canadian authors observed that scission of the presumed intermediate epihalonium species obeys the same regioselectivity as epoxide scission. In these cases also, no product was detected which would have arisen from attack of the nucleophile on the carbon atom adjacent to the methoxyl substituent in the *syn*-epihalonium species analogous to 135.

Table 1 shows further that the distribution of products is relatively unaffected by a change from acidic to basic reaction conditions, and by the size and nucleophilicity of the attacking group. Effects difficult to assess are (*1*) the presumed constancy of the inductive effect of the methoxyl group, in axial or equatorial orientation and *syn* or *anti* to the oxirane ring; and (*2*) any differences in repulsive interaction between the two oxygen atoms, which may be related to the anomeric effect in carbohydrates (110).

Extension of these studies to the cyclopentanoid analogues 146, 147, and 148 by the same authors (69) has served to emphasize the differences between the cyclopentanoid and cyclohexanoid series. The *syn*-epoxide 146 in ethereal

146   147   148

149   150

HCl gave about 17% of (1,4/5)-5-chloro product 149 and 83% of 150, whereas scission of 147 gave a high yield of a single product arising from attack of the nucleophile at the end of the oxirane ring more distant from the methoxyl substituent. Obviously, different conformational and steric effects are important in this series, and these are discussed in Section VII.B. Addition of

HOCl to **148** also gave a product mixture which did not resemble the distribution obtained from similar treatment of **134**.

B. Cyclopentanoid Epoxides

Our relatively clear understanding of the factors involved in scission of substituted cyclohexene oxides and pyranoid sugar epoxides rests on the well-developed principles of conformational analysis bearing on those systems. In contrast, the conformational analysis of cyclopentanoid and furanoid systems is not as highly developed (111); not only are the parameters less well understood, but unlike the cyclohexanes, in which one of the two chairs generally is the only reasonable conformation, in the cyclopentanes a large number of reasonable envelope and twist conformers, with different degrees of pseudorotation, may be written (112). As a result, although one can predict the conformation of most cyclohexanoid systems, specific conformational analysis is still necessary for most cyclopentanoid systems. Some examples may be useful. (*1*) It is impossible to obtain conformationally homogeneous cyclopentane derivatives with the aid of *t*-butyl groups. (*2*) In the 1,2-dihalogenocyclopentanes there are mainly two conformers (diaxial and diequatorial) differing in polarity; these are in dynamic equilibrium, with the diaxial form predominating, even in highly polar solvents. (*3*) The tribenzoate **151** of (1,3/2)-cyclopentanetriol is shown by its nmr spectrum (113) to exist largely in conformations that have the three bulky groups axial. The corresponding cyclohexanoid derivative would be expected to have the all-equatorial conformation. Another set of differences arises from the differing flexibility of the two ring systems. In epoxides the carbon atoms of the oxirane ring, and

TABLE 2
Relative Lability of Epoxydiols in
Aqueous Acid[a]

Compound	K_{hydrol}, M^{-1} sec^{-1}
156	0.477
157	0.257
158	0.0167
159	0.0113

[a] See Franks et al. (84).

the carbon atoms immediately adjacent to these, are coplanar. As a result, an epoxycyclohexane can exist in two half-chair conformations 152 and 153, and other conformations (twist or boat) are possible, but are generally likely to exist only in small amounts in the ground state. In cyclopentanoid epoxides, the requirement for coplanarity of four carbon atoms means that only the V_4 or V^4 conformations 154 and 155 are possible. In a considerable number of mono- and disubstituted cyclopentene oxides studied (113–115) the V_4 conformation was the predominant ($>95\%$) form, and in only one case was an appreciable amount of the V^4 conformation observed. Only recently have attempts been made (69) to apply the principles of conformational analysis to the scission of cyclopentanoid epoxides. If, as in the case of cyclohexanoid epoxides (63), the transition state for epoxide scission is indeed more product-like than oxidelike, not only 1,3-diaxial interactions, but also 1,2-eclipsing interactions in the transition state must be considered (69).

Recent studies on cyclopentanoid epoxides have been carried out by two research groups (69,84–86,113,114,116,117). The directive effects are most easily explained as a combination of polar and steric influences, the former effect predominating. The polar effect (see Sections III and VII.A) is a de-stabilization of one of the two possible transition states by a vicinal electro-negative group, and is well demonstrated by the different rates of hydrolysis of the epoxydiols 156–159, shown in Table 2. The two compounds 156 and 157 in which one end of the oxirane ring has a vicinal methylene group are attacked 30 times faster than the corresponding compounds 158 and 159 in which both ends of the oxirane ring have adjacent electronegative substitu-ents. The one and one-half to twofold differences in rates of 156 vs. 157 and of 158 vs. 159 are ascribed to steric hindrance of the single *trans* substituent in each case. The difference between 156 and 157 is in accord with the hypo-thesis of Langstaff et al. (69) that the transition state for scission is more productlike than oxidelike, for the following reasons. From the discussion of

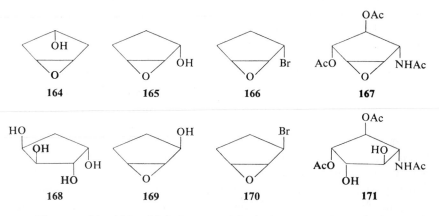

conformation given above, the ground state of **157** would be best represented by **154**, in which groups R and S are both OH, and X = H. The *trans*-hydroxyl at C-4 is equatorial, and is unlikely to impede the approach of the nucleophile. The productlike transition states leading to the two possible products of the reaction, **160** and **161**, however, can have many conformations in which the hydroxyl group could cause some hindrance to the attack at either end of the oxirane ring. The kinetic evidence of Table 2 is supported by the observed product ratio (85). Scission of **156** gives **162** and **163** in a ratio of about 100:1 (117), and scission of **157** gives **160** and **161** in a ratio of 10:1.

The epoxide **164**, which has no vicinal electronegative substituent, is hydrolyzed three times as rapidly as is **165** (113,114a). The effect of an electron-withdrawing group in slowing the rate of acidic ring scission at both oxirane carbon atoms has been noted in simpler cases by Addy and Parker (26a), and is in contrast to behavior under neutral conditions (see Section III). The lower concentration of conjugate acid intermediate, due to the electronegative substituent, is held to be responsible. The nature of the substituent is important, as shown by the fivefold slower hydrolysis of **166** compared to **165**. An extension of this principle is seen in the hydrolysis of **167**, in which two different electronegative groups flank the epoxide (86b). The amide nitrogen atom is less electronegative than the ester oxygen, so the ester is more effective in destabilizing the positive charge which develops in the transition state. As a result the attack occurs adjacent to the acetamido group, and the principal hydrolytic product is **171**.

The second important effect is steric hindrance by a vicinal *trans* substituent. Hydrolysis of **159** gives a mixture of **161** and **168** in which the former predominates by 20:1, in contradistinction to the cyclohexanoid series (118) in which the corresponding epoxide opens preferentially to give the isomer corresponding to **168**. The influence of the *trans* substituent is also apparent from the slower rate of hydrolysis of **159** relative to **158** (Table 2). Similarly, **165** and **166** are hydrolyzed 1.3 and 2.8 times as rapidly, respectively, as their corresponding *trans* isomers **169** and **170**. The kinetic data are valuable in a qualitative way, but quantitative interpretation is difficult, because k_{hydrol} is the sum of the separate rate constants for attack at the two ends of the oxirane ring (69,113,114a). More refined and more extensive measurements may, however, permit quantitative use of the kinetic data. Such attempts are now being made by one group of investigators (69).

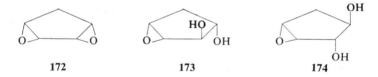

172   173   174

The hydrolysis of the diepoxide **172** to give **161** (58%) and **160** (42%) is unexpected (85). Since the presumed intermediates **173** and **174** are known to yield mainly **161** and **160**, respectively, on hydrolysis, the initial hydrolysis of **172** must be only slightly regioselective. The reason for this is not clear, but it may be significant that **173** and **174** do not accumulate during the reaction, i.e., the hydrolysis of **172** appears to be rate-controlling.

The regiospecificity and variation in k_{hydrol}, which has been ascribed above to steric hindrance, is susceptible to another explanation. Langstaff et al. (69) point out that since the transition states for epoxide scission are more productlike than oxidelike, ordinary conformational analysis may be applied to explain both the kinetic data and the product ratios. After extensive studies of the 3-methoxycyclohexene oxides (see Section VII.A), they investigated scission of the 3-methoxycyclopentene oxides **175** and **178**. Treatment of **175** with anhydrous HCl in ether gave **176** and **177** in a ratio of 87:13; similar treatment of **178** gave a high yield of **179**, none of the diastereoisomer **180** being detected. The difference between this system and the cyclohexanoid analogues (Table 1) is again apparent. In the cyclohexanoid case it is the *anti* isomer which gives about 10% of product derived from attack vicinal to the substituent, whereas the *syn* isomer gives no such products, directly opposite to what is observed in the scission of **175** and **178**. The productlike transition states **181–184** are proposed, leading to **176**, **177**, **179**, and **180**, respectively, and are contrasted with the corresponding transition states for the cyclohexanoid compounds (**142–145** in Section VII.A). The 1,3-diaxial interaction in transition state **143** does not occur in **182**, to which fact the significant amount of "abnormal" scission of **175** is ascribed, in contrast to the total absence of such scission in the cyclohexanoid case. In the case of the *anti* isomers, a similar 1,3-diaxial interaction is noted in the transition states for normal opening in both series, but more seriously in the cyclopentanoid case a semieclipsing interaction between chlorine and methoxyl in **184** strongly opposes "abnormal" scission. Alternative conformations are also relatively unfavorable, and thus the total absence of **180** is explained. The importance of 1,3-diaxiallike interactions in cyclopentane derivatives is uncertain in view of the examples cited earlier in this section, but the rest of the argument is undoubtedly correct.

C. Epoxides of Pyranoid Sugars

It was pointed out by Mills (119) that in the case of sugar epoxides containing fused rings the products of ring scission obeyed the Fürst-Plattner rule (44). Thus, the altroside **186** results from reaction of **185** or **187** with aqueous

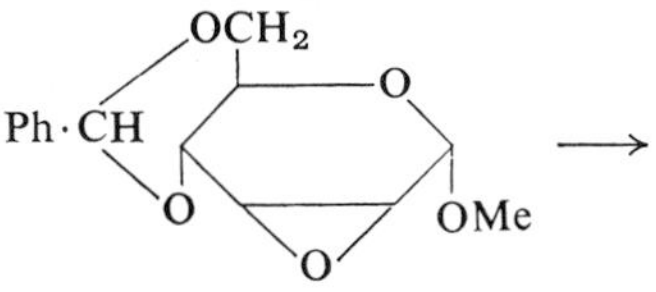

185

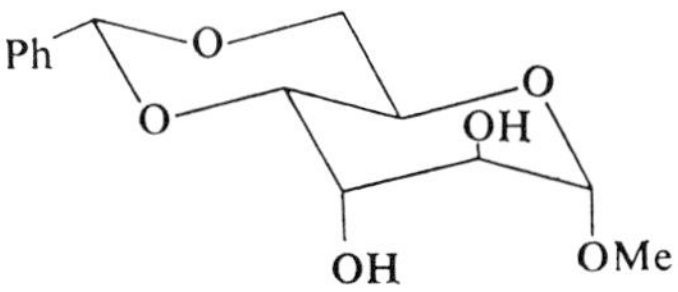

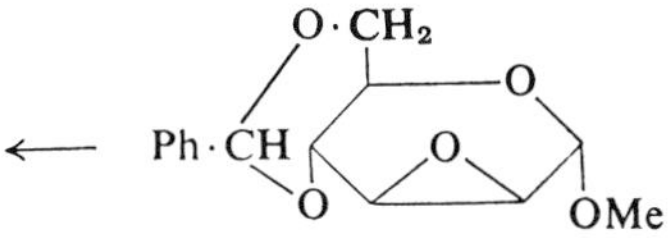

186   **187**

alkali (120,121). In addition, a small amount of diequatorial product is formed from **185** (121) and this must involve a boatlike transition state (see Section V). There are many other cases of diaxial opening in these compounds (122); a more recent example is the formation of *altro*-derivatives using azide ion (123). The epoxides **188** and **190**, derived from 1,6-anhydrohexoses, yield the diaxial products **189** and **191** on reaction with ammonia (124), another example of conformational control; there are other instances in this series (122,125).

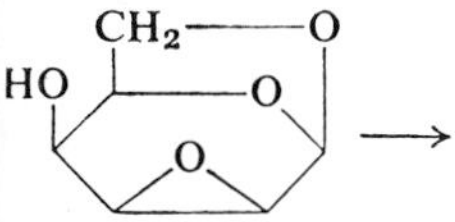

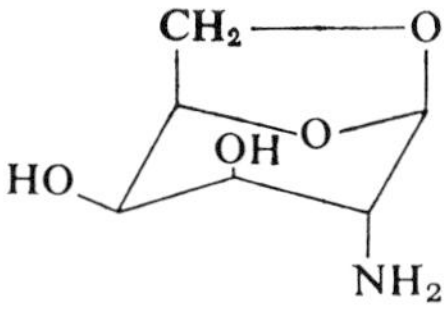

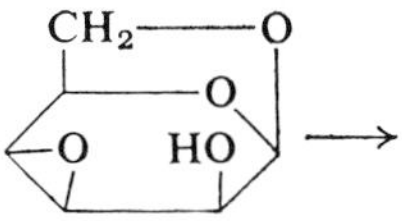

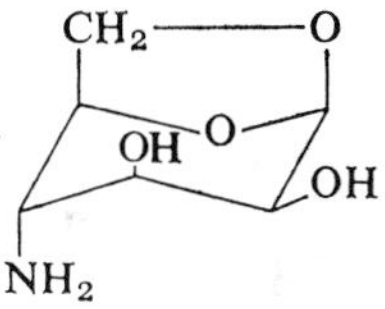

188   **189**   **190**   **191**

When the conformation of the pyranose ring is not stabilized, it becomes more difficult to predict the products, as shown by the number of attempts to do so (122,126,127). As an extreme case, let us consider the anhydropento-pyranosides, in which the lack of substituent at C-5 lowers the energy differences between the conformations (48).

The methyl 2,3-anhydropentopyranosides undergo ring-opening with a wide variety of reagents predominantly at C-3 (83,122,128). Thus, **194** and **196** are the major products resulting from treatment of **192** and **195**, respectively, with aqueous alkali (83,129). The regioselectivity is probably due to greater electron-withdrawal by the acetal group, compared with the C-4 hydroxyl group, first suggested by Cookson (51) and elaborated by Schaub and Weiss (130). It should be noted that the transition state **193** is not derived from the stable ground-state conformation of **192**, in which both substituents are pseudoaxial (131). Dick and Jones (132) have recently reported anomalous

HO — O — OMe → HO ... OH ... O ... OMe → HO ... OH ... O ... OMe ... OH

192   **193**   **194**

HO — O ... O ... OMe → HO ... HO ... O ... OMe ... OH

195   **196**

cases of ring scission in some methyl 2,3-anhydro-4-azido-4-deoxypento-pyranosides. Both **197** and **200** yield **199** as the major product when treated with aqueous potassium hydroxide. In these cases, and in the others cited, electron-withdrawal by the azido group (133) appears to be comparable to that by the acetal group, and conformational and steric effects become more

O — OMe ... N₃ ... O → O ... OMe ... N₃ ... OH → N₃ ... OH ... O ... OMe ... OH ← N₃ ... O ... OMe

197   **198**   **199**   **200**

important. Although the situation is by no means clear, the major product, in most cases, is derived from a transition state containing the anomeric methoxyl group in the more stable axial conformation, as in **198** (110a). It is interesting to note that the regioselectivity of the reaction **192** → **194** is opposite to that of the related sequence **197** → **199**. The methyl 3,4-anhydro-pentopyranosides, e.g., **201** (134), lack the electronic domination of C-1 and the regioselectivity should depend on conformational and steric effects. In the limited number of cases studied, there appears to be a preference for attack at C-4 (83,134). Detailed analysis of the results must await further studies.

In the methyl 2,3-anhydrohexopyranosides the carbon substituent on C-5 has an interesting effect. The anhydromannoside **202**, on treatment with hydrochloric acid in acetone, gives a high yield of chlorohydrin **204**, the isomer expected on polar and conformational grounds (135). The transition state **203** possesses an equatorial hydroxymethyl group, the anomeric disposition is favorable, and there is only a single unfavorable 1,3-diaxial interaction. In the case of the anhydroalloside **205**, however, the transition state **206** for attack at C_3 has two unfavorable 1,3-diaxial interactions, one of them

201   **202**   **203**   **204**

205   **206**   **207**   **208**

involving the axial hydroxymethyl group, in addition to an unfavorable anomeric effect. Both **207** and **208** are therefore formed in comparable amounts (136,82c). In the 4,6-*O*-benzylidene derivative of **205**, viz. **185**, the *trans* ring fusion causes conformational effects to be even more important. The only example in which a methyl 2,3-anhydrohexopyranoside is not attacked at C-3 is the reaction of **209a** with sodium methoxide in methanol, in which the sole product is the 2-methyl ether **210** (137). This is particularly surprising since similar treatment of **209b** gives mainly the expected product **211**. The 3,4-epoxides **212a** and **212b**, with the same reagents, also show a

209a,b   **210**   **211**   **212 a,b**
a; R = Me
b; R = H

difference in regioselectivity (137). Newth (122) has discussed this problem in terms of hydrogen bonding in **209b** and **212b**.

A further anomaly, in the 3,4-epoxides, is the preferential formation of the *gluco* isomer **216** from treatment of the anhydroalloside **213** with sodium

213   **214**   **215**   **216**

methoxide (138). The all-axial chairlike transition state **214** seems unlikely, in comparison with **215**, in spite of the unfavorable anomeric disposition in the latter. Perhaps a boatlike transition state may be involved (see Route B, Figure 2). Clearly, more experimental work is necessary in this field. The behavior of pyranoside epoxides towards Grignard reagents, briefly summarized by Newth (122), is similar to the reactions just described.

D. Epoxides of Furanoid Sugars

Epoxides of furanoid sugars have been of particular interest chiefly in studies concerned with the synthesis of nucleosides containing various natural and unnatural sugars. Such nucleosides are of interest for various biochemical purposes, and also for the designing of synthetic substances with medicinal properties. Because of the latter interest, various patents have been issued involving scission of furanoid epoxides [e.g., see (139)].

As in the case of cyclopentanoid epoxides, conformational factors are difficult to assess; in fact, the problem with furanoid sugars is even more complicated. Since the ring oxygen atom is unsubstituted, a number of possible eclipsing interactions that would be serious in cyclopentanoid compounds are absent here; however, there is the added factor of the anomeric effect (110), i.e., repulsion between the ring oxygen and the C_1 oxygen is greater when the latter is equatorial than when it is axial. The importance of the anomeric effect in furanoid compounds is shown (140) by the all-axial conformation adopted by 3,4-dibromo-2,5-dimethoxytetrahydrofuran (**217**). The reinforcement of the two anomeric effects, i.e., those due to the separate methoxyl groups, makes this the principal conformational influence. Most examples of scission of furanoid epoxides can be rationalized by reference to the polar effects described in Sections III, VII.A, and VII.B, and in view of the uncertainty of the conformational influences, the latter are largely omitted.

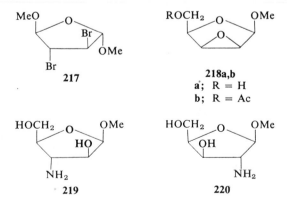

For example (141), ammonolysis of the anhydrolyxoside **218a** gives a mixture of the 3-amino (*arabino*) and 2-amino (*xylo*) derivatives **219** and **220** in which the former predominates by 100:1. The α-anomer of **218a** gives only the corresponding 3-amino product, due to steric hindrance by the methoxyl substituent. The transition state leading to **219** is destabilized by the inductive effect of the ring oxygen on C_4. However, the transition state leading to **220** is even more destabilized due to the inductive effect of the two electron-withdrawing substituents on C_1. Many other examples of predominant attack at

221   222

223   224

C_3 in 2,3-anhydrolyxofuranosides are known (142–147), particularly in the α-series. Two cases are known, however, in which attack occurs preferentially at C_2. With **218a**, sodium toluenethiolate in methanol gives **221** and **222** in a ratio of 3:2 (148), contrasting with exclusive attack at C_3 in the case of the α-anomer (149). Similarly, the acetate **218b** reacts with magnesium bromide in 1,2-dimethoxyethane (150) to yield the bromide **223**, together with **224** in the ratio 2:1; the α-anomer of **218b** is attacked normally at C_3. The basis of this reversal of selectivity is not apparent.

The anhydroribofuranosides have also been extensively studied. With ammonia (130,151,152), sodium ethanethiolate (153), and sodium azide (152) the β-anomer **225a** yields almost exclusively the C_3-substitution product, and **225b** reacts similarly with magnesium bromide (154). The importance of steric hindrance and nucleophilicity in this system is shown by the behavior of the trityl ethers **226**. Whereas **226a** is inert to ammonolysis (130), **226b** reacts smoothly at C_3 with sodium ethanethiolate (146). When **225a** is treated with

225a,b a; R = H
            b; R = Ac

226a,b a; R = OMe
            b; R = theophylline

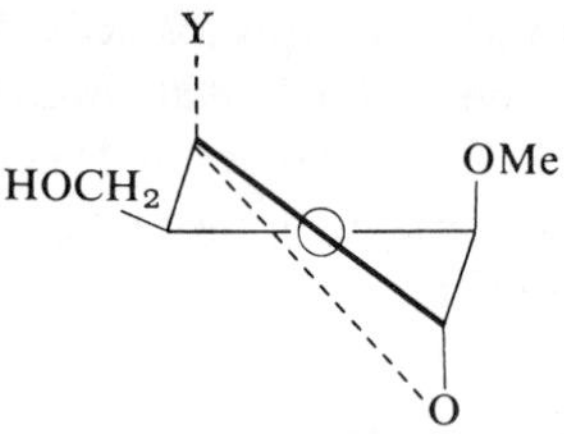

227 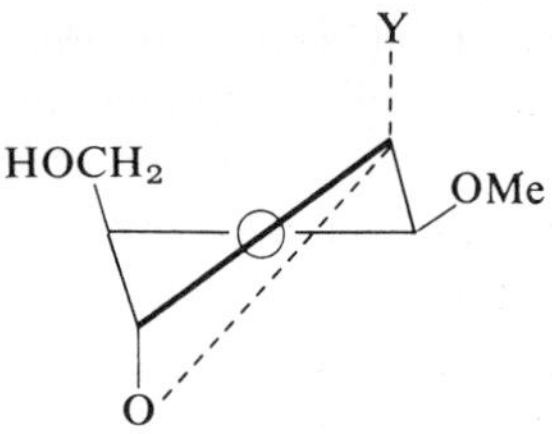   228

sodium methoxide the major product arises via epoxide migration [(155); see Section VIII.A], another indication of steric hindrance. The absence of attack at C_2 in these reactions may be ascribed to the combined effect of polarity, due to the two oxygen substituents on C_1 and steric hindrance [cf. the reaction of 178 with hydrogen chloride (Section VII.B)]. Reist and Holton (154) have suggested that the methoxyl group in 225b exhibits more steric influence than the acetoxymethyl group because of the anomeric effect, i.e., the methoxyl group tends to occupy an axial conformation. This would give a transition state of type 227 rather than 228 for these reactions.

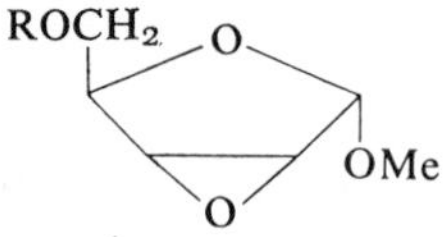

229a–c   230   231

a; R = H
b; R = Ts
c; R = Ac

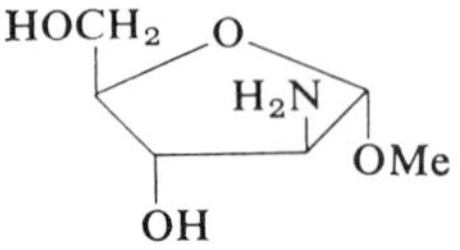

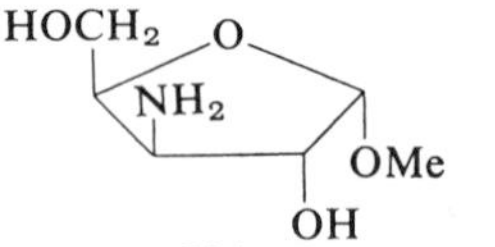

In the α-ribo series, as in 229, there is a preference for attack at C_2, shown by reaction of 229a with sodium methoxide (155), of 229b with lithium aluminum hydride or sodium toluenethiolate (156), and of 229c with magnesium bromide (154). Contrary to earlier reports (130,151), ammonolysis of 229a yields 230 and 231 in the ratio 2:1 (152). In these reactions the polar

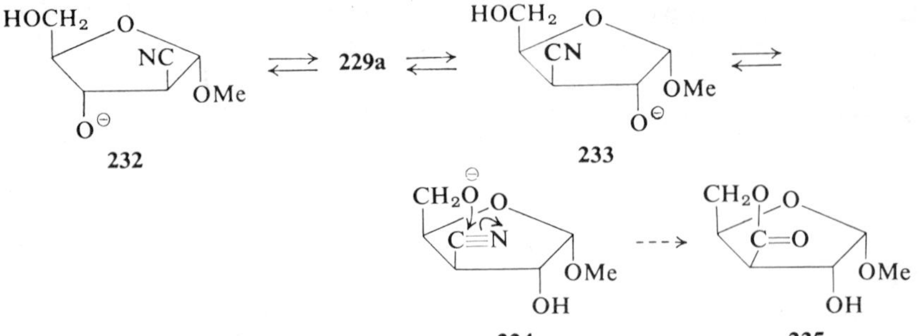

232   233

234   235

preference for attack at C_3 is inhibited by the substituent on C_4, whereas no such steric effect is present at C_2 [cf. the hydrolysis of **159** and the reaction of **175** with hydrogen chloride (Section VII.B)]. Cyanide ion attacks **229a** mainly at C_3 to give the lactone **235** (155,157). Reversibility of the cyanide addition, as in **232** and **233**, followed by intramolecular catalysis of the hydrolysis of the cyano group, as in **234**, have been proposed to account for this result (155).

E. Some Hydride Reductions

1. Some Anomalous Products of Lithium Aluminum Hydride Reduction of Epoxides

When the epoxycyclobutane **236** is reduced with lithium aluminum hydride in tetrahydrofuran the products are the expected alcohol **237** (64%),

| 236 | 237 | 238 | 239 |

| 240 | 241 | 242 | 243 |

together with **238** (36%), arising from carbon–carbon bond cleavage of the cyclobutane ring (158). Under the same conditions the analogue **239** reacted more rapidly to give **240** (54%) and **241** (46%). Considering first the normal product **240**, attack at C_6, and not C_7, is expected because of electron-withdrawal by the five-membered ring oxygen atom (see Section III). That it occurs more rapidly than with **236** is in keeping with the increased importance of bond-making in the transition state in these polar situations (26a).

On the basis of deuterium-labeling experiments, it was proposed that the abnormal product **241** arises by way of the intermediate **242** (158). The aluminum-containing species, after transference of the hydride ion, is still in the vicinity of C_6, and capable of assisting the cleavage of the C_6—C_7 bond, thereby releasing the angle strain of the cyclobutane ring. The resulting aldehyde **243** undergoes further reduction, and subsequent hydrolysis during isolation yields **241**.

77 →

244   245   246

A similar explanation has been advanced* (55) to account for the formation of 10% of the *cis*-alcohol **246** during the lithium aluminum hydride reduction of *trans*-4-*t*-butylcyclohexene oxide (**77**) (see Section V and Ref. 64). Hydride transfer to AlH_3 (alane) occurs through the intermediate **244**, with formation of the ketone **245**; reduction yields **246**.

2. Mixed Hydride Reductions

Elsewhere in this chapter many examples have been given of lithium aluminum hydride reduction of epoxides in which simple bimolecular attack by aluminum hydride ion has been assumed. When an excess of reagent is used, this is justifiable, but with a smaller quantity of reagent, intermediates of the type **247** and **248** will be formed from an epoxide such as **9**, and act as

$$RCHO\overset{\ominus}{AlH_3} \qquad (RCHO)_2\overset{\ominus}{AlH_2} \qquad LiAlH(OR)_3$$

Me   Me

247   248   **249a, b**
                                              a, R = Me
                                              b, R = CMe₃

the effective reducing agents (159). Such compounds will have a greater steric requirement and different regioselectivity might be expected. Fuchs and VanderWerf (160) found that in the case of 3,4-epoxy-1-butene (**16**) the amount of secondary attack changed from 30% (with 2.1 moles $LiAlH_4$) to 17% (0.26 mole). More recently Brown and Yoon (161) have found that **249a** and **249b** react more slowly than lithium aluminum hydride with some representative epoxides. In particular, **249b** gives no product of secondary attack on styrene oxide (**15**), compared with 4% in the case of lithium aluminum hydride. It would clearly be of interest to study the behavior of **249b** with more complex epoxides.

Several workers have studied the cleavage of epoxides by mixtures of lithium aluminum hydride and aluminum chloride, the so-called "mixed

---

* This "oxidative inversion" appears to be a very general side reaction in lithium aluminum hydride reduction of epoxides (64). It does not occur in epoxide reduction by alane(55,64*b*,264*b*) (see Section VII.E.2).

hydride" reagents (162–166). The results can be interpreted in terms of Lewis acid species present in the reaction mixture and are very dependent on the

relative amounts of aluminum chloride (164–167). Thus, in the case of tri-phenylethylene oxide (250) lithium aluminum hydride itself yields only 251, attack occurring at the least hindered carbon atom (162). In the presence of aluminum chloride (0.33 mole), which generates alane (164,165), the only product is 252. The regioselectivity in the latter reaction is determined by the Lewis acid character of alane, and the better conjugative stabilization of a δ^+ charge by two phenyl groups, as in 253; Ashby and Prather (165) have suggested a four-center transition state 254 (see Section IV, Formula 42), but a bimolecular reaction between 253 and a second molecule of alane may also occur (166). When an excess of aluminum chloride (4 moles) is present, the main product is the primary alcohol 255. The Lewis acid species, $AlHCl_2$ and $AlCl_3$, which are in this mixture (165), are powerful enough to cause rearrangement of 250 to triphenylacetaldehyde (256) before reduction. Similar general behavior has been reported for the reduction of 257 (164). Although a transition state analogous to 254 has been proposed for the reaction of 257 with alane (167), bimolecular attack by alane on the Lewis acid complex seems more likely in this case. The reaction of other epoxides with alane has been described (161,166).

3. Reductions with Diborane

Epoxides react rapidly with diborane in tetrahydrofuran in the presence of borohydride ion (168–170) to give products typical of an acid-catalyzed ring-opening (168,170). Thus, the trisubstituted epoxide 258 yields 259 (74%) and 260 (26%) in which attack has occurred predominantly at the more hindered carbon atom (170). It is interesting that when alane is the reductant,

258   259   260

the relative yield of **260** increases markedly, not more than 10% of **259** being formed (64b,161). Presumably this is due to the more stringent steric requirements of the reaction with alane.

Diborane itself in tetrahydrofuran reacts slowly with epoxides (169–171), to give only low yields of the expected alcohols. The various side-reactions are complex and appear to be a consequence of the weak nucleophilicity of diborane.

The effect of boron trifluoride on the diborane reduction of styrene oxides has been studied (171). 1-Phenylcyclopentene oxide (**261**), which reacts in a complex manner with diborane, yields **263** (82%) and **264** (18%) when treated with diborane and boron trifluoride (1:1) in tetrahydrofuran. Reduction is preceded by the boron trifluoride-catalyzed rearrangement of **261** to the

261   262   263   264

ketone **262** (the migrating group is not specified). These results parallel those with the aluminum hydride–aluminum chloride system at high aluminum chloride ratios (164).

F. Reduction of Certain Epoxyhydroxycyclohexanes and Cycloheptanes with Lithium Aluminum Hydride

The ability of lithium aluminum hydride to reduce cinnamyl alcohol to 3-phenylpropanol was reported by Hochstein and Brown (172). They showed convincingly that a five-membered cyclic intermediate **265** was formed, containing a carbon–aluminum bond. In order to explain the regioselectivity in the reduction of certain epoxy-compounds containing a hydroxyl group, Henbest (173–175) has postulated hydride transfer by way of a six-membered ring, e.g., **266**.

Although there is no direct evidence for the operation of this process, there are results which are most easily explained by neighboring group participation involving both 6- and 5-membered rings. In this discussion we

265   266

consider the polar influence in addition to the conformational and neighboring group effects noted by Henbest.

In their work on the epoxidation of cyclic allylic alcohols, Henbest and
Wilson (173) studied the behavior of the 1,2-epoxy-3-hydroxycyclohexanes
with lithium aluminum hydride in ether. On the basis of an extensive study of
the 1,2-epoxy-3-methoxycyclohexanes (63,98,99), it would be anticipated that
nucleophilic attack by the complex hydride ion would occur at C_1 in each case,
due to electron-withdrawal by the hydroxyl group. In keeping with this, the
cis-oxide **267** yielded the 1,2-diol **268** in about 90% yield. The *trans*-oxide
269, on the other hand, yielded the 1,3-diol **270** as the major product, a result
interpreted as axial attack at C_2 in the conformation **269** (173). In the light
of the Canadian work (63,98,99) it seems more probable that the major

267   268

269   270

271

product of reduction of the *trans*-isomer arises by participation, as in **271** →
270 (130b), and the minor product may also arise by participation as formulated by Henbest (173) in **266**. A similar explanation may be given of the
corresponding reactions of the 1,2-epoxy-3-hydroxycycloheptanes (176).

Henbest and Nicholls (174) have studied the reduction of the 1,2-epoxy-4-hydroxycyclohexanes with lithium aluminum hydride. Reduction in both cases led to a 1,4-diol. In the *cis*-isomer nucleophilic attack on the preferred conformation **272** (as the aluminum complex) yields the *cis*-1,4-diol **273**, whereas the *trans*-epoxide reacts by way of the intermediate **274**, to give the *trans*-1,4-diol **275**. In view of the studies of Rickborn and Lwo (57) in which

reduction of the 4-methylcyclohexene oxides yielded only 2% of equatorial alcohol, an intermediate **274** is strongly indicated.

Henbest and Nicholls (175) have also examined *trans*-4-hydroxymethyl-cyclohexene oxide (**276**), whose reduction follows that of the 4-methyl compound (57), forming the axial alcohol **277**. Participation would involve a seven-membered ring transition state. Reduction of the bicycloheptene oxide **278** yields the diol **280**, very probably by way of the aluminohydride complex **279** (175).

Neighboring group participation has been invoked (177) to explain the results of lithium aluminum hydride reduction of the alkaloid crinamidine and its derivatives (178). Crinamidine tetrahydropyranyl ether (**281**; R = tetrahydropyranyl) undergoes attack at C_1 with lithium aluminum hydride to give the axial alcohol **282**, in agreement with polar and conformational requirements. Crinamidine itself (**281**; R = H) gives the equatorial alcohol

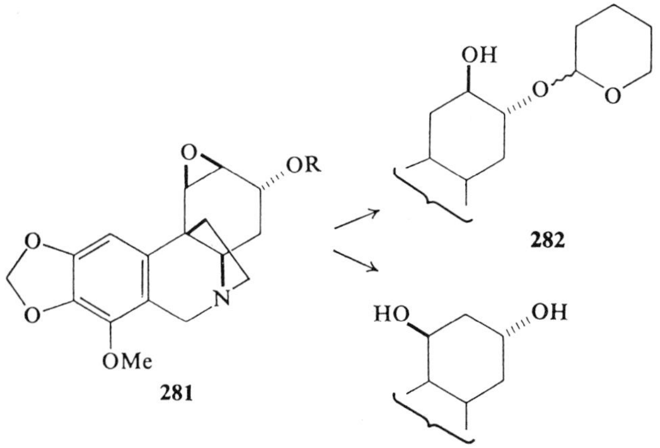

281

282

283

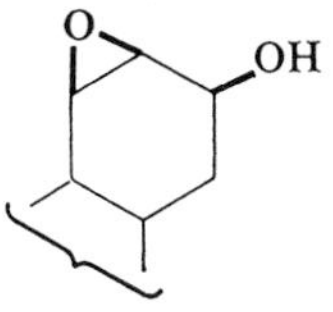

284

283, while under the same conditions epicrinamidine (**284**) is recovered unchanged. A five-membered ring transition state **285** seems probable (177) for the reduction of crinamidine.

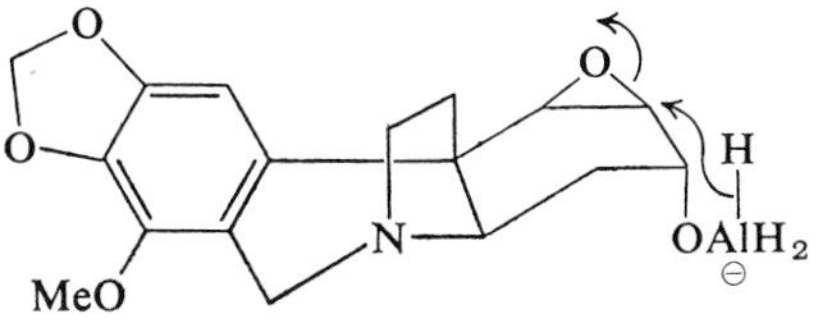

285

G. Reaction of Steroid Epoxides and Related Compounds

1. Reduction of Certain Steroid Epoxides with Lithium Aluminum Hydride

The steroid field is rich in examples of epoxide ring-opening and, because of the conformational restraints which exist, the contribution of various factors may be assessed. When the conformational effect in ring-opening is not complicated by steric or polar effects, as in the case of the 2,3- and 3,4-epoxides of 5α-steroids (47,179,180), lithium aluminum hydride gives a high

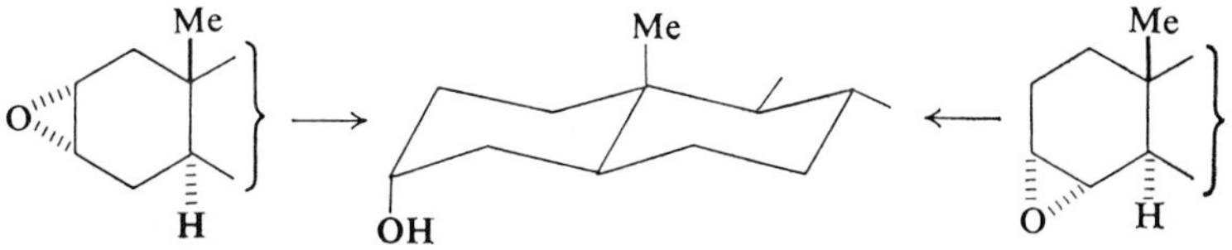

286

287

288

289   HO   290

yield of the axial alcohol, e.g., the same 3α-ol **287** results from **286** and **288**. Similarly, reduction of 3β,4β-epoxy-5β-cholestane **(289)** yields mainly the axial 3β-ol **290** (181). The susceptibility of lithium aluminum hydride to steric hindrance is seen in the case of 9α,11α-epoxides **291** (182,183) in which the epoxide ring is protected by the axial methyl groups from axial attack at the β-face of the molecule. Similar hindrance is encountered in the case of

291   292

7α,8α-epoxides **292**, but in this, and the previous instance, axial reduction may be achieved by lithium in ethylamine, in which the effective reagent, solvated electrons, has a lower steric requirement (184a). Brown et al. (184b) have recently described lithium in ethylenediamine as an improved reagent for this purpose. No difficulty is reported in the reduction of an 11α,12α-epoxide to give the axial 12α-ol (185).

The behavior of 4,5- and 5,6-epoxides shows an interesting balance of steric and conformational factors. When the epoxide ring is α, as in **293** and **295**, attack by aluminum hydride ion occurs at the secondary carbon to yield the axial 5α-ol **294** in both cases, due to a combination of steric and conformational factors (184,186,187). Reduction of 5β,6β-epoxycholestane **(296)** yields mainly (60%) the 5β-ol **297** (184) together with the 6β-ol **298**. In this example, steric and conformational factors compete. Attack at C_6 to give **297** requires a transition state in which Ring B is boatlike (see Section V, Route B); even though this transition state is of relatively high energy, it is obviously of lower energy than that which would be required if the attack occurred at the tertiary carbon atom, C_5. In the 4β,5β-epoxides Ring A, because of cis-fusion to Ring B, may adopt either of two half-chair forms of not widely different energies. 4β,5β-epoxycholestane **(299)** undergoes attack on the secondary carbon (C_4) yielding the 5β-ol **297** (186,187).

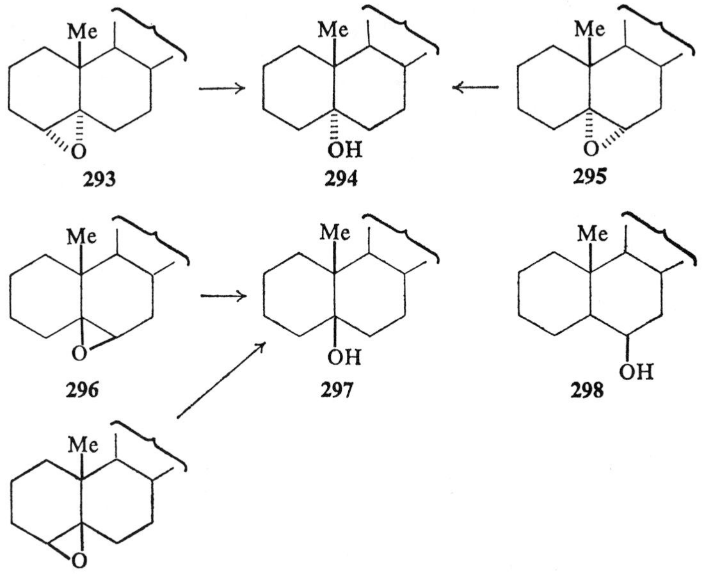

293 294 295

296 297 298

299

Reduction of 2β,3β-epoxy-4,4-dimethyl-5α-cholestane (**300**) with lithium aluminum hydride yields the equatorial alcohol **301** (188). Presumably steric interaction between the axial 19-methyl and 4β-methyl groups raises the energy of the chairlike transition state required for reaction at C_3, relative to the boatlike state required for reaction at C_2, in which this interaction is relieved. A similar regioselectivity is found in the reaction of **300** with hydrogen bromide.

In steroid 16α,17α-epoxides such as **302**, reduction of the oxide occurs regioselectively to yield the 17α-ol **303** (189). Because of the *trans*-fusion of

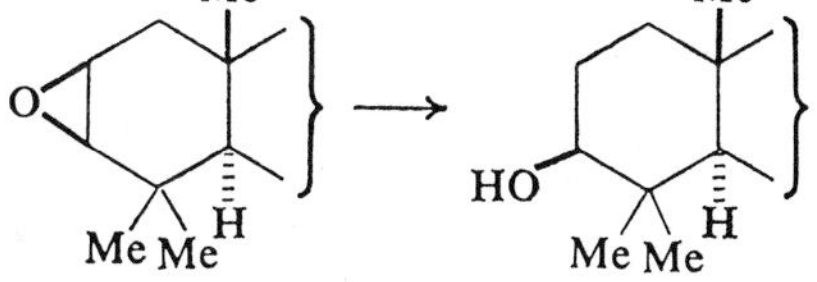

300 301

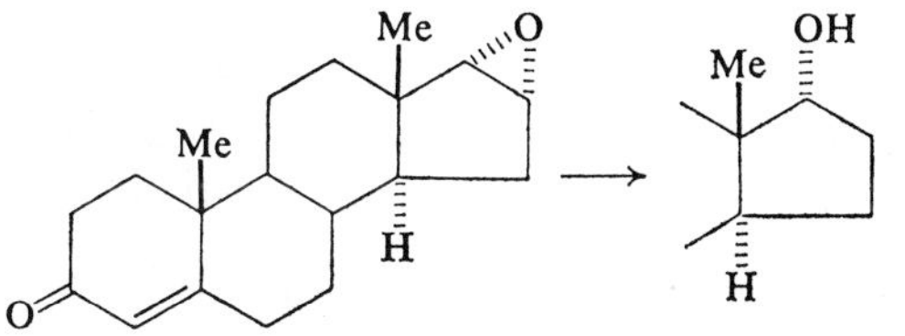

302 303

Rings C and D, axial attack at C_{17} would involve the attacking group in an eclipsing interaction with the adjacent axial methyl group, and reduction therefore occurs at C_{16}. There are several other cases in which this regioselectivity is reinforced by the presence of a carbon side-chain at C_{17} (190,191).

There are several instances of the reaction of hydroxy-steroid epoxides with lithium aluminum hydride that might be expected to exhibit polar or neighboring group effects. All of the 3-acetoxy-4,5-epoxycholestanes **304** (192) and some related model compounds (193) have been studied. The only

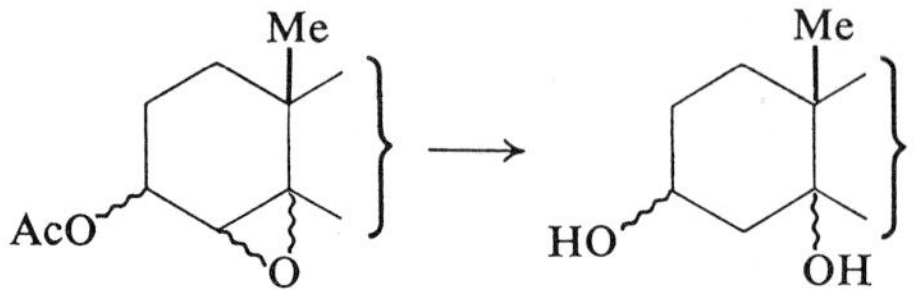

304   305

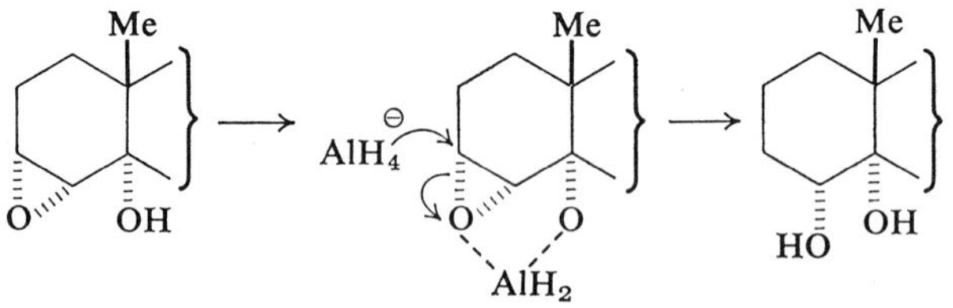

306   307   308

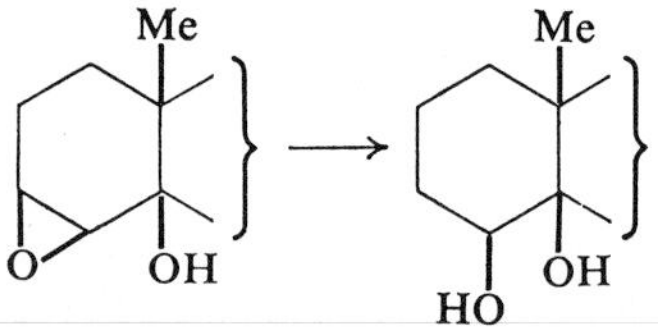

309   310

products isolated have arisen from attack at C_4 (i.e., **305**), in agreement with the unsubstituted epoxides (186). Polar effects should inhibit reduction at C_4, but apparently the steric hindrance to attack at C_5 is the overriding factor. Neighboring group participation, such as described in Section VII.F, may have been present in certain cases, but would not have been detected. Similarly, the 3-hydroxycholestane 5,6-epoxides resemble the parent epoxides **295** and **296** in that the α-epoxides yield the axial 5α-ol only, while the β-epoxides give a mixture of 5β- and 6β-ol, though in a different ratio (194,195). Again no conclusions regarding neighboring group participation can be drawn from the data available. The reduction of 3α,4α-epoxy-5α-cholestane-5-ol (**306**) yields the equatorial 4α-ol **308**, and the isomer **309** gives the 4β-ol **310** (196), in contrast to the behavior of the parent epoxides (179,181). In the case of **306** an intermediate of type **307** has been proposed. Alternatively, this may be an example of the destabilization of the normal chairlike transition state due to a polar effect of the C_5 hydroxyl group.

2. Reaction of Steroid Epoxides and Related Compounds with Hydrogen Bromide and other Protic Acids

The behavior of a large number of epoxides in this class towards hydrogen bromide and other acids has been studied and the results make an interesting comparison with the previous section. Conformational factors govern the formation of diaxial products from the 2,3-epoxy-5α-cholestanes (46) and other disecondary epoxides (179,181,197,198). During acidic ring-opening of epoxides the reaction mechanism becomes "borderline S_N2" (3) and there is a tendency for nucleophilic attack at a tertiary carbon, which is more capable of supporting a δ^+ charge in the transition state. The steric hindrance, which was evident in reactions with lithium aluminum hydride,

311   **312**   **313; R = H**   **315**
314; R = OAc

316   **317**

virtually disappears. In reaction of the 4α,5α-epoxide **311** with hydrogen bromide in acetic acid, conformational factors predominate and the 4β,5α-bromohydrin **312** results (186). The 4β,5β-epoxide **313**, however, yields the 5α-bromide **315**, the reaction showing the opposite regioselectivity to that with lithium aluminum hydride (Section VII.G.1). That the situation is finely balanced is shown by the formation of both **316** (major product) and **317** by hydrolysis of **314** with perchloric acid (whose anion is very weakly nucleophilic) (187).

With hydrogen halides the 5,6-epoxides **318** and **320** both yield diaxial 5α,6β products, **319** and **321**, respectively, (199–201). Recent papers by Lavie

318   319   320   321

and his colleagues (202,203) show an interesting variation in the regioselectivity of this reaction when a 4-acetoxy group is present. The results are
summarized in Formulas **322–330**. The conversion of **322** and **323** into **324**
and **325**, respectively, presents no mechanistic problem, being analogous to
318 → 319. It might be expected that the electron-withdrawing character of
the 4-acetoxy group would accentuate the tendency for nucleophilic attack

322; R_1 = H, R_2 = OAc   324; R_1 = H, R_2 = OAc
323; R_1 = OAc, R_2 = H   325; R_1 = OAc, R_2 = H

326   327

328   329   330

at C_6. The latter effect must be responsible for the regioselectivity of the
reaction **326 → 327** as compared to **320 → 321**. Lavie has pointed out that
Ring B in **326** must become boatlike in the transition state of ring cleavage,
with consequent change in the conformation of the 4β-acetoxy group from
axial to equatorial, but it is difficult to see why this, in itself, should inhibit

attack by bromide ion at C_5. Such a combination of polar, conformational, and steric effects could, however, explain the behavior of **328**. The 4α-acetoxy group becomes axial in the transition state required for the formation of **330**, thus inhibiting approach by bromide ion at C_6. Attack at C_5, although electronically unfavorable, requires a chairlike form of Ring B in the transition state, leaving the 4α-acetoxy group equatorial, and presenting no steric difficulty. Both **329** and **330** are therefore formed in comparable amounts.

More recently the behavior of 4β-acetoxy-$5\alpha,6\alpha$-epoxy-15-noreudesman-$8\beta,12$-olide (**331**) towards both hydrogen bromide in acetic acid and sulfuric acid in acetic acid has been examined (204). In both cases the product was **333**, arising via the acetoxonium ion **332**. Apparently the axial approach to C_6 by the nucleophile is prevented by the methyl group on C_{10} and by the

| 331 | 332 | 333 |

cis-fused lactone ring [the C_{11} epimer also gives this reaction (204)]. The C_{10} methyl group by itself offers considerably less hindrance (cf. **322** → **324**) and it is very interesting that no neighboring group participation is observed in the case of **322** or **328**. This may be due to the relatively high concentration of the nucleophilic bromide ions in these reactions (86c).

The addition of hydrogen bromide to the oxides **334** and **336**, derived from a D-homosteroid, shows regioselective attack at C_{16} in each case, with formation of **335** and **337** (205), a result best interpreted as a polar effect of the carbonyl group. In the case of **336** → **337** this is reinforced by conformational and steric factors, but for **334** → **335** polar and steric effects must override the conformational effect.

| 334 | 335 | 336 | 337 |

Methyl 3-keto-9α,11α-epoxy-5β-cholanate (**338**) reacts readily with hydrogen chloride to form **339**, the carbonyl group being necessary for chlorohydrin formation (206). The mechanism **340** → **341** elegantly accounts for this result, which is an example of internal acid catalysis of epoxide opening.

H. **Conformational and Solvent Effects in Ring Scission of the 1-Phenyl-4-*t*-butylcyclohexene Oxides**

Reference has already been made in Sections IV and V to the stereochemistry of ring-opening of styrene oxides (31–39,70). Berti and his colleagues have made a particularly thorough study of the ring scission of the epoxides **342** and **343** which are ideally constructed to examine conformational

effects (207). 1-Phenylcyclohexene oxide itself is very resistant to treatment with potassium hydroxide in ethanol or aqueous dioxan (208), but when **342** or **343** were heated with potassium hydroxide in dimethyl sulfoxide–water (85:15) the diaxial diol **344** was formed. This result is to be expected from S_N2 attack by hydroxyl ion, whose nucleophilicity is increased in dimethyl sulfoxide (97) by way of a chairlike transition state. Epoxide **343** undergoes reaction more slowly than **342**, a consequence of S_N2 attack at a tertiary center; bond-breaking of the benzylic C—O bond in the transition state appears to be relatively unimportant.

The stereochemical results of ring-opening under acidic conditions are very dependent on the type of acid used and on the solvent, but in all cases the benzylic C—O bond is cleaved as expected (see Section III). At one extreme, trichloroacetic acid in benzene causes exclusively *cis*-opening with both **342** and **343**. Thus, **342** yields the monoester **346** which arises by acyl migra-

tion from **345**. The *cis*-stereochemistry of the product is in agreement with other examples of ring-opening of 1-aryl-epoxides, especially by organic acids in nonpolar media (35). At the other extreme, aqueous sulfuric acid in dimethyl sulfoxide converts both **342** and **343** mainly into *trans*-diols. From **343**, the product is the diaxial diol **344**, in which the benzylic C—O cleavage and conformational effects augment each other; about 10% of the *cis*-diol **347** is formed. In the case of **342** the two effects are in opposition. The resulting product is the diequatorial diol **348**, showing that the cleavage of the benzylic C—O bond is the more important factor. The conformation of the transition state is probably boatlike, as in **349** (Route B, Section V). When **342** and **343** were treated with 98% formic acid in ether, both *cis*- and *trans*-isomers were produced, with the former in excess, and a similar result was found with 1-phenylcyclohexene oxide itself (208). Analogously, when 1-phenylcycloheptene was treated with performic acid, only the *cis*-diol was isolated (209).

VIII. INTRAMOLECULAR REARRANGEMENTS OF EPOXIDES

A. Acid and Base-Catalyzed Attack by an Intramolecular Hydroxyl Group. Epoxide Migration

There are many examples, particularly in the carbohydrate and inositol series, of intramolecular attack on an epoxide ring by a neighboring hydroxyl group or oxanion (122,210). Although these reactions are closely related to those dealt with in Section VI, we discuss them here separately because the immediate products are stable and capable of isolation. Two classes of reaction may be distinguished. The first yields cyclic ethers, containing a five- or six-membered ring, which are completely stable to the basic or acidic reaction conditions. The second yields oxiranes or oxetanes, under alkaline conditions, and an equilibrium is established with the original epoxide.

The anhydroalloside **350**, with aqueous alkali, yields the 3,6-anhydro-glucoside **352** (211), presumably by way of a transition state **351** derived from the C_6-oxanion, and there are similar examples in the earlier literature (210a).

a; R = Ac
b; R = H

Henbest and Nicholls (175) found that the acetate **354a**, prepared by epoxidation of the olefin **353a**, gave the ether **355** on treatment with alkali via the oxanion derived from **354b**. Epoxidation of the olefin **353b** with perbenzoic acid also yielded **355**, showing that **354b** is very easily converted to **355** under acidic as well as under basic conditions. The conversion of **350** and other 2,3-anhydroglycopyranosides into 3,6-anhydro derivatives under acidic conditions has been described (212), but the mechanistic interpretation is complicated by the lability of the glycosidic group and the possible formation of furanoid and acyclic forms of the 2,3-epoxide. A recent example of formation of a tetrahydrofuran is the conversion of 10-*epi*-γ-eudesmol (**356**) into

356   **357**   **358**

4-hydroxydihydroagarofuran (**358**) by *m*-chloroperbenzoic acid (213); the intermediate epoxide **357** was not isolated.

The stereochemical requirements for the formation of a tetrahydrofuran from an epoxide are interesting. Foster and his co-workers (211) found that the 3,4-epoxide **359** underwent normal ring-opening under alkaline conditions,

359   **360**   **361**   **362**   **363**

with no formation of 3,6-anhydride analogous to **352**, despite the *trans* relationship of the C_6-hydroxyl group to the epoxide ring. 1,2-Epoxy-4-butanol (**360**) yields only the triol **361** on alkaline hydrolysis (214). The anhydroaltritol **109a** gives very largely a mixture of hexitols (**110** and **111**) on treatment with dilute sulfuric acid (87,215). Study of Dreiding models shows that it is more difficult to form the transition state **362** than **363**, i.e., a 4-hydroxy-1,2-epoxide should form a tetrahydrofuran less readily than a 5-hydroxy-1,2-epoxide.

In favorable situations a six-membered ring may result from reaction of a 5-hydroxy-1,2-epoxide. Thus, 2,3-anhydro-D-iditol (**364**) yields a mixture of

364   **365**   **366**

367   368   369

1,4-anhydro-D-altritol (**365**) and 1,5-anhydro-L-glucitol (**366**), in the ratio 2:1, under acidic or basic conditions (87). Under alkaline conditions the epoxide **367** gives very largely **368**, together with **369** (216). The same products are formed in benzene in the presence of toluene-*p*-sulfonic acid, but more of **369** is produced owing to the increased tendency for epoxide ring scission at a tertiary carbon atom under acidic conditions.

Under basic conditions a 3-hydroxy-1,2-epoxide may undergo intramolecular ring-opening to yield another epoxide, as in the conversion of **370** into **371** by aqueous alcoholic alkali (217). Lake and Peat (218) described the formation of **373** from **372** when the latter was treated with sodium methoxide, and Peat later indicated the potential reversibility of the process (210a). Following renewed interest in this problem (219,220) Angyal and Gilham

370   371   372   373

374   375

(221a) isolated the inositol epoxides, **374** and **375**, and studied their interconversion ("epoxide migration") under alkaline conditions. Newth (220) had pointed out that in such equilibrations, provided the two conformations were similar, one isomer possessed an equatorial hydroxyl group and the other an axial hydroxyl group. He regarded the *anti*-axial group as more capable of attacking the epoxide ring, leading to a preponderance of the isomer with the equatorial hydroxyl group. Angyal correctly regarded the problem as a thermodynamic one, the isomer with the greater number of equatorial hydroxyl groups (or fewer axial groups) being the more stable. This is borne out in some inositol epoxides (221), where the symmetrical properties of the system make conformational effects most important, in accord with the behavior of inositol epoxides towards normal ring scission (see Section V). Thus, **375** is preferred over **374** by 9:1 (221a). Epoxides

CH₂——O   OH   →   CH₂——O

376   **377**

OH   OMe   ⇌   MeO   OH

378   **379**

derived from the conformationally stable 1,6-anhydrohexopyranoses have been studied by Czechoslovakian workers (222). The dianhydroaltrose **376** is strongly preferred in equilibrium with **377** (220,222b); this may be due to the presence of an axial hydroxyl group in **377** (220), but a strong destabilizing interaction between the epoxide oxygen and the C_6 methylene group in **377** has also been suggested (131).

The prediction of the position of epoxide equilibrium in the case of glycopyranoside epoxides (e.g., **372** and **373**) is more complex, due to the incursion of other factors (81c,82a,83,131). Only one example will be given. In the equilibrium between **378** and **379** in aqueous alkali, the isomer **378** is preferred (2:1) despite the axial C_4-hydroxyl group. The latter probably has a very small 1,3-diaxial interaction with the lone-pair of electrons on the pyranose ring oxygen. Other examples are discussed in Ref. 131 and earlier papers.

Payne has studied epoxide migration of 3-hydroxy-1,2-epoxides in 0.5 M sodium hydroxide solution (223) and his results are shown in Formulas **380–389**; the amounts of each isomer at equilibrium are shown in parentheses. Two structural effects were noted. The more highly substituted epoxide (or less-substituted alcohol) was preferred (cf. **370** → **371**) and, in disubstituted epoxides, a *trans* arrangement was favorable, while a *cis* arrangement was unfavorable. The first effect is seen in **380** and **381**; the two effects reinforce each other in two of the other examples, making **382** more stable than **383**, and **389** than **388**. In other cases the two effects compete. The epoxide ring in **384** is disubstituted, but has a *cis* arrangement of large groups, so comparable quantities of **384** and **385** are present at equilibrium. **387** is a trisubstituted epoxide, but **386** is *trans*-disubstituted, so again the equilibrium is evenly balanced. The reason for the influence of the degree of substitution of the epoxide is not clear. It is reminiscent of the rate-enhancing effect of methyl

$$H_2C-\overset{O}{\overset{|}{\underset{\underset{OH}{|}}{CH}}}CHCMe_2 \rightleftarrows HOCH_2CH-CMe_2$$

380 (8%)   **381** (92%)

382 (93%)   **383** (7%)

384 (58%)   **385** (42%)

386 (44%)   **387** (56%)

388 (5%)   **389** (95%)

substitution on epoxide formation from ethylene chlorohydrin, for which steric factors have been suggested (224), but electronic stabilization by alkyl substituents should not be ruled out (225). Payne has suggested (223) that the acid strength of the alcohol is important, and it would be interesting to re-examine the equilibria at lower hydroxyl ion concentrations. The two influences described by Payne are also evident in two recent examples in the anhydrohexitol series. The anhydroaltritol **390** is converted into the *trans*

390   **391**   **392**   **393**

epoxide **391** under alkaline conditions (87) and **392** is similarly converted into **393** (226).

It is interesting that when **380** was converted into its sodium salt in tetrahydrofuran by means of sodium hydride, it could be recovered with little or no rearrangement into **381** (223). This again emphasizes the importance of solvation in epoxide ring-opening (9–11).

When the oxirane **394** is treated with potassium *t*-butoxide in *t*-butanol, the oxetane **395** results (175). Since oxetanes undergo ring scission with alkoxide ion (227), the conversion of **394** into **395** should be reversible, but

this was not observed. The epoxide **397**, on treatment with aqueous alkali, gave the oxetane **396** in 12% yield (228) together with **398** (229) and **399**. When **396** itself was subjected to longer alkaline treatment **398** and **399** resulted (228), showing the reversibility of the epoxide migration. The susceptibility of **397** to attack at C_6 by hydroxyl ion prevented the study of the equilibrium between **396** and **397**. In the case of the furanosides **400** and **401**, however, the epoxide **400** is hindered towards nucleophilic attack (see Section VII.D) and equilibrium was established in favor of **401** ($> 20:1$) in M aqueous sodium hydroxide at 100° (155).

B. Acid-Catalyzed Carbon and Hydrogen Migrations, Including Transannular Reactions

In most of the examples described in the earlier sections of this review, openings of the oxirane ring are caused by or followed by attack of a nucleophile on the carbon atom from which the C—O bond has departed. Contrasted with these are a number of species in which epoxide-scission is accompanied by varying degrees of rearrangement, giving products which bear little or no superficial resemblance to the usual products of epoxide-opening. Some of these classes of rearrangements have been discussed in detail in recent reviews, but because of their inherent interest to the topic of this review, some mention of them is desirable.

1. Medium Ring Epoxides

The best-studied group of reactions of this class is the solvolysis of epoxides of medium-ring (C_8–C_{12}) cycloalkanes. The products obtained closely resemble the products obtained from solvolysis of the corresponding cycloalkyl tosylates, and very similar mechanistic explanations apply. These reactions have been studied principally in the laboratories of Cope, Prelog, and Traynham, and have been the subject of recent reviews (230,231) which should be consulted for excellent discussions and complete references. Most of the principles are demonstrated by the products obtained from formolysis of the *cis* and *trans* cyclooctene oxides **402** and **412**. The *cis* oxide (232–234) gives about 20% of normal product, the *trans*-1,2-diol **403**, and about 30% of *cis*-1,4-diol **404**. This product arises by a 1,3-hydride shift from C_4 to C_2, with attack of the nucleophile at C_4. A 1,5-hydride shift from C_6 to C_2 gives the same product, indicated by **405**; in the other medium rings the products corresponding to **404** and **405** would not be identical. The unsaturated alcohols **406** (11%) and **407** (4%) are also formed by hydride shifts, as are both 9-*oxa*-bicyclononanes **408** and **409** (0.1% total). A small yield (0.1%) of *endo-cis*-bicyclo[4.2.0]octan-7-ol (**410**) and a trace of cyclooctanone (**411**) are also formed. The *trans* oxide **412** (235,236) gives no 1,2 diol; *trans*-1,4-diol **413** forms to the extent of 33%, *trans*-1,3-diol **414**, 1%, and the products **415–417** resulting from carbon migration represent 55% of the total. Studies with specifically deuterated and ^{14}C-labeled cycloalkene oxides have led to quantitative evaluation of the relative amounts of 1,3 and 1,5 hydride shifts and other rearrangements (230,231).

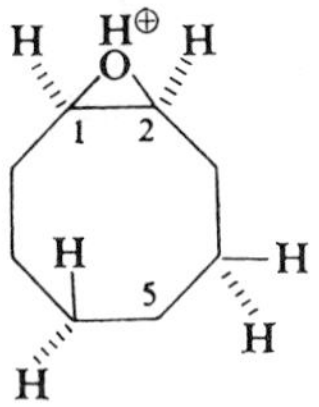

402 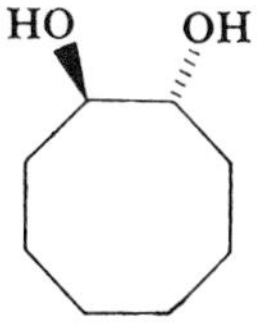   403 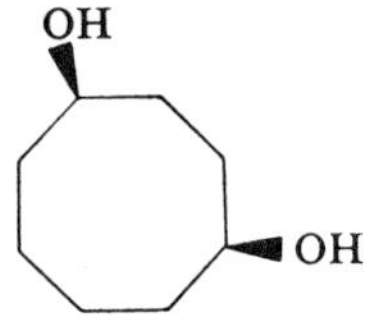   404 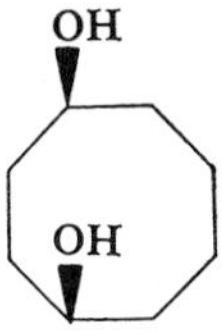   405

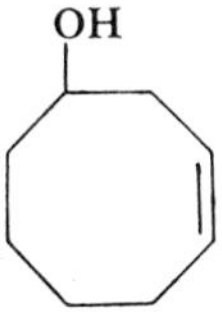

406 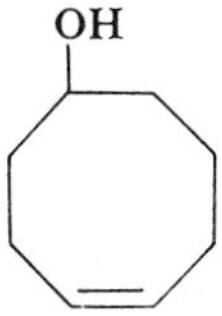   407 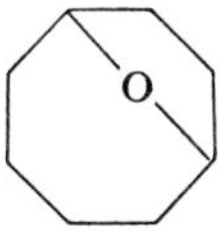   408 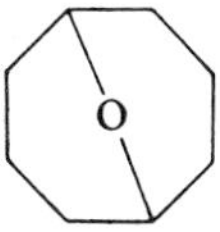   409

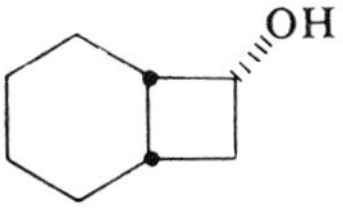

410 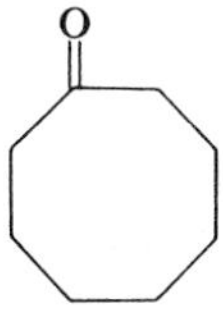   411 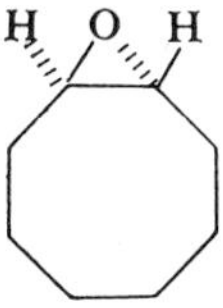   412 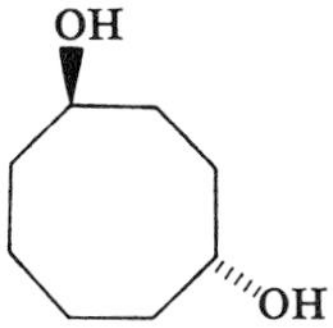   413

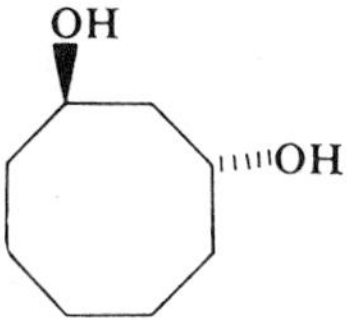

414 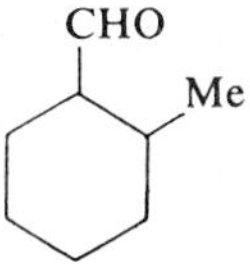   415 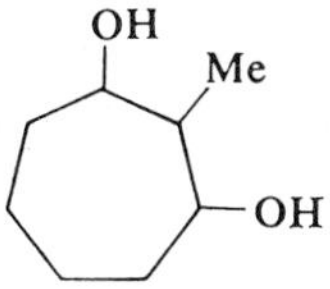   416 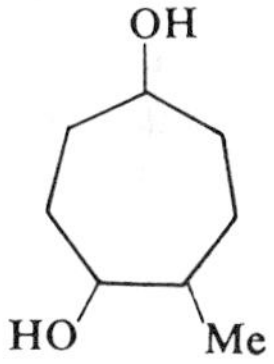   417

2. Ring Expansion of Vinylcyclohexanols

Cheer and Johnson have described a ring expansion of 1-vinylcyclo-alkanol epoxides and a corresponding ring expansion on halogenation of the corresponding alkenes (237–239). Under catalysis by BF_3–etherate or acidic alumina, the epoxyalcohol **418** is converted into the ring-expanded ketone **419**. A similar reaction occurs when alkenes like **420** are treated with aqueous bromine or *t*-butyl hypochlorite, and the authors postulate the intervention of an epihalonium ionic intermediate. Study of models shows that one of the ring carbon atoms (asterisk in structures **421–423**) in the hydrogen-bonded conformer is in a position favorable for migration to the positive center generated by scission of the epoxide. Although these compounds are conformationally mobile, rearrangement of the diastereoisomeric epoxides occurs in a largely stereospecific manner under the influence of acidic alumina (239); to explain this finding the authors postulate a conformationally immobile, surface-adsorbed transition state.

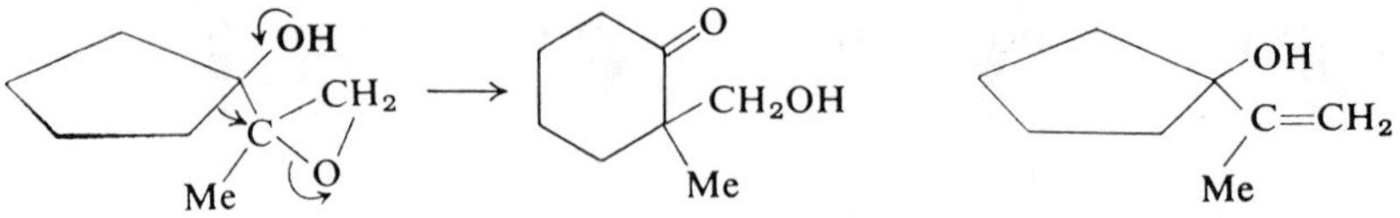

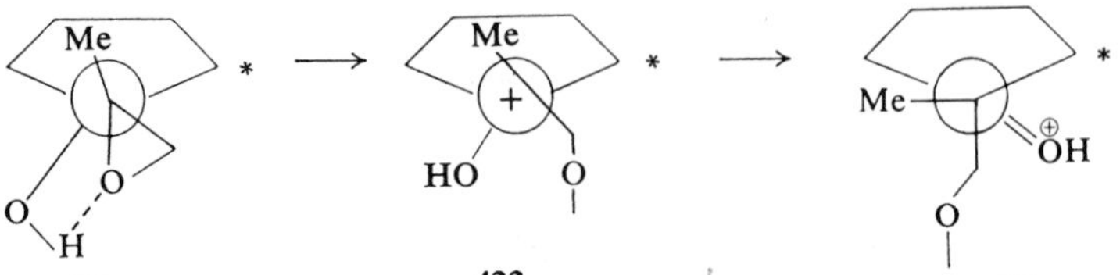

3. Terpenoid Epoxides and Related Systems

Many rearrangements of epoxides are observed in terpenoid compounds, a series whose chemistry abounds in rearrangements in general (240). One of the most remarkable reactions known is the cyclization of the acyclic triterpenoid epoxide, 2,3-oxidosqualene 424, to form lanosterol 425 and related sterols (241–244). Although such reactions are catalyzed by highly specific enzymes, they are properly considered under the heading of acid-catalyzed

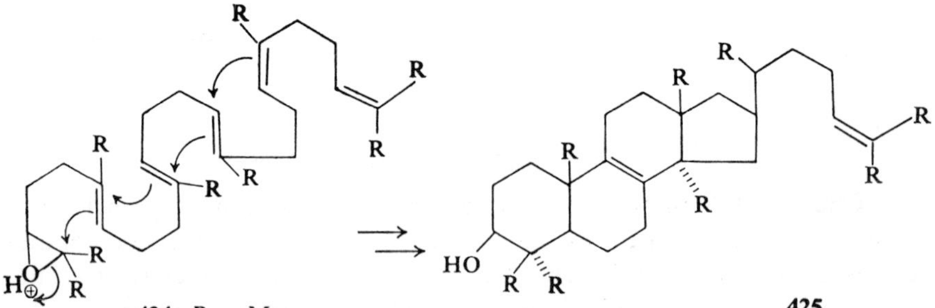

424; R = Me   425

rearrangements, since protonation of the oxirane ring is a necessary first step (241,242) in the cyclization. The concerted reaction indicated yields an intermediate species, not shown here, which loses a proton and undergoes ad-

ditional rearrangement, in the form of methyl migrations, to give lanosterol
(see Ref. 245 for additional details and references). Similar cyclizations of
olefinic epoxides are catalyzed by protic and Lewis acids (242b,253) (see
Section VIII.B.3 for additional details).

The caryophyllene oxides undergo numerous rearrangements during
acid-catalyzed epoxide scission. Since the skeleton contains a medium ring and
a small ring, both of which are subject to rearrangement even in the absence
of epoxides (246), the diversity of products is not surprising. When hydrolyzed
in dilute mineral acid, caryophyllene oxide (**426**) and isocaryophyllene oxide-*a*

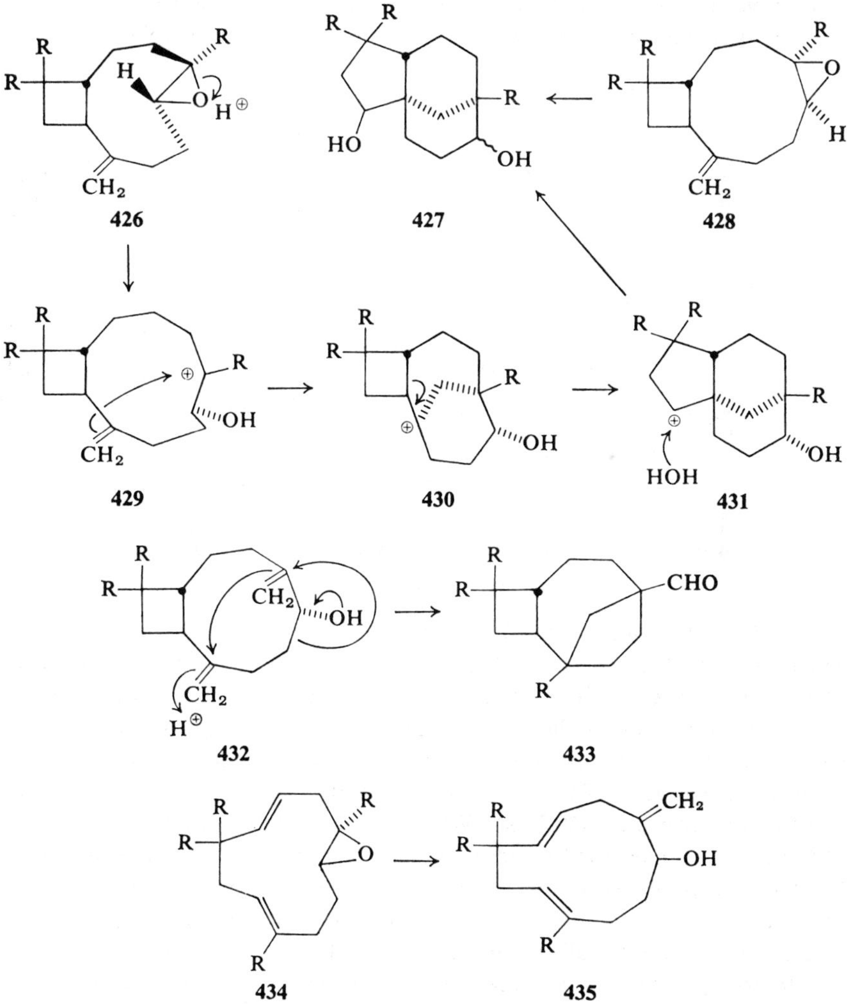

426   427   428

429   430   431

432   433

434   435

R = Me

(428) are converted (247,248) by transannular ring closure and a ring expansion into the epimeric diols 427, each oxide yielding only one of the epimers. The concerted or sequential bond-migrations indicated by structures 429–431 rationalize the transformation. In addition 426 may rearrange to the allylic alcohol 432 which is subject to further rearrangements, giving aldehydic products 433, etc. (248).

In related studies, Nigam and Levi (249,250) have shown that humulene monoxide 434 which is present in the essential oil of wild ginger is converted into the allylic alcohol 435 during chromatography on activated alumina. They have also proposed (251) that the biosynthesis of *cis*- and *trans*-dihydrocarvone 438 from limonene (436) may occur via the oxide 437, presumably

436   437   438

439   440   441   442

R = Me

by a hydride shift. Royals and Leffingwell (252) observed that α-pinene oxide 439 is converted into at least three products, 440–442, during solvolysis in acetic–sodium acetate. In each case skeletal rearrangement has occurred.

Goldsmith et al. (253a–c,253e) reported the cyclization of epoxyolefins to form bornane derivatives and related bridged compounds under the influence of Lewis acids. The reaction is also catalyzed by protic acids, as found by van Tamelen et al. (242b,253d). When a mixture of the campholene epoxide 443 and its 7-epimer was treated with SnCl₄ in benzene (253a), 444–446 were the principal products obtained. The yield of bornane products was greater than 50%. The formation of a bridged cation and intramolecular hydride migration accounts for all the products formed.

Rearrangements of a large number of steroid epoxides with boron trifluoride have been studied. A review of this field is beyond the scope of

443 → **444** + **445** + **446**

this chapter, but the work has been summarized by Wendler (245) and more recently by Hartshorn and Kirk (254,255).

C. Base-Catalyzed Rearrangements

The strong base lithium diethylamide causes rearrangement of aryl-substituted epoxides, in ether or benzene solution, into carbonyl compounds (256). The epoxide **447** is converted to **450** presumably by way of the anionic intermediates **448** and **449**. *trans*-Stilbene oxide (**451**), on the other hand, yields **454**, the carbanion **452** undergoing phenyl group migration to yield **453**.

447 → **448** → **449** → $R_2CH \cdot COR'$ **450**

451 → **452** → **453** → → $OCH \cdot CHR_2$ **454**

R = Ph
R' = p-MeC$_6$H$_4$

The behavior of alicyclic and aliphatic epoxides towards lithium diethylamide has been studied by Cope (230,257–260), Crandall (261–263), and Rickborn (264) and their co-workers. Three main types of reaction may be discerned: formation of allylic alcohols by β-elimination, skeletal rearrangement via a carbenoid intermediate produced by α-elimination, and formation of ketones.

α-Pinene oxide (**455**) rearranges to the allylic alcohol **456**, and **457** gives **458**, both in high yield (262a). Recent work has shown that the formation of allylic alcohols is highly selective, involving proton abstraction by the bulky lithium diethylamide from the least substituted carbon atom (264a), and proceeds by a *syn*-elimination (262b,264a,b); cyclohexene oxide, for example, is converted into **460** via the adduct **459** (262a,264b). Other examples, involving aliphatic epoxides (260,264a), epoxides having 1-methyl substituents

455   456   457   458

459   460

(262a,263,264a,265) and epoxides of normal and large-ring carbocycles (262a, 264b,266), have been reported.

Carbenoid intermediates were proposed by Cope (230,257–259) to account for rearrangement products arising from medium-ring epoxides. *cis*-Cyclodecene oxide (**461**) is converted into **462** (83%), **463** (9%), and **464** (8%) by lithium diethylamide in benzene, while the *trans*-isomer **465** yields **466** (36%) and **464** (64%). The formation of the bicyclic alcohols **462** and **466** is completely stereospecific. A mechanism involving concerted formation of a

461   462   +   463

464   465   466

467   468   469   470

$$461 \longrightarrow \text{[471]} \longrightarrow 462$$

471

carbenoid intermediate and insertion into a C—H bond was proposed (258) and confirmed by deuterium labeling at C_5 and C_6 or at C_1 and C_2 (230,259). Crandall (261) has shown that norbornene oxide (**467**) is converted into **470** in high yield, showing that the C—H bond approaches the carbenoid carbon from the side opposite the departing oxygen atom, as in **468** and **469**. In agreement with this we formulate the conversion of **461** into **462** as taking place via **471**; **463** will be formed in similar fashion via a transition state analogous to **471**. The stereochemistry of carbene insertion in alicyclic compounds has been discussed by Crandall (262a). The allylic alcohol **464** arises by β-elimination in both cases.

Ketones are sometimes produced from aliphatic or alicyclic epoxides by base treatment, especially in cases where β-elimination is prohibited and a *trans*-annular hydrogen is not available (262a). The ketones **473** and **474** are the major products from the epoxide **472**. A mechanism such as that described for aryl epoxides (e.g., **448 → 450**) may operate, but rearrangement of a carbenoid intermediate, e.g., **475 → 476**, is also a possible route (262a).

472   473   474

475   476

Base-catalyzed rearrangements of α-epoxyketones have been extensively studied. Two main types of reaction are found, depending on the nature of the proton abstracted by the base. In the first a methylene or methine group is adjacent to the carbonyl group; these reactions were studied by Treibs (267) and later by House and Gilmore (268), who formulated reaction mechanisms. When piperitone oxide (**477**) is treated with potassium *t*-butoxide in 1,2-dimethoxyethane, the major product (70%) is the lactone **480** (268).

Under these relatively nonpolar conditions a stereospecific Favorskii-type rearrangement is observed, involving intramolecular attack in the anion **478** to form the cyclopropanone **479** (269). On the other hand, aqueous potassium

hydroxide converts **477** into the hydroxyketone **483** (88%). It is envisaged that under these polar conditions the enolate ion **481** is formed, which promotes allylic opening of the epoxide to yield **482** (see Section III); elimination of hydroxyl ion from **482** gives **483**. With methanolic sodium methoxide, a system of intermediate polarity, both the Favorskii route and the allylic route are in operation. References 268 and 269 should be consulted for full details. The configurations of the two pulegone oxides have recently been established by Favorskii rearrangement (270,271).

The second type of α-epoxyketone rearrangement involves formation of a diketone, which then undergoes a benzilic acid rearrangement. The epoxide **484** (R = Ph), on treatment with sodium hydroxide in aqueous ethanol, yields the carboxylate **488** by way of **486** as intermediate (272). The migration of the benzyl group in **486** → **487** → **488** was shown by C^{14} labeling (273), and **485** seems a plausible intermediate for the formation of **486**. Recently the behavior of **484** (R = cycloalkyl) towards base has been studied (274). For cyclohexyl, cyclopentyl, and cyclobutyl derivatives the reaction involves a Favorskii rearrangement. In the case of **484** (R = cyclopropyl), however, the reaction follows the same course as when R = Ph. The cyclopropyl carbanion, required for the Favorskii route, cannot be stabilized by resonance with the carbonyl group, and is less easily formed (274).

D. Miscellaneous Rearrangements

1. Migration of α-Substituents in Epoxide-Carbonyl Rearrangements

McDonald et al. (275–278) have investigated the migration of chlorine in the rearrangement of α-chloroepoxides to α-chloroketones. On treating *trans*-α-chlorostilbene **489** with peroxybenzoic acid, they obtained only the α-chloroketone **490** instead of the expected chlorostilbene oxide **491**. They

assumed that **491** was formed as an intermediate which immediately underwent an epoxide-carbonyl rearrangement (see Ref. 3 for earlier references); **490** could then arise by ring-opening to give a carbonyl group, accompanied by either hydrogen or chlorine migration. By studying the same reaction in a series of unsymmetrically substituted α-chlorostilbenes they found that only the chlorine migrated, and they present a reasonable argument for believing that the migration is intramolecular (275). The migration is considered to proceed by way of an ion-pair formed *before* opening of the oxirane ring, as shown by the sequence **492–494**. The same authors (276) have proposed that, under basic conditions and in nonpolar solvent, the epoxides presumed to be intermediates in the Darzens condensation undergo an identical rearrangement.

In none of the reactions just described were epoxides proved to be involved, although their role as intermediates seems reasonable. In further studies (277) the alicyclic α-chloroepoxide **495** was converted into *trans*-2-chloro-4-methylcyclohexanone **496** by neat thermal rearrangement, none of the 5-methyl isomer **497** being formed. Here again only chlorine migration has occurred. The starting material is actually a mixture of **495a** and **495b**,

495a,b   496   497

a; Me, Cl *trans*
b; Me, Cl *cis*

but both epimers appear to give rise to **496**; this finding is rationalized by the formation of the same α-ketocarbonium ion–chloride ion pair from both epimers. When solid $ZnCl_2$ was added to the thermolysis flask, both **496a** and **496b** were formed. It is proposed that the catalysis by $ZnCl_2$ can occur by coordination at chlorine rather than at oxygen, as in **498**, leading to the

498   499

formation of the enol hypochlorite **499**. The latter could give either an enolate anion–chloronium ion pair or an α-ketocarbonium ion–chloride ion pair, the latter being favored on theoretical grounds.

The rearrangements of 2-chlorobicyclo[2.2.1]hept-2-ene *exo*-oxide **500** have also been studied (278). The principal products of neat thermolysis were the chloroketones **501** and **502** and several minor products, all of which are rationalized by some combination of ion-pair formation, Wagner-Meerwein rearrangements, and proton elimination. Solvolysis of **500** in glacial acetic acid led to **503–505** along with several of the same products as were obtained on thermolysis. Undoubtedly the mechanism is similar to that involved in the solvolysis of the nonchlorinated epoxide (279,280).

A related rearrangement is the conversion of an α-acetoxyepoxide into an α-acetoxyketone (245). When heated above its melting point 3β-acetoxy-2,3α-epoxycholestane (**506**) yields the 2β-acetoxyketone **508** (281), probably by an intramolecular process (282), e.g., **507** (283). Under acidic conditions, however, **506** is converted into **512** without **508** being an intermediate (283).

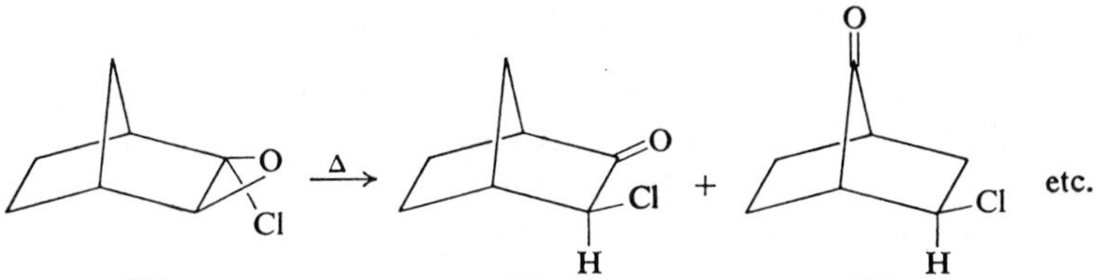

500 **501** **502** etc.

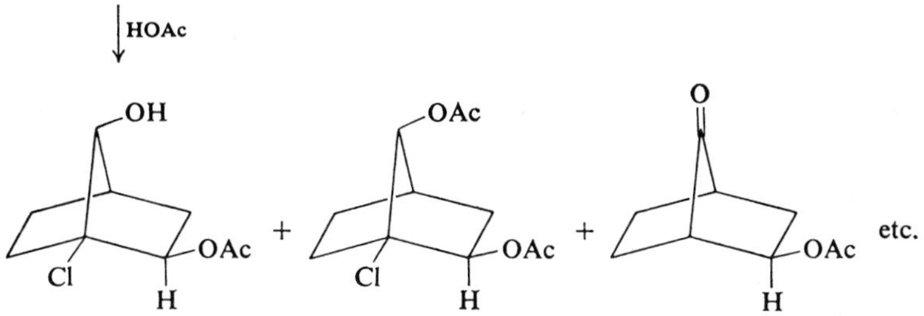

503 **504** **505** etc.

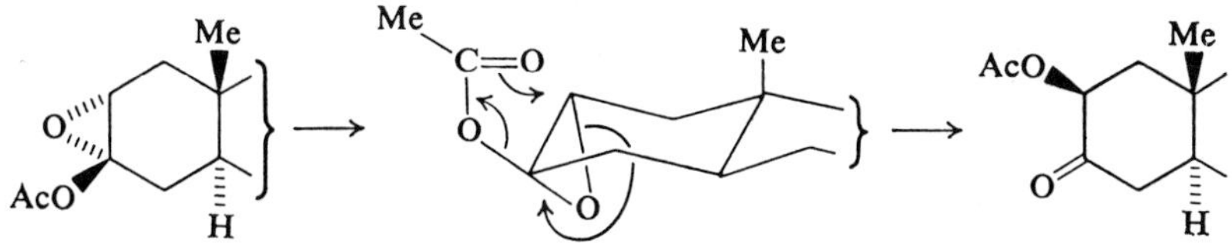

506 **507** **508**

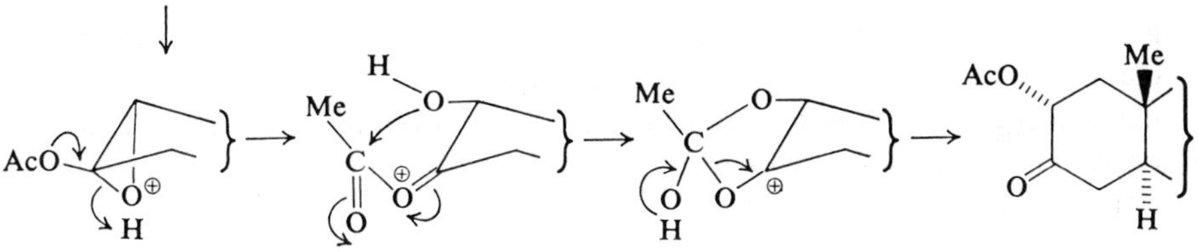

509 **510** **511** **512**

Several mechanisms have been proposed for the acid-catalyzed rearrangement, of which the most plausible involves **509–511** as intermediates (245,283).

In a related study by Churi and Griffin (284) α,β-epoxyvinylphosphonates **513** were rearranged thermally or by BF_3 to the corresponding α-formylalkyl-

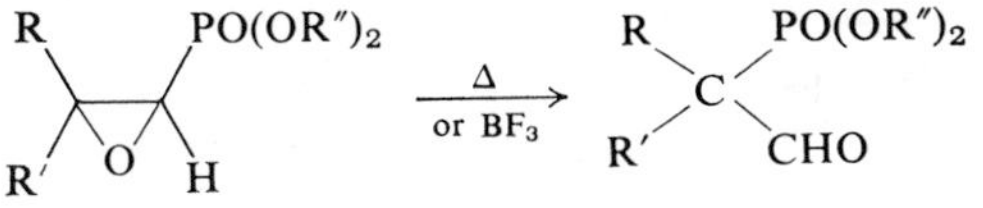

513 **514**

phosphonates **514**. The authors comment that the high migratory capacity of the phosphono group was previously unrecognized.

2. *Reactions of Epoxides with Ylids*

Epoxides react with phosphorus ylids to give derivatives of cyclopropane (285–287). Thus, the ylid **515** reacts with cyclohexene oxide (**517**) at 200° to give ethyl 7-norcaranecarboxylate (**518**) and triphenylphosphine oxide (**519**) (285). The intermediates **520–522** are probably involved. When the ylid is

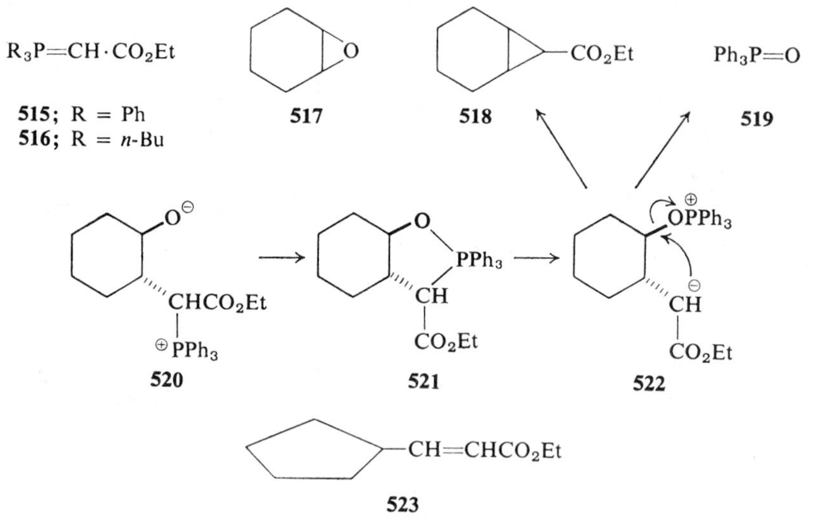
$R_3P{=}CH\cdot CO_2Et$
515; R = Ph
516; R = *n*-Bu
O
517
CO_2Et
518
$Ph_3P{=}O$
519
$O^{\ominus}$
$CHCO_2Et$
$^{\oplus}PPh_3$
520
O
PPh_3
CH
CO_2Et
521
$\overset{\oplus}{O}PPh_3$
$\overset{\ominus}{C}H$
CO_2Et
522
$CH{=}CHCO_2Et$
523

generated from a phosphonium salt by means of butyllithium, rearranged olefins result (288), e.g., **523** from **516** and **517**. This is probably due to prior rearrangement of **517** into cyclopentanecarboxyaldehyde by the lithium halide present (289). Mechanistic studies have also been carried out by McEwen, VanderWerf, and their co-workers (290).

3. *Small-Ring Epoxides*

In connection with studies on small-ring compounds in general, Crandall and Paulson (291,292) have investigated derivatives of oxaspiropentane **524** and 1,5-dioxaspiroheptane **525**. By gas-phase thermolysis **524** is converted into the ring-expanded cyclobutanone **528** and the acyclic ketone **529**. The thermolytic process appears to proceed via the diradicals **526** and **527**, the first step being a homolytic cleavage of one of the bonds of the cyclopropane ring. By solvolysis in aqueous dioxane or acetic acid the same products are formed; in this case the first step is considered to be cleavage of a C—O bond of the protonated oxirane ring to give **530** which rearranges to give **528** directly, or **529** by way of **531**.

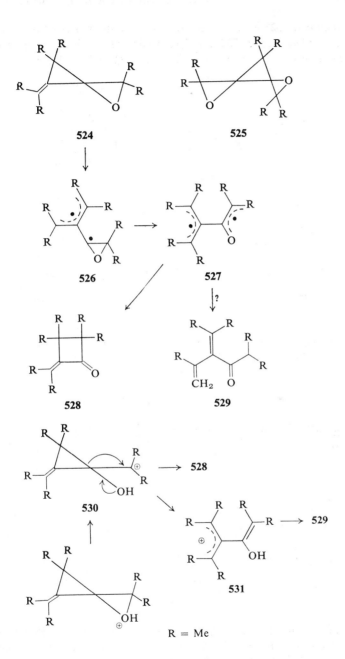

524

525

526

527

528

529

530

531

R = Me

The diepoxide **525** is converted by gas-phase thermolysis (292) into the two ring-expanded ketones **534** and **536**. Again the reaction is considered to proceed by homolytic cleavage into diradicals **532**, **533**, and **535**. Solvolysis

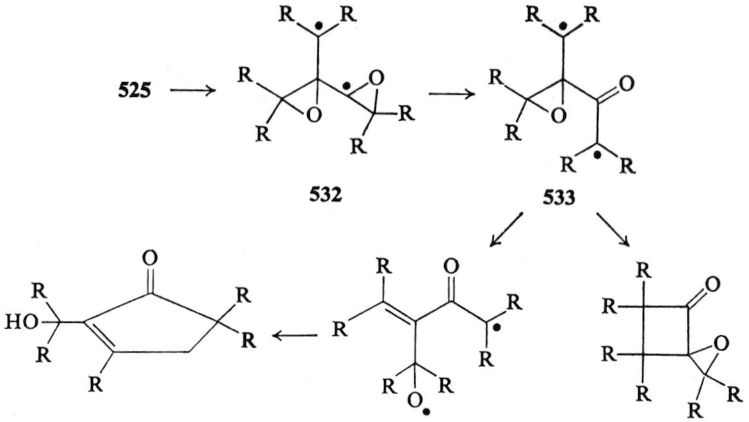

532 **533** **536** **535** **534**

R = Me

in acetic acid, containing H_2SO_4, gave both of the possible hexamethylcyclopentanediones **539** and **540**. The cyclobutanone **534** is probably formed from

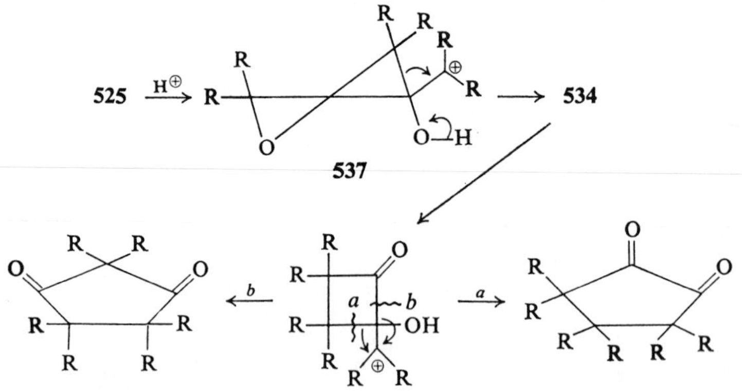

537 **539** **538** **540**

R = Me

537 as an intermediate in the process; rearrangement of the ion **538** then leads to the products. An entirely different product was obtained by a base-promoted isomerization of **525**; a single substance **543** was obtained on treatment with lithium diethylamide. The authors propose that the initial product is

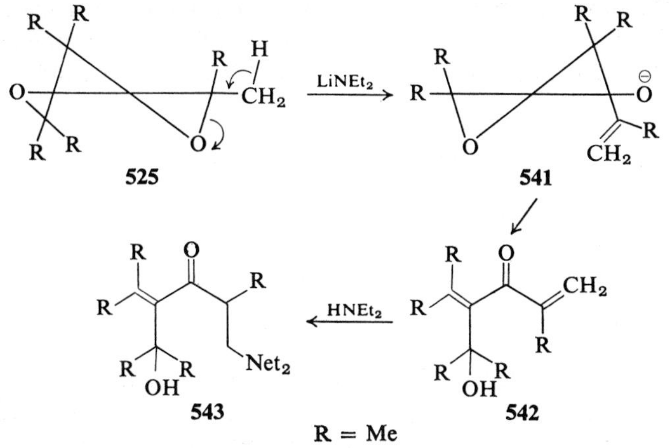

525   541

543   542

R = Me

542, formed by a rearrangement of **541** which is produced by a β-elimination; **543** is then formed by a Michael-type addition of diethylamine.

IX. ADDITION OF HYPOHALOUS ACIDS TO ALKENES, AN ANALOGOUS SYSTEM

Closely related in mechanism to epoxide scission is a group of reactions exemplified by the electrophilic addition of the elements of hypohalous acids to alkenes. In these reactions, as in polar addition of halogen, there is a diaxial transition state (293), as indicated by the sequence **544–546**. Although

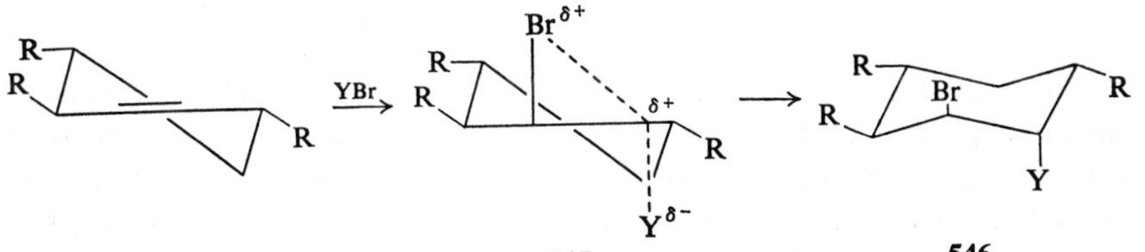

544   545   546

a bromine derivative is used as the example, the discussion applies equally well to the chloro and iodo analogs. Y, the other part of the reagent, may be an OH, alkoxy, or acyloxy group, and it is assumed that the reagent is polarized or completely dissociated, so that the reaction may be considered to occur stepwise; electrophilic addition of a halonium ion forms the epihalonium species, which is then attacked by $Y^{\ominus}$. Also included are cases in which Y is part of the substrate molecule, and neighboring group participation is involved. The formation of iodolactone **548** from **547** is an example of such reactions (294), and other examples are known in the steroids and sapogenins

547   548

(295). It has been pointed out (230,296) that the cyclic halonium ion is formally analogous to protonated epoxides, and the analogy is extended by the finding that, in the addition of bromine to medium-sized cycloolefins, *trans*-annular reactions give products analogous to those obtained from the solvolysis of medium-ring epoxides (see Section VIII.B.1). In some recently reported cases of addition of HOBr to medium-ring terpenoid triolefins (297), the products were 1,4-bromohydrins, and one or more new ring-junctions were produced. The products are easily rationalized by transition states such as those postulated in VIII.B.1 for solvolysis of related terpenoid epoxides.

Although there is general agreement that the reaction occurs in two steps, there has been some uncertainty about the nature of both the attacking species and the intermediate formed in the first step. Equally logical arguments can be presented which agree with the postulation that the adduct derived from **549** is a halocarbonium ion **550** and that the reagent is a hydrated halonium ion, rather than hypohalous acid (34b,298). Gould (299)

$$Me_2C{=}CH_2 + H_2\overset{\oplus}{O}Cl \longrightarrow Me_2\overset{\oplus}{C}{-}CH_2Cl$$

549   550

pointed out that the products of bromination of maleic and fumaric acids can be explained by the epibromonium intermediate, and are inconsistent with a carbonium ion; however, he questioned the validity of extending this concept to include chloronium ions. More recently Bannard, Moir, and one of the present authors have independently observed that the distribution of products from addition of both HOCl and HOBr is analogous to the distribution from epoxide scission (see below), and this lends much weight to the idea that epihalonium ions always intervene in these reactions. Poutsma (300) has shown that for polar chlorination of unsymmetrically substituted alkenes also the product ratio is explicable by the epichloronium and not by the carbonium ion intermediate.

Two well-known procedures for converting alkenes into glycols (74,301,

302) involve similar intermediates and merit discussion at this point. In both the Prévost and Woodward reactions the reagent is a silver halogenocarboxylate complex. In the former case the complex $Ag(C_6H_5CO_2)_2X$ is prepared first and the alkene is then added; in the latter the complex is generated *in situ* in the presence of the alkene. Because of small but important differences in the conditions, the Prévost reaction leads to derivatives of *trans*-glycols (**551** → **553**) whereas the Woodward reagent leads to a *cis*-glycol monoacetate

| | 551 | | 552 | | 553 |

(**554** → **558**). In both cases the displacement of the halogen involves participation of the neighboring *trans*-acyloxy functional group. That one gives a *cis* and the other a *trans* glycol is due to the profound influence of water on the mechanism of participation by the neighboring group (73,303); the proposed mechanism is supported by ^{18}O-tracer studies (74) of both the "wet" and "dry" reactions. In both sequences the reagent that adds to the

| | 554 | | 555 | | 556 |
| | 558 | | 557 | | |

double bond is formulated as acetyl (or benzoyl) hypohalite. In the formation of **553** the first intermediate is presumed to be the 1,2-epichloronium ion, which is then attacked by the benzoate ion to give **552**; there is no doubt about the nature of **552**, since Gaoni (301d) has isolated and identified this substance. The acetylated iodohydrin **556** has not been isolated, but it is most reasonably formulated (74,302) as the diaxial product (293). It is assumed that the epiiodonium species **555** is formed by addition to the less hindered side of the double bond. In some cases (302b) the Woodward procedure leads to a mixture of *cis* and *trans* glycols.

Recent studies in several laboratories have clarified considerably both the stereochemistry and mechanism of addition of hypohalous acids (49,69, 86b,86c,100–102). Studies on the 3-methoxycycloalkenes (69,100–102) have been particularly valuable in this connection, because the carefully executed experiments permit direct comparisons with work by the same authors on scission of the corresponding epoxides. Addition of HOBr and HOCl to the cycloalkene **559** leads to the sets of bromohydrins **562a–565a** and chloro-hydrins **562b–565b**. The pair **562–563** is derived from the *syn*-adduct **560**, and the other pair from the *anti*-adduct **561**. The same set of halohydrins is ob-

a; X = Br
b; X = Cl

tained from scission of the corresponding epoxides with HBr or HCl; how-ever, in this case **562** and **563** arise from the *anti*-epoxide and **564** and **565** from the *syn*-epoxide. Table 3 shows the distribution of products obtained as well as the results obtained with 3-methoxycyclopentene. In the addition of HOBr the vicinal methoxyl group (presumably in the quasi-equatorial orientation shown) seems to offer no steric hindrance in the addition step; indeed, there is an apparent preference for *syn*-addition. No such preference is seen in the addition of HOCl. It has been suggested (102) that this apparent difference may be due to two factors: (*1*) the formation of **560** and **561** from **559** is rapidly reversible, and (*2*) the second step occurs at markedly different rates for the various adducts. Once formed, the epihalonium species is sus-ceptible to nucleophilic scission by the H_2O or OH^-. As in the opening of

TABLE 3
Formation of Halohydrins from 3-Methoxycycloalkenes[a]

Cycloalkene	Hypohalite	Products from			
		syn-Adduct		anti-Adduct	
		562	**563**	**564**	**565**
Cyclohexene	HOBr	66.7	0.5	25.6	6.7
	HOCl	51.3	0	38.8	10.1
Cyclopentene[b]	HOCl	37.2	7.0	55.8	0

[a] Adapted from the data of Bannard, Moir, and their associates (see text). The values are percent of total halohydrin product recovered.
[b] The products are not **562–565**, but cyclopentanoid analogs of the same configurations.

the oxides (see Table 1, Section VII.A) there is preference for attack at the carbon atom more distant from the electronegative substituent, and the same factors appear to control the regioselectivity in both situations. As with the oxides, the syn-adduct gives only minute amounts of products arising from vicinal attack, whereas substantial amounts of such products are formed from the anti-adduct. The addition of HOCl to methoxycyclopentene leads to a different product distribution, due first to a small preference for formation of the anti-adduct and second to the different steric and conformational effects in the cyclopentanoid system. The product distribution derived from these adducts, however, closely resembles the scission of the corresponding oxides. From all of these results it is therefore reasonable to conclude that the transition states for oxide scission and for hypohalite addition are nearly identical. Langstaff et al. (69) refer to unpublished work that shows the transition states for oxide scission are more like the halohydrins than the oxides. The probable similarities of the transition states, coupled with the product ratios, strengthen the assumption that the cyclic epihalonium ion is involved in the addition, rather than an α-carbonium ion.

In studies of the synthesis of cyclopentanoid aminocyclitols (86b,86c) several examples involved the addition of HOBr to more hindered precursors than those described above, and the stepwise nature of the addition became more clearly defined. In all cases the addition of bromine showed a high degree of selectivity for the less hindered side of the plane of the cycloalkene, thus effectively diminishing the number of products from four to two. In this study all the substrates had acetoxyl or acetamido groups flanking the double bond, so the addition of OH$^-$ in the second step was indirect, involving the

566   →   567   ⇒   568

569   →   570   ⇒   571

participation of neighboring *trans* acetyl groups. For example, **568** and **571** were obtained in 97% and 70% yield, respectively, from **566** and **569** by way of the bromonium ions **567** and **570**. Similar selectivity has been observed in cyclohexanoid cases (49,80a,80d,304, and Section VI). Because of the high degree of selectivity of the second step, the mode of addition of the halogen controls the over-all stereochemistry. In all of the examples studied, the regioselectivity was exactly the same as that observed for opening of the corresponding epoxide, which again supports the postulation that an epihalonium species is involved, and that the transition state for hypohalite addition resembles that for epoxide scission.

X. EPOXIDES IN BIOLOGICAL SYSTEMS

Most studies of epoxides are carried out in a strictly organic chemical context. There is, however, an increasing number of cases in which epoxides are biologically active, either as intermediates in enzyme-catalyzed reactions, as end products of metabolism, as enzyme inhibitors, or as toxic (e.g., carcinogenic) agents. In addition, epoxides have been used, in reactions similar to those discussed in the main part of this chapter, as intermediates for the preparation of deuterated and tritiated compounds of known chirality, for use in studies of enzymic stereospecificity. Some examples of each of these cases are given in the succeeding paragraphs.

The cyclization of 2,3-oxidosqualene (**424**) to lanosterol (**425**) has been

572   573   574

noted in Section VIII.B.3. Retey et al. (305) have proposed that the epoxy alcohol **573** is an intermediate in the formation of propionaldehyde **574** from propan-1,2-diol **572** catalyzed by the cobamide-containing enzyme diol dehydrase.

A recently discovered broad-spectrum antibiotic, phosphonomycin (**575**) is a small molecular compound, (−)-(1R,2S)-*cis*-1,2-epoxypropylphosphonate (306). Phosphonomycin is produced by various strains of *Streptomyces*, and

575

inhibits cell-wall synthesis. The inhibition is irreversible and is due to the alkylation of pyruvate-uridine diphospho-*N*-acetylglucosamine transferase, the enzyme which catalyzes the first step in a biosynthetic pathway leading to the synthesis of cell-wall components. Other enzymes, carbohydrases, and a peptidase are also inhibited by epoxides and it is obvious that in each case one is dealing with a substrate analog which is a highly selective alkylating agent. The carbohydrases in question are lysozyme and β-glucosidase. Thomas et al. (307) have found that 2′,3′-epoxypropyl-β-D-glycosides of 2-acetamido-2-deoxy-D-glucose (**576**) and its oligomers are specific and irreversible in-

576   **577**

hibitors of hen's egg-white lysozyme; 1 mole of inhibitor is covalently bound to 1 mole of enzyme when 100% inactivation is achieved. Legler (308a,308b) has reported that conduritol B epoxide (**577**) reacts with β-glucosidases from various sources, causing specific, irreversible inactivation, Similarly, a substrate analog, 1,2-epoxy-3-phenoxypropane, inhibits chymotrypsin (308c) by alkylating a methionine residue in the active site.

The cytotoxic nature of diepoxy compounds has been known for many years (309) and they have been studied as potential carcinostatic agents. Some diepoxides have also been found to be carcinogenic (309b,310). Structural factors that may be related to the biological activity have been evaluated, and the possibility that the bifunctional reagents may be able to crosslink the strands of double helical DNA are discussed (310b,310c).

Some interesting examples of the use of epoxides as intermediates in the chemical synthesis of compounds of biological interest have been reported recently. Fried et al. have been concerned with the synthesis of prostaglandins and compounds related to them. Recently they have reported (311) excellent yields in the alkynylation of alicyclic epoxides with alkynyldiethylalanes **578**

578   **579**

to give products **579**. The elegant studies of Cornforth, Popják, and their associates on the stereochemistry of carbon–carbon bond formation in terpenoid biosynthesis have required optically pure intermediates with stereo-chemically defined substitution of deuterium or tritium (see, e.g., Ref. 312). This was accomplished by a sequence of reactions in which the regioselective

580   **581**

reduction of an epoxide by a group-selective reagent, e.g., **580** → **581**, was the crucial step (313). It is noteworthy in this connection that $LiBH_4$ is a group-selective reductant, attacking epoxides and esters but not affecting amides. In a different problem concerning the stereochemistry of the malate synthase reaction, monodeuterio-monotritio-acetic acid of known chirality was required (314), and again the regioselective reduction of an epoxide was the key reaction.

XI. ADDENDUM

The literature review for this chapter was completed in the spring of 1969. In this Addendum we describe important work published since that time.

Brief reference has been made in Section V to a complete analysis of the effect of conformation on epoxide ring opening undertaken by Rickborn and his colleagues (64b). 1-Methylcyclohexene oxide (**258**) reacts with lithium aluminum hydride in ether to yield the tertiary alcohol **260** in 99.4% yield, together with small amounts of **259** and its epimer [arising by a process similar to **244** → **246** (55)]. Since chairlike transition states will be favored in all these reactions, Rickborn has calculated a difference in free energy of activation between secondary and tertiary attack of about 3.1 kcal/mole. Similarly the reaction with the octalin oxide **582** has been examined. Because of the *trans*-ring fusion, only one chairlike transition state is possible, that requiring attack at the tertiary carbon to yield **583**. The other possible reaction, in which the

582

LiAlH₄ LiAlH₄

OH OH Me

Me

H

H Me

583 **584** **585**

product is **585**, involves a boatlike transition state related to **584**. These two pathways correspond to Routes A and B, respectively, in Figure 1 (Section V). Since 21.2% of **585** is formed, corresponding to a difference in activation free energy of 0.80 kcal/mole, Rickborn calculates that the energy difference due purely to conformational effects is $0.8 + 3.1 = 3.9$ kcal/mole. These calculations have enabled Rickborn to estimate the relative importance of Routes B and C (Figure 2, Section V) in reactions of flexible systems.

The role of electrophilic assistance to ring opening, by solvation of the epoxide oxygen atom, has already been noted (Section II, Refs. 9–11; and Section VI, Ref. 96). Of particular interest is the work of Cruickshank and Fishman (315) who have studied the reaction of ω-bromo-1,2-epoxyalkanes

$$\text{PhOCH}_2\text{CH(OH)(CH}_2)_2\text{Br} \longleftarrow \text{CH}_2\text{—CH(CH}_2)_2\text{Br} \longrightarrow \text{CH}_2\text{—CH(CH}_2)_2\text{OPh}$$

586 **587** **588**

with nucleophiles. In this competitive situation they found that sodium phenoxide in ethanol gave mainly the ether **586** from **587** by epoxide opening, whereas in an aprotic solvent, e.g., dimethylformamide or dimethyl sulfoxide, displacement of bromide ion occurred to give the epoxyether **588**.

Additional studies on rearrangements of allene oxides and dioxides have been reported (316–318) (see Section VIII.D.3).

Fukushima and his colleagues have published two interesting papers (319) on neighboring group participation in epoxide opening (see Sections VI and

VII.G.2). In the androstane derivatives **589** the regioselectivity of ring-opening by dilute sulfuric acid is as expected. When the acetamide and epoxide groups are *cis* to each other, attack occurs at C_5 because of both the polar influence of the acetamido group and the tendency for reaction at a tertiary carbon; when they are *trans*, neighboring group participation causes attack at C_4. In a similar series of reactions on analogous acetyl ureido compounds the only

589   **590**

591   **592**

discordant note is the formation of the $4\beta,5\beta$–diol **592** from **590**. The authors suggest that a *trans* intermediate **591** undergoes *trans* ring-opening at C-4 to restore the 4,5-*cis* relationship.

It is remarkable that the tumor inhibitor crotepoxide (**593**) shows no evidence of neighboring group participation when treated with acidic reagents (320). With hydrochloric acid, for example, the chlorohydrins **594** and **595** are produced, despite the presence of suitably situated acetoxy and benzoyloxy groups.

The conversion of the carbanion **596** into **597** by intramolecular attack on the epoxide group has been described by McMurry (321).

593   **594**   **595**

The solvolytic behavior of *exo*-2,3-epoxybicyclo[3.3.1]nonane (**598**) makes an interesting comparison with that of *cis*-cyclooctene oxide (**402**; Section VIII.B.1), because it can be regarded as a conformationally restricted form of the latter (322,323).

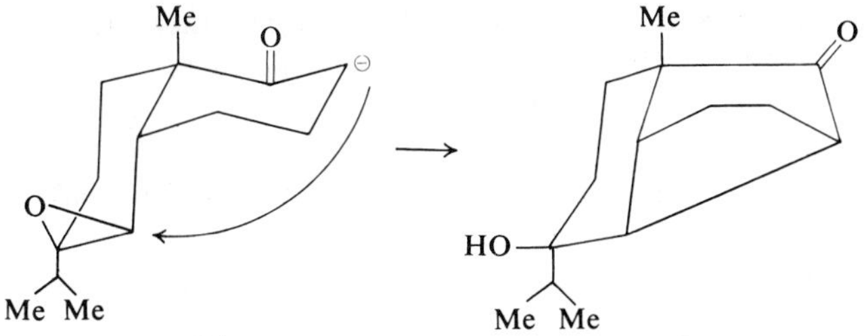

596   **597** 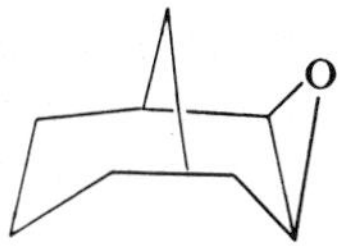   **598**

There have been some examples quoted of nucleophilic participation by epoxide oxygen in solvolytic reactions. Epichlorohydrin (**599**), for example, undergoes acetolysis at about one hundred times the rate of allyl chloride, and one of the products is 3-acetoxyoxetane (**601**) (324). Ion-pairs derived from **600** have been proposed in this and other reactions (324,325).

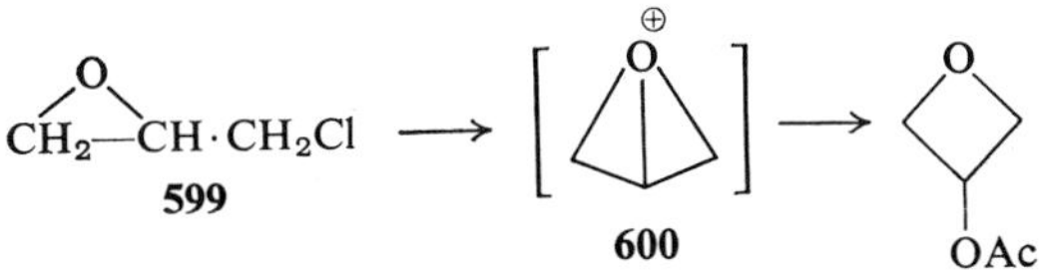

599   **600**   **601**

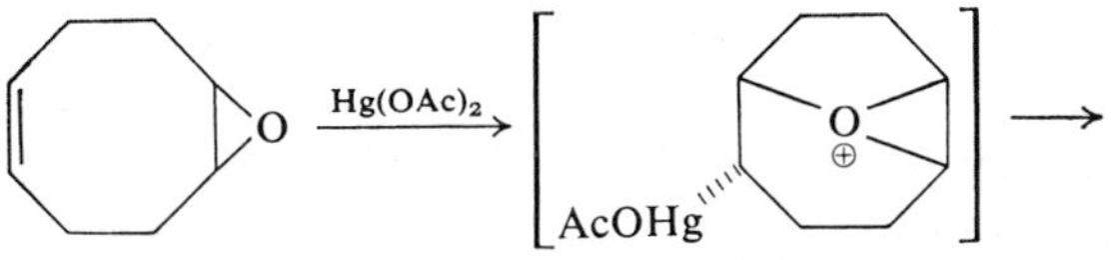

602   **603**

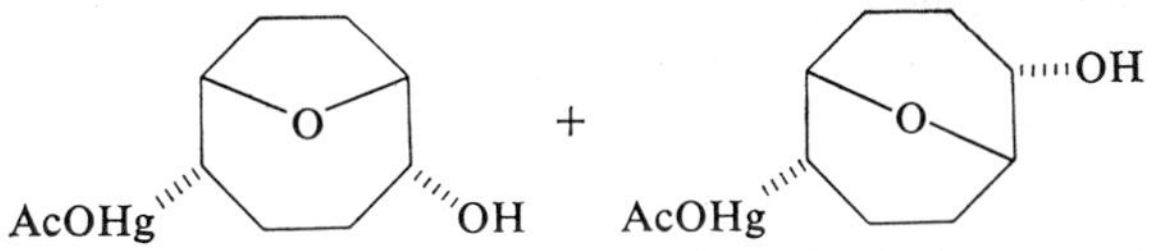

604   +   **605**

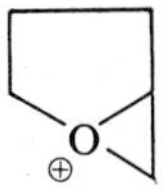

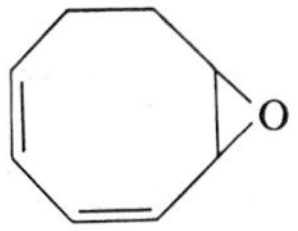

606   **607**   **608**   **609**

Related observations are those by Closson and his colleagues on oxymercuration of certain 1,5-diene monoxides (326). The monoepoxide **602** of *cis,cis*-1,5-cyclooctadiene, for example, when treated with mercuric acetate in aqueous tetrahydrofuran, followed by demercuration with sodium borohydride, yields the alcohols **608** and **609** in the ratio 1:3. The most likely mercuration products are **604** and **605**, formed by hydrolysis of the oxonium intermediate **603**. A similar intermediate **606** may be involved in solvolytic reactions of tetrahydrofurfuryl tosylate (327,328). Transannular participation by an epoxide group is also observed in acid treatment of the epoxydiene **607** (329).

There have been two recent papers on hydrolysis and ring-opening of allylic epoxides (329,330).

McDonald and Steppel (331) have reinvestigated the epoxide-carbonyl rearrangement (Section VIII.D.1) with an optically pure form of 2-chloro-norbornene-2,3-*exo*-oxide. The data confirm the proposed intramolecular migration of Cl$^-$, oppose a hydride shift, and strongly support the earlier suggestion of the intermediacy of a tight carbonium ion–chloride ion pair.

610   **611**

Goldsmith and Joines (332) have extended earlier work (see Section VIII.B.3) on cyclization of epoxyolefins to include diastereoisomers (and homologs) of **610**, which are constitutional isomers of the compounds **443** studied earlier (253). It was anticipated that cyclization would occur involving a participating π-bond and a cation. However, no bridged bicyclic products were detected, the only product formed from **610** being the ketone **611**. The authors conclude that in these cyclization reactions steric factors play a larger role than had been previously assumed.

More examples have been reported in which epoxides are formed in the course of metabolic interconversions (Section X). Arene oxides, e.g., benzene oxide, toluene-3,4-oxide, and naphthalene-1,2-oxide, have been implicated as intermediates in the metabolism of aromatic compounds (333–334). It is suggested (334c) that the cationoid transition states required for the "NIH shift" (335) could arise from the acid-catalyzed opening of the appropriate epoxide. These oxides also seem to be involved in the biosynthesis of premer-capturic acids by an enzyme-catalyzed attack of the —SH group of gluta-thione on the oxirane ring (334c). The oxides can be converted to phenols, apparently nonenzymatically (334c). An enzyme present in liver, epoxide hydrolase (336), hydrolyzes aliphatic epoxides to 1,2-glycols, and the same or a closely related enzyme hydrolyzes benzene oxide, naphthalene oxide, styrene oxide, and indene oxide to the corresponding 1,2-*trans* glycols (334c). The occurrence of epoxides as obligatory intermediates in the enzymic oxidation of alkenes to *vic*-diols has been demonstrated (336). These epoxides are then hydrolyzed by the epoxide hydrolase; if extraneous epoxides are present during the oxidation, the hydrolase is inhibited and the epoxy intermediates accumulate.

Acknowledgments

We thank Professors D. Lavie, W. D. Closson, and J. E. McMurry for sending manuscripts (204, 326, and 321, respectively) in advance of publication. We thank Mrs. A. B. Sakami, Mrs. Patricia Mensch, and Miss Margarethe H. Hagemann for valuable assistance in the preparation of the manuscript of this chapter.

H.Z.S. thanks the United States Public Health Service for support of his research through grant AM-07719.

Note Added in Proof

Williams (337) has reviewed oxirane derivatives of aldoses.

An important review of Akhrem et al. (30b), which came to the authors' attention after the manuscript of this chapter had been set into type, is a comprehensive and useful treatment of epoxide-opening reactions that proceed with retention of configuration, a subject with which we have dealt briefly in Sections IV and VII.H. In this connection, one of us has repeated the work of Boeseken (338), and we substantiate his findings that hydrolysis of indene oxide gives both *cis* and *trans*-1,2-indanediol. The products were analyzed by vapor phase chromatography. At reflux temperature, in dioxane-water solutions, the *trans/cis* ratios varied with pH: 1 N H$_2$SO$_4$ gave a ratio of 1.6:1; 0.1 N, 0.82:1; no added acid, 1.15:1; and 0.1 N NaOH, no *cis* glycol detected. When indene oxide was added to pure water and left at room temperature for two weeks, the ratio was 0.68:1.

An epoxide hydrase has been purified and its properties studied (339). The juvenile hormones of the Cecropia silk moth have been shown to be terpenoid epoxides (340).

References

1. S. Winstein and R. B. Henderson, in R. C. Elderfield, Ed., *Heterocyclic Compounds* Vol. 1, Wiley, New York, 1950, p. 1ff.
2. E. L. Eliel, in M. S. Newman, Ed., *Steric Effects in Organic Chemistry*, Wiley, New York, 1956, p. 61ff.
3. R. E. Parker and N. S. Isaacs, *Chem. Rev.*, **59**, 737 (1959).
4. (a) A. Rosovsky, in A. Weissberger, Ed., *The Chemistry of Heterocyclic Compounds*, Vol. 19, *Heterocyclic Compounds with Three and Four-Membered Rings*, Part I, Wiley-Interscience, New York, 1964, p. 1. (b) M. S. Malinovskii, *Epoxides and Their Derivatives*, Israel Program for Scientific Translations, Jerusalem, 1965.

5. A. Hassner, *J. Org. Chem.*, **33**, 2684 (1968).
6. H. G. Viehe, *Angew. Chem. Int. Ed. Eng.*, **6**, 767 (1967).
7. H. E. Zimmerman, L. Singer, and B. S. Thyagarajan, *J. Amer. Chem. Soc.*, **81**, 108 (1959).
8. E. L. Eliel, *Stereochemistry of Carbon Compounds*, McGraw-Hill, New York, 1962, p. 436.
9. W. H. Horne and R. L. Shriner, *J. Amer. Chem. Soc.*, **54**, 2925 (1932).
10. R. E. Parker and B. W. Rockett, *J. Chem. Soc.*, **1965**, 2569.
11. N. N. Lebedev and Y. I. Baranov, *Kinet. Katal.*, **7**, 619 (1966); through *Chem. Abstr.*, **65**, 16812a (1966).
12. F. A. Long and M. A. Paul, *Chem. Rev.*, **57**, 935 (1957).
13. C. F. Wilcox, Jr., and J. S. McIntyre, *J. Org. Chem.*, **30**, 777 (1965).
14. J. K. Addy and R. E. Parker, *J. Chem. Soc.*, **1963**, 915.
15. C. A. Stewart and C. A. VanderWerf, *J. Amer. Chem. Soc.*, **76**, 1259 (1954).
16. C. G. Swain and C. B. Scott, *J. Amer. Chem. Soc.*, **75**, 141 (1953).
17. P. D. Bartlett and S. D. Ross, *J. Amer. Chem. Soc.*, **70**, 926 (1948).
18. (a) W. E. McEwen, W. E. Conrad, and C. A. VanderWerf, *J. Amer. Chem. Soc.*, **74**, 1168 (1952). (b) C. A. VanderWerf, R. Y. Heisler, and W. E. McEwen, *J. Amer. Chem. Soc.*, **76**, 1231 (1954).
19. A. Feldstein and C. A. VanderWerf, *J. Amer. Chem. Soc.*, **76**, 1626 (1954).
20. (a) R. R. Russell and C. A. VanderWerf, *J. Amer. Chem. Soc.*, **69**, 11 (1947). (b) G. Van Zyl and E. E. van Tamelen, *J. Amer. Chem. Soc.*, **72**, 1357 (1950).
21. (a) C. H. De Puy, F. W. Breitbeil, and K. L. Eilers, *J. Org. Chem.*, **29**, 2810 (1964). (b) P. M. G. Bavin, D. P. Hansell, and R. G. W. Spickett, *J. Chem. Soc.*, **1964**, 4535.
22. (a) E. E. van Tamelen, G. Van Zyl, and G. D. Zuidema, *J. Amer. Chem. Soc.*, **72**, 488 (1950). (b) J. K. Crandall, D. B. Banks, R. A. Colyer, R. J. Watkins, and J. P. Arrington, *J. Org. Chem.*, **33**, 423 (1968).
23. C. L. Stevens and T. H. Coffield, *J. Amer. Chem. Soc.*, **80**, 1919 (1958).
24. C. L. Stevens and C. H. Chang, *J. Org. Chem.*, **27**, 4392 (1962).
25. R. Fuchs and C. A. VanderWerf, *J. Amer. Chem. Soc.*, **76**, 1631 (1954).
26. (a) J. K. Addy and R. E. Parker, *J. Chem. Soc.*, **1965**, 644. (b) E. A. S. Cavell, R. E. Parker, and A. W. Scaplehorn, *J. Chem. Soc.*, **1965**, 4780.
27. Y. Okamoto and H. C. Brown, *J. Amer. Chem. Soc.*, **80**, 4976 (1958).
28. Y. Liwschitz, Y. Rabinsohn, and D. Perera, *J. Chem. Soc.*, **1962**, 1116.
29. (a) H. B. Henbest, M. Smith, and A. Thomas, *J. Chem. Soc.*, **1958**, 3293. (b) B. Rickborn and D. K. Murphy, *J. Org. Chem.*, **34**, 3209 (1969).
30. (a) H. Meerwein, *Justus Liebigs Ann. Chem.*, **542**, 123 (1939); (b) A. A. Akhrem, A. M. Moiseenkov, and V. N. Dobrynin, *Russian Chemical Reviews*, **37**, 448 (1968). (Pagination is that of the English translation.)
31. G. Berti, F. Bottari, P. L. Ferrarini, and B. Macchia, *J. Org. Chem.*, **30**, 4091 (1965).
32. P. A. Levene and A. Rothen, *J. Biol. Chem.*, **127**, 237 (1939).
33. H. Hart and H. S. Eleuterio, *J. Amer. Chem. Soc.*, **76**, 1379 (1954).
34. (a) M. J. S. Dewar and R. C. Fahey, *J. Amer. Chem. Soc.*, **85**, 3645 (1963). (b) R. C. Fahey, in E. L. Eliel and N. L. Allinger, Eds., *Topics in Stereochemistry*, Vol. 3, Wiley-Interscience, New York, 1968, p. 237.
35. J. H. Brewster, *J. Amer. Chem. Soc.*, **78**, 4061 (1956).
36. D. Y. Curtin, A. Bradley, and Y. G. Hendrickson, *J. Amer. Chem. Soc.*, **78**, 4064 (1956).
37. H. H. Wasserman and N. E. Aubrey, *J. Amer. Chem. Soc.*, **78**, 1726 (1956).
38. C. C. Tung and A. J. Speziale, *J. Org. Chem.*, **28**, 2009 (1963).

39. S. O. Chan and E. J. Wells, *Can. J. Chem.*, **45**, 2123 (1967).
40. M. Korach, D. R. Nielsen, and W. H. Rideout, *J. Amer. Chem. Soc.*, **82**, 4328 (1960).
41. H. Z. Sable and T. Posternak, *Helv. Chim. Acta*, **45**, 370 (1962).
42. R. Criegee, *Justus Liebigs Ann. Chem.*, **481**, 263 (1930).
43. P. A. Mayor and G. D. Meakins, *J. Chem. Soc.*, **1960**, 2792.
44. A. Fürst and P. A. Plattner, Abstracts of Papers of the 12th International Congress on Pure and Applied Chemistry, New York, 1951, p. 409.
45. D. H. R. Barton, *J. Chem. Soc.*, **1953**, 1027.
46. G. H. Alt and D. H. R. Barton, *J. Chem. Soc.*, **1954**, 4284.
47. A. Fürst and P. A. Plattner, *Helv. Chim. Acta*, **32**, 275 (1949).
48. S. J. Angyal, *Chem. Ind.* (London), **1954**, 1230.
49. F. G. Bordwell, R. R. Frame, and J. G. Strong, *J. Org. Chem.*, **33**, 3385 (1968).
50. E. L. Eliel, N. L. Allinger, S. J. Angyal, and G. A. Morrison, *Conformational Analysis*, Wiley, New York, 1965, p. 482.
51. R. C. Cookson, *Chem. Ind.* (London), **1954**, 223, 1512.
52. Ref. 50, p. 101.
53. J. Sicher, F. Šipoš, and M. Tichý, *Collect. Czech. Chem. Commun.*, **26**, 847 (1961).
54. N. A. Le Bel and R. F. Czaja, *J. Org. Chem.*, **26**, 4768 (1961).
55. B. Rickborn and J. Quartucci, *J. Org. Chem.*, **29**, 3185 (1964).
56. D. Y. Curtin and R. J. Harder, *J. Amer. Chem. Soc.*, **82**, 2357 (1960).
57. B. Rickborn and S.-Y. Lwo, *J. Org. Chem.*, **30**, 2212 (1965).
58. T. Posternak, *Les Cyclitols*, Hermann, Paris, 1962.
59. (a) Ref. 50, p. 358. (b) S. J. Angyal, V. Bender, and J. H. Curtin, *J. Chem. Soc., C*, **1966**, 798.
60. M. Nakajima and N. Kurihara, *Chem. Ber.*, **94**, 515 (1961).
61. M. Nakajima, N. Kurihara, and A. Hasegawa, *Chem. Ber.*, **95**, 141 (1962).
62. B. Rickborn and W. E. Lamke, II, *J. Org. Chem.*, **32**, 537 (1967).
63. R. A. B. Bannard, A. A. Casselman, E. J. Langstaff, and R. Y. Moir, *Can. J. Chem.*, **46**, 35 (1968).
64. (a) N. A. Le Bel and G. G. Ecke, *J. Org. Chem.*, **30**, 4316 (1965). (b) D. K. Murphy, R. L. Alumbaugh, and B. Rickborn, *J. Amer. Chem. Soc.*, **91**, 2649 (1969).
65. J. C. Leffingwell and E. E. Royals, *Tetrahedron Lett.*, **1965**, 3829.
66. E. E. Royals and J. C. Leffingwell, *J. Org. Chem.*, **31**, 1937 (1966).
67. Ref. 50, p. 28.
68. Ref. 50, p. 359.
69. E. J. Langstaff, R. Y. Moir, R. A. B. Bannard, and A. A. Casselman, *Can. J. Chem.*, **46**, 3649 (1968).
70. G. Berti, F. Bottari, B. Macchia, and F. Macchia, *Tetrahedron*, **22**, 189 (1966).
71. (a) R. S. Tipson, *J. Biol. Chem.*, **130**, 55 (1939). (b) R. K. Ness and H. G. Fletcher, Jr., *J. Amer. Chem. Soc.*, **80**, 2007 (1958).
72. H. S. Isbell, *Ann. Rev. Biochem.*, **9**, 65 (1940).
73. (a) S. Winstein and R. E. Buckles, *J. Amer. Chem. Soc.*, **64**, 2780, 2787 (1942). (b) S. Winstein, H. V. Hess, and R. E. Buckles, *J. Amer. Chem. Soc.*, **64**, 2796 (1942).
74. K. B. Wiberg and K. A. Saegebarth, *J. Amer. Chem. Soc.*, **79**, 6256 (1957).
75. (a) E. J. Hedgley and H. G. Fletcher, Jr., *J. Amer. Chem. Soc.*, **84**, 3726 (1962); **85**, 1615 (1963); **86**, 1576 (1964). (b) C. Pedersen, *Tetrahedron Lett.*, **1967**, 511.
76. (a) S. J. Angyal, P. A. J. Gorin, and M. E. Pitman, *Proc. Chem. Soc.*, **1962**, 337 (1962). (b) S. J. Angyal, P. A. J. Gorin, and M. E. Pitman, *J. Chem. Soc.*, **1965**, 1807.

77. A. Hassner and C. Heathcock, *J. Org. Chem.*, **29**, 3640 (1964).

78. B. R. Baker and T. L. Hullar, *J. Org. Chem.*, **30**, 4038, 4045, 4049 (1965).

79. (a) W. S. Johnson and E. N. Schubert, *J. Amer. Chem. Soc.*, **72**, 2187 (1950). (b) G. E. McCasland and D. A. Smith, *J. Amer. Chem. Soc.*, **72**, 2190 (1950).

80. (a) S. Winstein, L. Goodman, and R. Boschan, *J. Amer. Chem. Soc.*, **72**, 2311 (1950). (b) S. Winstein and L. Goodman, *J. Amer. Chem. Soc.*, **76**, 4368, 4373 (1954). (c) L. Goodman and S. Winstein, *J. Amer. Chem. Soc.*, **79**, 4788 (1957). (d) L. Goodman, S. Winstein, and R. Boschan, *J. Amer. Chem. Soc.*, **80**, 4312 (1958).

81. (a) J. G. Buchanan, *J. Chem. Soc.*, **1958**, 995. (b) J. G. Buchanan, *J. Chem. Soc.*, **1958**, 2511. (c), J. G. Buchanan and R. Fletcher, *J. Chem. Soc.*, **1965**, 6316.

82. (a) J. G. Buchanan and J. C. P. Schwarz, *J. Chem. Soc.*, **1962**, 4770. (b) J. G. Buchanan and R. M. Saunders, *J. Chem. Soc.*, **1964**, 1791. (c) J. G. Buchanan and R. M. Saunders, *J. Chem. Soc.*, **1964**, 1796.

83. J. G. Buchanan and R. Fletcher, *J. Chem. Soc.*, C, **1966**, 1926.

84. J. A. Franks, Jr., B. Tolbert, R. Steyn, and H. Z. Sable, *J. Org. Chem.*, **30**, 1440 (1965).

85. B. Tolbert, R. Steyn, J. A. Franks, Jr., and H. Z. Sable, *Carbohyd. Res.*, **5**, 62 (1967).

86. (a) A. Hasegawa and H. Z. Sable, *J. Org. Chem.*, **31**, 4149 (1966). (b) A. Hasegawa and H. Z. Sable, *J. Org. Chem.*, **31**, 4154 (1966). (c) A. Hasegawa and H. Z. Sable, *J. Org. Chem.*, **31**, 4161 (1966).

87. J. G. Buchanan and A. R. Edgar, *Carbohyd. Res.*, **10**, 295 (1969).

88. J. M. Coxon, M. P. Hartshorn, and D. N. Kirk, *Tetrahedron*, **20**, 2547 (1964).

89. J. G. Buchanan and A. R. Edgar, *Chem. Commun.*, **1967**, 29.

90. (a) S. Julia and J.-P. Lavaux, *Bull. Soc. Chim. Fr.*, **1963**, 1238. (b) S. Julia and B. Fürer, *C. R. Acad. Sci., Paris*, **257**, 710 (1963).

91. J. F. King and A. D. Allbutt, *Tetrahedron Lett.*, **1967**, 49; *Can. J. Chem.*, **48**, 1754 (1970).

92. P. H. Gross, K. Brendel, and H. K. Zimmerman, *Justus Liebigs Ann. Chem.*, **680**, 159 (1964).

93. M. Nakajima, N. Kurihara, and T. Ogino, *Chem. Ber.*, **96**, 619 (1963).

94. H. I. Schlesinger, H. C. Brown, H. R. Hoekstra, and L. R. Rapp, *J. Amer. Chem. Soc.*, **75**, 199 (1953).

95. S. J. Angyal, Personal Communication.

96. S. J. Angyal and T. S. Stewart, *Aust. J. Chem.*, **20**, 2117 (1967).

97. (a) A. J. Parker, *Quart. Rev.* (London), **16**, 163 (1962); (b) L. Goodman, *Advan. Carbohyd. Chem.*, **22**, 109 (1967).

98. R. A. B. Bannard and L. R. Hawkins, *Can. J. Chem.*, **36**, 1241 (1958).

99. R. U. Lemieux, R. K. Kullnig, and R. Y. Moir, *J. Amer. Chem. Soc.*, **80**, 2237 (1958).

100. R. A. B. Bannard and L. R. Hawkins, *Can. J. Chem.*, **39**, 1530 (1961).

101. R. A. B. Bannard, A. A. Casselman, and L. R. Hawkins, *Can. J. Chem.*, **43**, 2398 (1965).

102. E. J. Langstaff, E. Hamanaka, G. A. Neville, and R. Y. Moir, *Can. J. Chem.*, **45**, 1907 (1967).

103. A. Streitwieser, Jr., *Chem. Rev.*, **56**, 571 (1956).

104. F. H. Westheimer and J. G. Kirkwood, *J. Chem. Phys.*, **6**, 513 (1938).

105. H. D. Holtz and L. M. Stock, *J. Amer. Chem. Soc.*, **86**, 5188 (1964).

106. H. D. Holtz and L. M. Stock, *J. Amer. Chem. Soc.*, **87**, 2404 (1965), and references cited therein.

107. R. L. Golden and L. M. Stock, *J. Amer. Chem. Soc.*, **88**, 5928 (1966).

108. P. E. Peterson, R. J. Bopp, D. M. Chevli, E. L. Curran, D. E. Dillard, and R. J. Kamat, *J. Amer. Chem. Soc.*, **89**, 5902 (1967).
109. M. J. S. Dewar and A. P. Marchand, *J. Amer. Chem. Soc.*, **88**, 3318 (1966).
110. (a) Ref. 50, p. 376. (b) R. O. Hutchins, L. D. Kopp, and E. L. Eliel, *J. Amer. Chem. Soc.*, **90**, 7174 (1968). (c) E. L. Eliel and C. A. Giza, *J. Org. Chem.*, **33**, 3754 (1968).
111. Ref. 50, p. 200ff.
112. H. R. Buys, Ph.D. Thesis, University of Leiden, 1968.
113. R. Steyn, Ph.D. Thesis, Case Western Reserve University, Cleveland, 1968.
114. (a) R. Steyn and H. Z. Sable, *Tetrahedron*, **25**, 3579 (1969). (b) R. Steyn and H. Z. Sable, *Tetrahedron*, in press, 1971.
115. J. J. McCullough, H. B. Henbest, R. J. Bishop, G. M. Glower, and L. E. Sutton, *J. Chem. Soc.*, **1965**, 5496.
116. H. Z. Sable, T. Adamson, B. Tolbert, and T. Posternak, *Helv. Chim. Acta*, **46**, 1157 (1963).
117. H. Z. Sable and K. A. Powell, Unpublished Experiments (1969).
118. M. Nakajima, I. Tomida, N. Kurihara, and S. Takei, *Chem. Ber.*, **92**, 173 (1959).
119. J. A. Mills, addendum to F. H. Newth and R. F. Homer, *J. Chem. Soc.*, **1953**, 989.
120. G. J. Robertson and C. F. Griffith, *J. Chem. Soc.*, **1935**, 1193.
121. N. K. Richtmyer and C. S. Hudson, *J. Amer. Chem. Soc.*, **63**, 1727 (1941).
122. F. H. Newth, *Quart. Rev.* (London), **13**, 30 (1959).
123. R. D. Guthrie and D. Murphy, *J. Chem. Soc.*, **1963**, 5288.
124. S. P. James, F. Smith, M. Stacey, and L. F. Wiggins, *J. Chem. Soc.*, **1946**, 625.
125. M. Černý and J. Pacák, *Collect. Czech. Chem. Commun.*, **27**, 94 (1962).
126. W. G. Overend and G. Vaughan, *Chem. Ind.* (London), **1955**, 995.
127. G. Huber and O. Schier, *Helv. Chim. Acta*, **43**, 129 (1960).
128. N. R. Williams, *Chem. Commun.*, **1967**, 1012.
129. J. Honeyman, *J. Chem. Soc.*, **1946**, 990.
130. (a) R. E. Schaub and M. J. Weiss, *J. Amer. Chem. Soc.*, **80**, 4683 (1958). (b) See also F. Sweet and R. K. Brown, *Can. J. Chem.*, **46**, 707 (1968).
131. J. G. Buchanan, R. Fletcher, K. Parry, and W. A. Thomas, *J. Chem. Soc., B*, **1969**, 377.
132. A. J. Dick and J. K. N. Jones, *Can. J. Chem.*, **45**, 2879 (1967).
133. P. A. S. Smith, J. H. Hall, and R. O. Kan, *J. Amer. Chem. Soc.*, **84**, 485 (1962).
134. (a) P. W. Kent and P. F. V. Ward, *J. Chem. Soc.*, **1953**, 416. (b) N. A. Hughes, R. Robson and S. A. Saeed, *Chem. Commun.*, **1968**, 1381.
135. F. H. Newth and R. F. Homer, *J. Chem. Soc.*, **1953**, 989.
136. F. H. Newth, W. G. Overend, and L. F. Wiggins, *J. Chem. Soc.*, **1947**, 10.
137. G. Charalambous and E. Percival, *J. Chem. Soc.*, **1954**, 2443.
138. S. Peat and L. F. Wiggins, *J. Chem. Soc.*, **1938**, 1088, 1810.
139. British Patent 762,540, cited in *Chem. Abstr.*, **52**, 11912g (1958).
140. D. Gagnaire and P. Vottero, *Bull. Soc. Chim. Fr.*, **1963**, 2779.
141. J. M. Anderson and E. Percival, *J. Chem. Soc.*, **1956**, 819.
142. B. R. Baker, R. E. Schaub, and J. H. Williams, *J. Amer. Chem. Soc.*, **77**, 7 (1955).
143. B. R. Baker and R. E. Schaub, *J. Amer. Chem. Soc.*, **77**, 5900 (1955).
144. E. E. Percival and R. Zobrist, *J. Chem. Soc.*, **1953**, 564.
145. I. L. Doerr, J. F. Codington, and J. J. Fox, *J. Org. Chem.*, **30**, 467 (1965).
146. J. Davoll, B. Lythgoe, and S. Trippett, *J. Chem. Soc.*, **1951**, 2230.
147. E. J. Reist, A. Benitez, L. Goodman, B. R. Baker, and W. W. Lee, *J. Org. Chem.*, **27**, 3274 (1962).
148. G. Casini and L. Goodman, *J. Amer. Chem. Soc.*, **86**, 1427 (1964).
149. J. E. Christensen and L. Goodman, *J. Org. Chem.*, **28**, 2995 (1963).

150. E. J. Reist and S. L. Holton, *Carbohyd. Res.*, **2**, 181 (1966).
151. C. D. Anderson, L. Goodman, and B. R. Baker, *J. Amer. Chem. Soc.*, **80**, 5247 (1958).
152. J. G. Buchanan and D. R. Clark, Unpublished Results.
153. C. D. Anderson, L. Goodman, and B. R. Baker, *J. Amer. Chem. Soc.*, **81**, 898 (1959).
154. E. J. Reist and S. L. Holton, *Carbohyd. Res.*, **9**, 71 (1969).
155. P. W. Austin, J. G. Buchanan, and E. M. Oakes, *Chem. Commun.*, **1965**, 374, 472.
156. L. Goodman, *J. Amer. Chem. Soc.*, **86**, 4167 (1964).
157. P. W. Austin, J. G. Buchanan, and R. M. Saunders, *J. Chem. Soc.*, C, **1967**, 372.
158. L. A. Paquette, A. A. Youssef, and M. L. Wise, *J. Amer. Chem. Soc.*, **89**, 5246 (1967).
159. L. W. Trevoy and W. G. Brown, *J. Amer. Chem. Soc.*, **71**, 1675 (1949).
160. R. Fuchs and C. A. VanderWerf, *J. Amer. Chem. Soc.*, **74**, 5917 (1952).
161. H. C. Brown and N. M. Yoon, *J. Amer. Chem. Soc.*, **88**, 1464 (1966).
162. E. L. Eliel and D. W. Delmonte, *J. Amer. Chem. Soc.*, **80**, 1744 (1958).
163. E. L. Eliel and M. N. Rerick, *J. Amer. Chem. Soc.*, **82**, 1362 (1960).
164. M. N. Rerick and E. L. Eliel, *J. Amer. Chem. Soc.*, **84**, 2356 (1962).
165. E. C. Ashby and J. Prather, *J. Amer. Chem. Soc.*, **88**, 729 (1966).
166. P. T. Lansbury, D. J. Scharf, and V. A. Pattison, *J. Org. Chem.*, **32**, 1748 (1967).
167. B. Cooke, E. C. Ashby, and J. Lott, *J. Org. Chem.*, **33**, 1132 (1968).
168. H. C. Brown and B. C. SubbaRao, *J. Amer. Chem. Soc.*, **82**, 681 (1960).
169. D. J. Pasto, C. C. Cumbo, and J. Hickman, *J. Amer. Chem. Soc.*, **88**, 2201 (1966).
170. H. C. Brown and N. M. Yoon, *J. Amer. Chem. Soc.*, **90**, 2686 (1968).
171. H. C. Brown and N. M. Yoon, *Chem. Commun.*, **1968**, 1549.
172. F. A. Hochstein and W. G. Brown, *J. Amer. Chem. Soc.*, **70**, 3484 (1948).
173. H. B. Henbest and R. A. L. Wilson, *J. Chem. Soc.*, **1957**, 1958.
174. H. B. Henbest and B. Nicholls, *J. Chem. Soc.*, **1957**, 4608.
175. H. B. Henbest and B. Nicholls, *J. Chem. Soc.*, **1959**, 221.
176. A. C. Cope, J. K. Heeren, and V. Seeman, *J. Org. Chem.*, **28**, 516 (1963).
177. Ref. 50, p. 297.
178. H. M. Fales and W. C. Wildman, *J. Org. Chem.*, **26**, 181 (1961).
179. A. Fürst and R. Scotoni, Jr., *Helv. Chim. Acta*, **36**, 1332 (1953).
180. A. R. Davies and G. H. R. Summers, *J. Chem. Soc.*, C, **1967**, 1227.
181. C. W. Davey, E. L. McGinnis, J. M. McKeown, G. D. Meakins, M. W. Pemberton, and R. N. Young, *J. Chem. Soc.*, C, **1968**, 2674.
182. L. F. Fieser and S. Rajagopalan, *J. Amer. Chem. Soc.*, **73**, 118 (1951).
183. C. Djerassi, E. Batres, M. Velasco, and G. Rosenkranz, *J. Amer. Chem. Soc.*, **74**, 1712 (1952).
184. (a) A. S. Hallsworth and H. B. Henbest, *J. Chem. Soc.*, **1957**, 4604. (b) H. C. Brown, S. Ikegami, and J. H. Kawakami, *J. Org. Chem.*, **35**, 3243 (1970).
185. R. Hirschmann, C. S. Snoddy, Jr., C. F. Hiskey, and N. L. Wendler, *J. Amer. Chem. Soc.*, **76**, 4013 (1954).
186. C. W. Shoppee, M. E. H. Howden, R. W. Killick, and G. H. R. Summers, *J. Chem. Soc.*, **1959**, 630.
187. A. R. Davies and G. H. R. Summers, *J. Chem. Soc.*, C, **1966**, 1012.
188. D. H. R. Barton, D. A. Lewis, and J. F. McGhie, *J. Chem. Soc.*, **1957**, 2907.
189. H. Heusser, M. Feurer, K. Eichenberger, and V. Prelog, *Helv. Chim. Acta*, **33**, 2243 (1950).
190. P. A. Plattner, H. Heusser, and M. Feurer, *Helv. Chim. Acta*, **31**, 2210 (1948).
191. J. Elks, G. H. Phillipps, and W. F. Wall, *J. Chem. Soc.*, **1958**, 4001.

192. (a) P. A. Plattner, H. Heusser, and A. B. Kulkarni, *Helv. Chim. Acta*, **31**, 1885 (1948). (b) P. A. Plattner, H. Heusser, and A. B. Kulkarni, *Helv. Chim. Acta*, **32**, 1070 (1949).
193. H. B. Henbest and J. McEntee, *J. Chem. Soc.*, **1961**, 4478.
194. P. A. Plattner, H. Heusser, and M. Feurer, *Helv. Chim. Acta*, **32**, 587 (1949).
195. P. A. Plattner, A. Fürst, F. Koller, and H. H. Kuhn, *Helv. Chim. Acta*, **37**, 258 (1954).
196. E. Glotter, S. Greenfield, and D. Lavie, *Tetrahedron Lett.*, **1967**, 5261.
197. A. Fürst and R. Scotoni, Jr., *Helv. Chim. Acta*, **36**, 1410 (1953).
198. J. W. Cornforth, J. M. Osbond, and G. H. Phillipps, *J. Chem. Soc.*, **1954**, 907.
199. D. H. R. Barton, E. Miller, and H. T. Young, *J. Chem. Soc.*, **1951**, 2598.
200. C. W. Shoppee, R. H. Jenkins, and G. H. R. Summers, *J. Chem. Soc.*, **1958**, 1657.
201. C. W. Shoppee and R. Lack, *J. Chem. Soc.*, **1960**, 4864.
202. D. Lavie, Y. Kashman, and E. Glotter, *Tetrahedron*, **22**, 1103 (1966).
203. S. Greenfield, E. Glotter, D. Lavie, and Y. Kashman, *J. Chem. Soc.*, *C*, **1967**, 1460.
204. P. Vita-Finzi, Y. Kashman, E. Glotter, and D. Lavie, *Tetrahedron*, **24**, 5847 (1968).
205. N. L. Wendler, D. Taub, S. Dobriner, and D. K. Fukushima, *J. Amer. Chem. Soc.*, **78**, 5027 (1956).
206. H. Heymann and L. F. Fieser, *J. Amer. Chem. Soc.*, **73**, 5252 (1951).
207. G. Berti, B. Macchia, and F. Macchia, *Tetrahedron*, **24**, 1755 (1968).
208. G. Berti, F. Bottari, B. Macchia, and F. Macchia, *Tetrahedron*, **21**, 3277 (1965).
209. H. Riviere, *Bull. Soc. Chim. Fr.*, **1964**, 97.
210. (a) S. Peat, *Advan. Carbohyd. Chem.*, **2**, 37 (1946). (b) R. U. Lemieux, in P. de Mayo, Ed., *Molecular Rearrangements*, Wiley-Interscience, New York, 1963, chap. 12.
211. A. B. Foster, M. Stacey, and S. Vardheim, *Acta Chem. Scand.*, **12**, 1819 (1958).
212. J. G. Buchanan and J. Conn, *J. Chem. Soc.*, **1965**, 201.
213. J. A. Marshall and M. T. Pike, *J. Org. Chem.*, **33**, 435 (1968).
214. F. C. Hartman and R. Barker, *J. Org. Chem.*, **28**, 1004 (1963).
215. P. Bladon and L. N. Owen, *J. Chem. Soc.*, **1950**, 604.
216. M. Mousseron-Canet, C. Levallois, and H. Huerre, *Bull. Soc. Chim. Fr.*, **1966**, 658.
217. E. P. Kohler, N. K. Richtmyer, and W. F. Hester, *J. Amer. Chem. Soc.*, **53**, 205 (1931).
218. W. H. G. Lake and S. Peat, *J. Chem. Soc.*, **1939**, 1069.
219. J. G. Buchanan, *Chem. Ind.* (London), **1954**, 1484.
220. F. H. Newth, *J. Chem. Soc.*, **1956**, 441.
221. (a) S. J. Angyal and P. T. Gilham, *J. Chem. Soc.*, **1957**, 3691. (b) S. J. Angyal, V. Bender, and J. H. Curtin, *J. Chem. Soc.*, **1966**, 798.
222. (a) M. Černý, I. Buben, and J. Pacák, *Collect. Czech. Chem. Commun.*, **28**, 1569 (1963). (b) M. Černý, J. Pacák, and J. Staněk, *Collect. Czech. Chem. Commun.*, **30**, 1151 (1965).
223. G. B. Payne, *J. Org. Chem.*, **27**, 3819 (1962).
224. E. S. Gould, *Mechanism and Structure in Organic Chemistry*, Holt, Rinehart and Winston, New York, 1959, p. 567.
225. M. T. Rogers, *J. Amer. Chem. Soc.*, **69**, 2544 (1947).
226. M. Jarman and W. C. J. Ross, *Carbohyd. Res.*, **9**, 139 (1969).
227. S. Searles and C. F. Butler, *J. Amer. Chem. Soc.*, **76**, 56 (1954).
228. J. G. Buchanan and E. M. Oakes, *Carbohyd. Res.*, **1**, 242 (1965).
229. A. S. Meyer and T. Reichstein, *Helv. Chim. Acta*, **29**, 152 (1946).
230. A. C. Cope, M. M. Martin, and M. A. McKervey, *Quart. Rev.*, **20**, 119 (1966).

231. V. Prelog and J. G. Traynham, in P. de Mayo, Ed., *Molecular Rearrangements*, Wiley-Interscience, New York, 1963, chap. 9.

232. A. C. Cope and A. Fournier, Jr., *J. Amer. Chem. Soc.*, **79**, 3896 (1957).

233. A. C. Cope, A. H. Keough, P. E. Peterson, H. E. Simmons, Jr., and G. W. Wood, *J. Amer. Chem. Soc.*, **79**, 3900 (1957).

234. A. C. Cope and R. W. Gleason, *J. Amer. Chem. Soc.*, **84**, 1928 (1962).

235. A. C. Cope, A. Fournier, Jr., and H. E. Simmons, Jr., *J. Amer. Chem. Soc.*, **79**, 3905 (1957).

236. A. C. Cope, P. Scheiner, and M. J. Youngquist, *J. Org. Chem.*, **28**, 518 (1963).

237. C. R. Johnson, C. J. Cheer, and D. J. Goldsmith, *J. Org. Chem.*, **29**, 3320 (1964).

238. C. J. Cheer and C. R. Johnson, *J. Org. Chem.*, **32**, 428 (1967).

239. C. J. Cheer and C. R. Johnson, *J. Amer. Chem. Soc.*, **90**, 178 (1968).

240. J. F. King and P. de Mayo, in P. de Mayo, Ed., *Molecular Rearrangements*, Wiley-Interscience, New York, 1963, chap. 13.

241. (a) E. J. Corey, W. E. Russey, and P. R. Ortiz de Montellano, *J. Amer. Chem. Soc.*, **88**, 4750 (1966). (b) E. J. Corey and W. E. Russey, *J. Amer. Chem. Soc.*, **88**, 4751 (1966).

242. (a) E. E. van Tamelen, J. D. Willett, R. B. Clayton, and K. E. Lord, *J. Amer. Chem. Soc.*, **88**, 4752 (1966). (b) E. E. van Tamelen, J. D. Willett, M. Schwartz, and R. Nadeau, *J. Amer. Chem. Soc.*, **88**, 5937 (1966).

243. (a) E. J. Corey and P. R. Ortiz de Montellano, *J. Amer. Chem. Soc.*, **89**, 3362 (1967). (b) E. J. Corey, K. Lin, and H. Yamamoto, *J. Amer. Chem. Soc.*, **91**, 2132 (1969).

244. E. E. van Tamelen, J. D. Willett, and R. B. Clayton, *J. Amer. Chem. Soc.*, **89**, 3371 (1967).

245. N. L. Wendler, in P. de Mayo, Ed., *Molecular Rearrangements*, Wiley-Interscience, New York, 1963, chap. 16.

246. R. Breslow, in P. de Mayo, Ed., *Molecular Rearrangements*, Wiley-Interscience, New York, 1963, chap. 4.

247. A. Aebi, D. H. R. Barton, and A. S. Lindsey, *J. Chem. Soc.*, **1953**, 3124.

248. E. W. Warnhoff, *Can. J. Chem.*, **42**, 1664 (1964).

249. I. C. Nigam and L. Levi, *Can. J. Chem.*, **41**, 1726 (1963).

250. I. C. Nigam and L. Levi, *J. Org. Chem.*, **29**, 2803 (1964).

251. M. C. Nigam, I. C. Nigam, and L. Levi, *Can. J. Chem.*, **43**, 521 (1965).

252. E. E. Royals and J. C. Leffingwell, *J. Org. Chem.*, **29**, 2098 (1964).

253. (a) D. J. Goldsmith, B. C. Clark, Jr., and R. C. Joines, *Tetrahedron Lett.*, **1966**, 1149. (b) D. J. Goldsmith, B. C. Clark, Jr., and R. C. Joines, *Tetrahedron Lett.*, **1967**, 1211. (c) D. J. Goldsmith and B. C. Clark, Jr., *Tetrahedron Lett.*, **1967**, 1215. (d) E. E. van Tamelen, A. Storni, E. J. Hessler, and M. Schwartz, *J. Amer. Chem. Soc.*, **85**, 3295 (1963). (e) D. J. Goldsmith and C. F. Phillips, *J. Amer. Chem. Soc.*, **91**, 5862 (1969).

254. M. P. Hartshorn and D. N. Kirk, *Tetrahedron*, **21**, 1547 (1965).

255. D. N. Kirk and M. P. Hartshorn, *Steroid Reaction Mechanisms*, Elsevier, Amsterdam, 1969, chap. 8.

256. A. C. Cope, P. A. Trumbull, and E. R. Trumbull, *J. Amer. Chem. Soc.*, **80**, 2844 (1958).

257. A. C. Cope, H.-H. Lee, and H. E. Petree, *J. Amer. Chem. Soc.*, **80**, 2849 (1958).

258. A. C. Cope, M. Brown, and H.-H. Lee, *J. Amer. Chem. Soc.*, **80**, 2855 (1958).

259. A. C. Cope, G. A. Berchtold, P. E. Peterson, and S. H. Sharman, *J. Amer. Chem. Soc.*, **82**, 6370 (1960).

260. A. C. Cope and J. K. Heeren, *J. Amer. Chem. Soc.*, **87**, 3125 (1965).

261. J. K. Crandall, *J. Org. Chem.*, **29**, 2830 (1964).

262. (a) J. K. Crandall and L.-H. Chang, *J. Org. Chem.*, **32**, 435 (1967). (b) J. K. Crandall and L.-H. Chang, *J. Org. Chem.*, **32**, 532 (1967).
263. J. K. Crandall and L.-H. C. Lin, *J. Org. Chem.*, **33**, 2375 (1968).
264. (a) B. Rickborn and R. P. Thummel, *J. Org. Chem.*, **34**, 3583 (1969). (b) R. P. Thummel and B. Rickborn, *J. Amer. Chem. Soc.*, **92**, 2064 (1970).
265. C. C. Price and D. D. Carmelite, *J. Amer. Chem. Soc.*, **88**, 4039 (1966).
266. H. Nozaki, T. Mori, and R. Noyori, *Tetrahedron*, **22**, 1207 (1966).
267. W. Treibs, *Ber.*, **66**, 610, 1483 (1933), and earlier papers.
268. H. O. House and W. F. Gilmore, *J. Amer. Chem. Soc.*, **83**, 3972 (1961).
269. H. O. House and W. F. Gilmore, *J. Amer. Chem. Soc.*, **83**, 3980 (1961), and references cited therein.
270. W. Reusch and P. Mattison, *Tetrahedron*, **23**, 1953 (1967).
271. G. W. K. Cavill and C. D. Hall, *Tetrahedron*, **23**, 1119 (1967).
272. W. Baker and R. Robinson, *J. Chem. Soc.*, **1932**, 1798.
273. C. J. Collins and O. K. Neville, *J. Amer. Chem. Soc.*, **73**, 2471 (1951).
274. G. R. Treves, H. Stange, and R. A. Olofson, *J. Amer. Chem. Soc.*, **89**, 6257 (1967).
275. R. N. McDonald and P. A. Schwab, *J. Amer. Chem. Soc.*, **85**, 4004 (1963).
276. R. N. McDonald and P. A. Schwab, *J. Org. Chem.*, **29**, 2459 (1964).
277. R. N. McDonald and T. E. Tabor, *J. Amer. Chem. Soc.*, **89**, 6573 (1967).
278. R. N. McDonald and T. E. Tabor, *J. Org. Chem.*, **33**, 2934 (1968).
279. H. M. Walborsky and D. F. Loncrini, *J. Amer. Chem. Soc.*, **76**, 5396 (1954).
280. H. M. Walborsky and D. F. Loncrini, *J. Org. Chem.*, **22**, 1117 (1957).
281. K. L. Williamson and W. S. Johnson, *J. Org. Chem.*, **26**, 4563 (1961).
282. A. L. Draper, W. J. Heilman, W. E. Schaefer, H. J. Shine, and J. N. Shoolery, *J. Org. Chem.*, **27**, 2727 (1962).
283. K. L. Williamson, J. I. Coburn, and M. F. Herr, *J. Org. Chem.*, **32**, 3934 (1967).
284. R. H. Churi and C. E. Griffin, *J. Amer. Chem. Soc.*, **88**, 1824 (1966).
285. (a) D. B. Denney and M. J. Boskin, *J. Amer. Chem. Soc.*, **81**, 6330 (1959). (b) D. B. Denney, J. J. Vill, and M. J. Boskin, *J. Amer. Chem. Soc.*, **84**, 3944 (1962).
286. S. Trippett, *Quart. Rev.* (London), **17**, 406 (1963).
287. I. Tömöskösi, *Tetrahedron*, **19**, 1969 (1963).
288. R. M. Gerkin and B. Rickborn, *J. Amer. Chem. Soc.*, **89**, 5850 (1967).
289. B. Rickborn and R. M. Gerkin, *J. Amer. Chem. Soc.*, **90**, 4193 (1968).
290. W. E. McEwen, A. P. Wolf, C. A. VanderWerf, A. Blade-Font, and J. W. Wolfe, *J. Amer. Chem. Soc.*, **89**, 6685 (1967).
291. J. K. Crandall and D. R. Paulson, *J. Org. Chem.*, **33**, 991 (1968).
292. J. K. Crandall and D. R. Paulson, *J. Org. Chem.*, **33**, 3291 (1968).
293. D. H. R. Barton and R. C. Cookson, *Quart. Rev.* (London), **10**, 44 (1956).
294. (a) J. Meinwald, S. S. Labana, and M. S. Chadha, *J. Amer. Chem. Soc.*, **85**, 582 (1963). (b) P. B. D. de la Mare, in P. de Mayo, Ed., *Molecular Rearrangements*, Wiley-Interscience, New York, 1963, chap. 2.
295. (a) A. Winterstein and G. Stein, *Z. Physiol. Chem.*, **202**, 217 (1931). (b) H. Wieland and G. Hanke, *Z. Physiol. Chem.*, **241**, 93 (1936).
296. A. C. Cope and G. W. Wood, *J. Amer. Chem. Soc.*, **79**, 3885 (1957).
297. (a) J. M. Greenwood, J. K. Sutherland, and A. Torre, *Chem. Commun.*, **1965**, 410. (b) E. D. Brown, M. D. Solomon, J. K. Sutherland, and A. Torre, *Chem. Commun.*, **1967**, 111.
298. (a) P. B. D. de la Mare and A. Salama, *J. Chem. Soc.*, **1956**, 3337. (b) J. G. Traynham and O. S. Pascual, *J. Amer. Chem. Soc.*, **79**, 2341 (1957). (c) P. B. D. de la Mare and R. Bolton, *Electrophilic Addition to Unsaturated Systems*, Elsevier, Amsterdam, 1966.

299. E. S. Gould, *Mechanism and Structure in Organic Chemistry*, Holt, Rinehart and Winston, New York, 1959, pp. 523ff.
300. M. L. Poutsma, *Science*, **157**, 997 (1967).
301. (a) C. Prévost, *C. R. Acad. Sci., Paris*, **196**, 1129 (1933). (b) *C. R. Acad. Sci., Paris*, **197**, 1661 (1933). (c) C. Prévost and J. Wiemann, *C. R. Acad. Sci., Paris*, **204**, 700 (1937). (d) Y. Gaoni, *Bull. Soc. Chim. Fr.*, **1959**, 701.
302. (a) R. B. Woodward and F. V. Brutcher, Jr., *J. Amer. Chem. Soc.*, **80**, 209 (1958). (b) C. A. Bunton and M. D. Carr, *J. Chem. Soc.*, **1963**, 770.
303. S. Winstein and R. M. Roberts, *J. Amer. Chem. Soc.*, **75**, 2297 (1953).
304. M. Nakajima, A. Hasegawa, and F. W. Lichtenthaler, *Justus Liebigs Ann. Chem.*, **680**, 21 (1964).
305. J. Retey, A. Umani-Ronchi, J. Seibl, and D. Arigoni, *Experientia*, **22**, 502 (1966).
306. B. G. Christensen, W. J. Leanza, T. R. Beattie, A. A. Patchett, B. H. Arison, R. E. Ormond, F. A. Kuehl, Jr., G. Albers-Schonberg, and O. Jardetzky, *Science*, **166**, 123 (1969).
307. (a) E. W. Thomas, J. F. McKelvy, and N. Sharon, *Nature*, **222**, 485 (1969). (b) E. W. Thomas, *Carbohyd. Res.*, **13**, 225 (1970).
308. (a) G. Legler, *Biochim. Biophys. Acta*, **151**, 728 (1968). (b) G. Legler, *Z. Physiol. Chem.*, **349**, 767 (1968). (c) K. J. Stevenson and L. B. Smillie, *J. Mol. Biol.*, **12**, 937 (1965).
309. (a) J. L. Everett and G. A. R. Kon, *J. Chem. Soc.*, **1950**, 3131. (b) J. A. Hendry, R. F. Homer, F. L. Rose, and A. L. Walpole, *Brit. J. Pharmacol.*, **6**, 235 (1951). (c) L. A. Elson, M. Jarman, and W. C. J. Ross, *Eur. J. Cancer*, **4**, 617 (1968).
310. (a) B. L. Van Duuren, N. Nelson, L. Orris, E. D. Palmes, and F. L. Schmitt, *J. Nat. Cancer Inst.*, **31**, 41 (1963). (b) B. L. Van Duuren, L. Orris, and N. Nelson, *J. Nat. Cancer Inst.*, **35**, 707 (1965). (c) B. L. Van Duuren and B. M. Goldschmidt, *J. Med. Chem.*, **9**, 77 (1966).
311. J. Fried, C. H. Lin, and S. H. Ford, *Tetrahedron Lett.*, **1969**, 1379.
312. J. W. Cornforth, R. H. Cornforth, G. Popják, and L. Yengoyan, *J. Biol. Chem.*, **241**, 3970 (1966).
313. J. W. Cornforth, R. H. Cornforth, C. Donninger, and G. Popják, *Proc. Roy. Soc., Ser. B*, **163**, 492 (1966).
314. J. W. Cornforth, J. W. Redmond, H. Eggerer, W. Buckel, and C. Gutschow, *Nature*, **221**, 1212 (1969).
315. P. A. Cruickshank and M. Fishman, *J. Org. Chem.*, **34**, 4060 (1969).
316. J. K. Crandall and D. R. Paulson, *Tetrahedron Lett.*, **1969**, 2751.
317. J. K. Crandall et al., *J. Amer. Chem. Soc.*, **90**, 7292, 7346, 7347 (1968).
318. R. L. Camp and F. D. Greene, *J. Amer. Chem. Soc.*, **90**, 7349 (1968).
319. (a) D. K. Fukushima, M. Smulovitz, J. S. Liang, and G. Lukacs, *J. Org. Chem.*, **34**, 2702 (1969). (b) G. Lukacs and D. K. Fukushima, *J. Org. Chem.*, **34**, 2707 (1969).
320. S. M. Kupchan, R. J. Hemingway, and R. M. Smith, *J. Org. Chem.*, **34**, 3898 (1969).
321. J. E. McMurry, *Tetrahedron Lett.*, **1970**, 3731.
322. R. A. Appleton, J. R. Dixon, J. M. Evans, and S. H. Graham, *Tetrahedron*, **23**, 805 (1967).
323. E. N. Marvell, J. Seubert, D. Sturmer, and W. Federici, *J. Org. Chem.*, **35**, 396 (1970).
324. H. Morita and S. Oae, *Tetrahedron Lett.*, **1969**, 1347.
325. H. G. Richey, Jr., and D. V. Kinsman, *Tetrahedron Lett.*, **1969**, 2505.
326. J. L. Jernow, D. Gray, and W. D. Closson, Abstracts, 159th National Meeting of the American Chemical Society, Houston, Texas, February, 1970, ORGN 130.

327. D. Gagnaire, *Bull. Soc. Chim. Fr.*, **1960**, 1813.
328. G. T. Kwiatkowski, S. J. Kavarnos, and W. D. Closson, *J. Heterocycl. Chem.*, **2**, 11 (1965).
329. N. Heap, G. E. Green, and G. H. Whitham, *J. Chem. Soc.*, C, **1969**, 160.
330. G. O. Pierson and O. A. Runquist, *J. Org. Chem.*, **34**, 3654 (1969).
331. R. N. McDonald and R. N. Steppel, *J. Amer. Chem. Soc.*, **92**, 5664 (1970).
332. D. J. Goldsmith and R. C. Joines, *J. Org. Chem.*, **35**, 3572 (1970).
333. J. Holtzman, J. R. Gillette and G. W. A. Milne, *J. Amer. Chem. Soc.*, **89**, 6341 (1967).
334. (a) D. M. Jerina, J. W. Daly, and B. Witkop, *J. Amer. Chem. Soc.*, **90**, 6523 (1968). (b) D. M. Jerina, J. W. Daly, B. Witkop, P. Zaltman-Nirenberg, and S. Udenfriend, *J. Amer. Chem. Soc.*, **90**, 6525 (1968). (c) D. M. Jerina, J. W. Daly, B. Witkop, P. Zaltman-Nirenberg, and S. Udenfriend, *Arch. Biochem. Biophys.*, **128**, 176 (1968).
335. D. M. Jerina, J. W. Daly, and B. Witkop, *J. Amer. Chem. Soc.*, **89**, 5488 (1967).
336. E. W. Maynert, H. L. Foreman, and T. Watabe, *J. Biol. Chem.*, **245**, 5234 (1970).
337. N. R. Williams, *Adv. Carbohyd. Chem. Biochem.*, **25**, 109 (1970).
338. (a) J. Boeseken, *Rec. Trav. Chim.*, **47**, 683 (1928); (b) D. Lutkus, J. Pipa, and H. Z. Sable, unpublished experiments (1971).
339. (a) F. Oesch, D. M. Jerina, and J. Daly, *Biochem. Biophys. Acta*, **227**, 685 (1971); (b) F. Oesch and J. Daly, *ibid.*, **227**, 692 (1971).
340. A. S. Meyer, E. Hanzmann, H. A. Schneiderman, L. I. Gilbert, and M. Boyette, *Arch. Biochem. Biophys.*, **137**, 190 (1970).

Alcohol Oxidation by Lead Tetraacetate

MIHAILO LJ. MIHAILOVIĆ

Department of Chemistry, Faculty of Sciences
of the University of Belgrade,
and
Institute for Chemistry, Technology and Metallurgy,
Belgrade, Yugoslavia

RICHARD E. PARTCH

Department of Chemistry, Clarkson College of Technology,
Potsdam, New York

I. INTRODUCTION

Oxidation of organic molecules to yield useful, reactive intermediates and important end products is one of five reaction classes. The other four are reduction, addition, elimination, and substitution. Of those types of molecular groups oxidized, the alcohol group is possibly the most often involved. Lead tetraacetate, hereafter referred to as LTA, has been a much used oxidant.

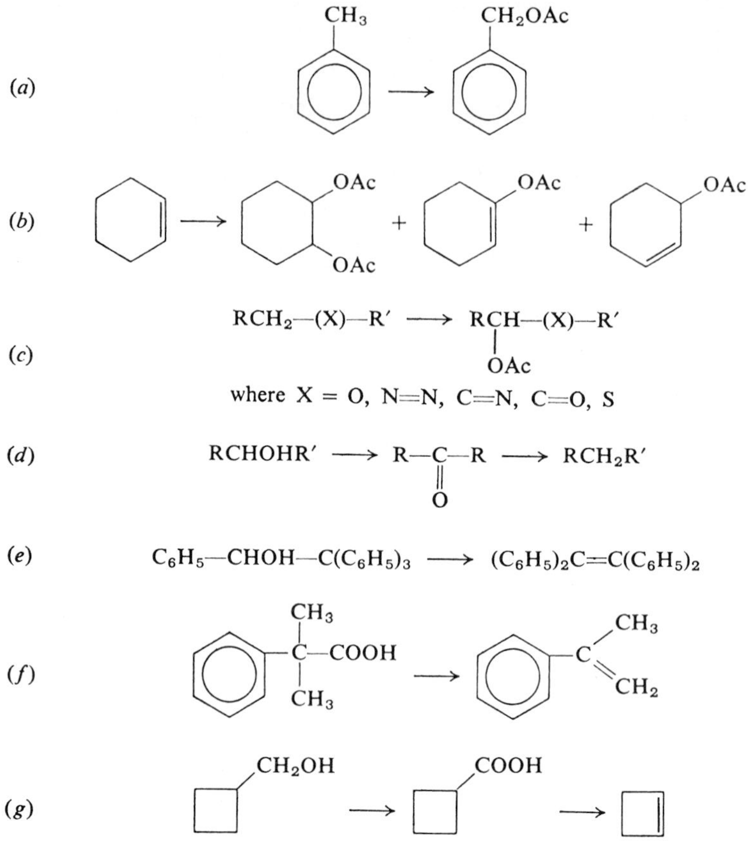

(a)

$$CH_3 \longrightarrow CH_2OAc$$

(b)

(c)

$$RCH_2\text{---}(X)\text{---}R' \longrightarrow \underset{\underset{OAc}{|}}{RCH\text{---}(X)\text{---}R'}$$

where X = O, N=N, C=N, C=O, S

(d)

$$RCHOHR' \longrightarrow R\text{---}\underset{\underset{O}{\|}}{C}\text{---}R \longrightarrow RCH_2R'$$

(e)

$$C_6H_5\text{---}CHOH\text{---}C(C_6H_5)_3 \longrightarrow (C_6H_5)_2C\text{=}C(C_6H_5)_2$$

(f)

(g)

Figure 1

Historically, this oxidant has served organic chemists in a reversible fashion. It has been used to prepare alcohols, in the state of acetate esters, from active hydrocarbons such as alkylbenzenes and alkenes, and from activated carbon–hydrogen bonds (1); it has been used alone, or in conjunction with other oxidizing and reducing agents, to convert alcohols back into hydrocarbons (Figure 1).

Approximately twenty different nonbiological reagents are in popular use as oxidants of organic compounds. They range in complexity from molecular oxygen to potassium ferricyanide, phenyl iodosoacetate, and ceric ammonium nitrate. For a given organic reducing agent each oxidant is capable of generating its own set of products and in relative yields depending on the reaction medium (Figure 2). In most cases all oxidants yield some of the same product from a given organic reducing agent, and the optimization of *desired* product yield has stimulated research in how to use each one. Studies by molecular structure, kinetics, and product analysis to date have revealed the unfortunate fact that for the different metal ion oxidants, e.g., CrO_3, $KMnO_4$, and $Pb(OAc)_4$, no correlation can be made between the oxidation potential of the metal atom and the organic products obtained. Additionally, the role of reaction solvent too often remains anonymous in that coordination and solvation characteristics cannot always be related to changes in product yield and ratios.*

LTA has one of the highest redox potentials of the metal ion oxidants.†

(a) $(C_6H_5)_2CH_2 \xrightarrow{\text{[OX]}} (C_6H_5)_2CHOAc + (C_6H_5)_2C{=}O$

[OX]	Yield (%)	
Pb(IV) + C_6H_6	71	—
Ce(IV) + H_2O	—	95
Cr(VI) + H_2SO_4	—	95
Mn(VII) + NaOH	—	65
Cr(VI) + C_5H_5N	—	71

(b)

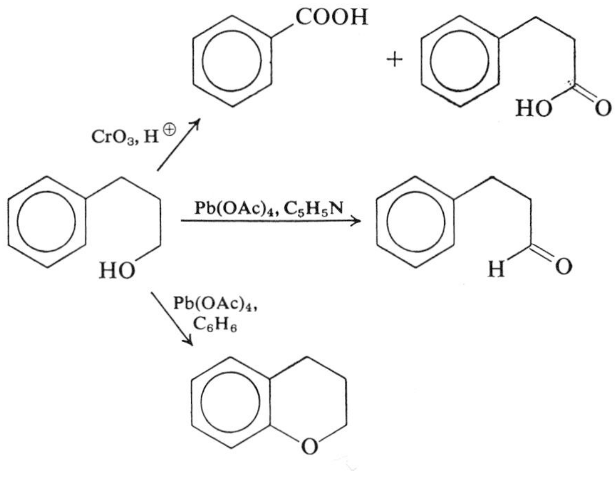

Figure 2. (2)

* Little is known of the magnitude of solvent effects on various oxidation reactions, at least in the mechanism interpretation. Difficulties are exemplified by the variation of Mn(VII)–Mn(IV) potentials as a function of medium: 1.70 V in H_2SO_4; 0.59 V in NaOH (3).

† Pb(IV)–Pb(II), 1.6 V in $HClO_4$ (anion dependent); Ce(IV)–Ce(II), 1.87–1.28 V (anion dependent); and Mn(VII)–Mn(IV), 1.70 V in acid, 0.59 V in base. Other potentials are found in standard references (4).

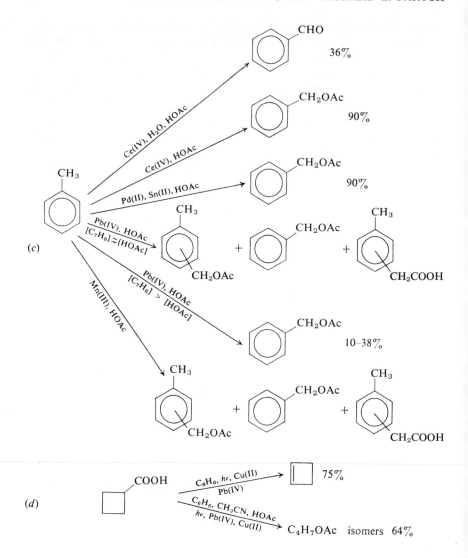

Figure 2. (2)—*Continued*

This implies that electron transfer to the lead atom should be extremely
facile. Thermodynamically, this is borne out, as the reagent reacts with some
substrates inert to oxidants with lesser potentials (Figure 3). On the other
hand, reaction kinetics have demonstrated LTA to be equal to or less rapid
an oxidant than its competitors. Interpretation difficulties arise, however,

(e)

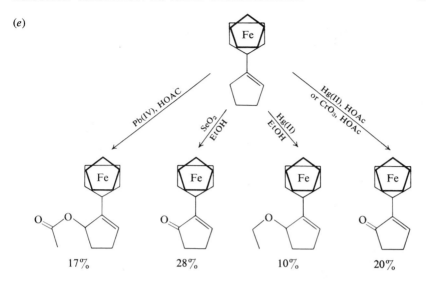

Figure 2. (2)—*Continued*

(a)

(b)

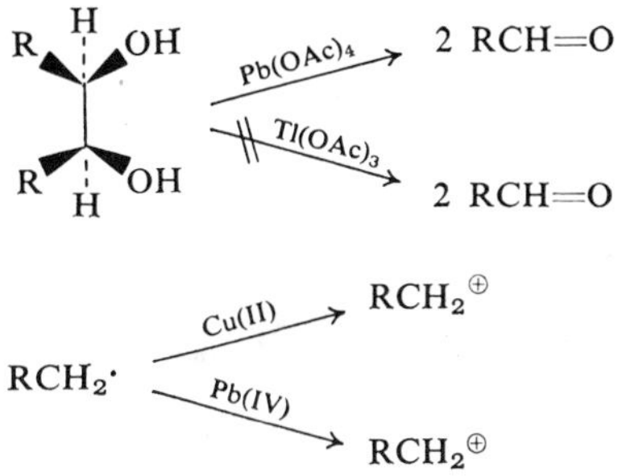

$$k_{(Cu(II))} \gg k_{(Pb(IV))}$$

Figure 3. (5)

since the media are often different and for each oxidant multiple equilibria are involved prior to the rate determining step.*

* To date no examples of these initial steps being rate determining have been reported. See, however, Figure 31. When oxidation is carried out in complexing solvents, such as acetic acid or pyridine, other equilibria involving species as $H_n[Pb(OAc)_{4+n}]$ and $[Pb(OAc)_4 \cdot mC_5H_5N]$, respectively, are prevalent (11). The process of alcoholysis can take place by two methods: concerted metathesis or by ionic dissociation of LTA to $+Pb(OAc)_3$ and $-OAc$ followed by capture of the cation by alcohol. Hydrolysis and dissociation equilibrium constants for Pb(IV) species seem to indicate that this latter type of alcoholysis of LTA may be relatively unfavorable (11g).

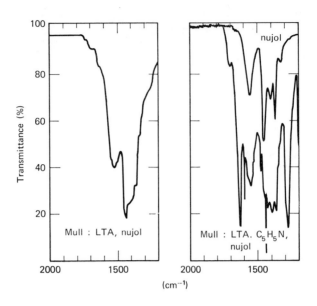

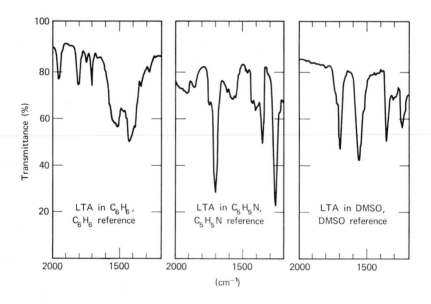

Figure 4. (7)

Structurally, LTA has been reported as being monomeric in the solid and in benzene and acetic acid solution states, although there is some evidence of dimerization in concentrated solutions (6). Their spectra in Nujol or in benzene solution show no "ester type" carbonyl groups and suggest each oxygen of any of the four acetate groups to be equivalent (Figure 4) (7). In this respect the acetate groups in LTA are ionically associated with the lead atom and resemble those found in sodium acetate. The possibility that one or more of the acetate anions in LTA bridge between lead atoms, as is the case in trisalkyltin and trisalkyllead acetates, is not likely. Crystal structure determination, though incomplete (8), suggests LTA to be of distorted cubic geometry (1–2) (6). That the acetate groups are labile to rapid exchange in acetic acid supports the concept of minor covalency in the lead-oxygen bonds (9). Contraindicating loosely paired ions are conductance measurements in solvents capable of separating ion pairs. These show LTA is not much

Cubic	Square antiprism
1	**2**

different than a non- or weak-electrolyte but very much less ionized than the electrolyte tetraphenyl arsonium chloride (10).

Fragmentary evidence defines the method of initial interaction of functional groups with LTA. When used as an alcohol oxidant, the oxidant does not dissociate into free radicals *prior* to other reactions. For the most part the electron acceptor properties of the lead atom dictate a sequence such as : (*1*) primary coordination of the functional group to the lead atom via an electron pair, and (*2*) secondary exchange of an acetate group for the group with the electron pair. The initial coordination can be interpreted as quite strong if the coordinating electron pair is of the nonbonded type (11). For example, an alcohol coordinates first via an n electron pair on oxygen. This is followed by loss of a molecule of acetic acid to give a mixed acetate–alcoholate of lead. The process of alcoholysis can take place by two methods: concerted metathesis or by ionic dissociation of LTA to $+Pb(OAc)_3$ and $-OAc$ followed by capture of the cation by alcohol. Subsequent alcoholysis reactions, as well as the first, must be reversible unless the generated acetic acid is eliminated or an alcoholate (A and/or B, Figure 5) decomposes (the tertiary step).* Evidence for alcoholysis comes from the isolated half alcoholate $(CH_3O)(HO)Pb(OAc)_2$ (1a), and the generation of nonequivalent acetate oxygen atoms in the ir spectrum of LTA when placed in alcohol solution (Figure 4) (7). Unfortunately, the lability of the supposed alcoholates of many alcohols, especially the cyclic bisalcoholates from *cisoid* vicinal diols, has precluded their isolation.

When the coordinating electron pair is less free, interaction with the lead atom is not enough to alter the benzene solution ir spectrum of LTA (12). More energetic conditions are needed in these cases to effect the primary and

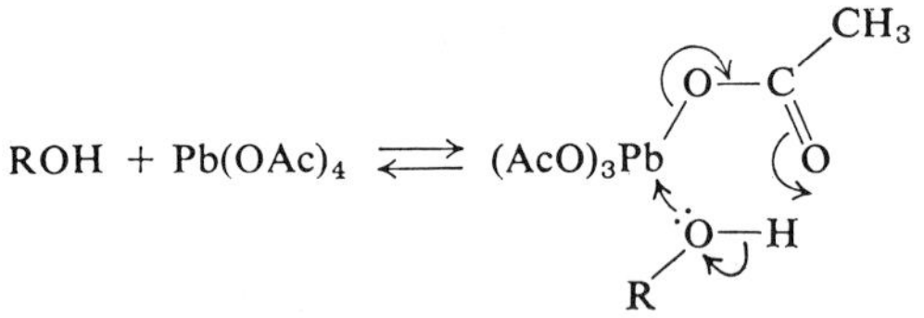

$$ROH + Pb(OAc)_4 \rightleftharpoons (AcO)_3Pb \begin{array}{c} O-C \diagdown CH_3 \\ | \qquad \diagup C \diagdown O \\ O-H \\ | \\ R \end{array}$$

$$ROH + Pb(OAc)_4 \rightleftharpoons \underset{A}{ROPb(OAc)_3} + HOAc$$

$$ROPb(OAc)_3 \rightarrow Pb(OAc)_2 + Organic\ products$$

$$ROPb(OAc)_3 + ROH \rightleftharpoons \underset{B}{(RO)_2Pb(OAc)_2} + HOAc$$

$$(RO)_2Pb(OAc)_2 \rightarrow Pb(OAc)_2 + Organic\ products$$

$$(RO)_2Pb(OAc)_2 + ROH \rightleftharpoons (RO)_3PbOAc + HOAc$$

$$(RO)_3PbOAc + ROH \rightleftharpoons Pb(OR)_4 + HOAc$$

$$Primary\ step = ROH + Pb(OAc)_4 \rightleftharpoons \begin{array}{c} R \\ \diagdown \\ O: \rightarrow Pb(OAc)_4 \\ \diagup \\ H \end{array}$$

$$Secondary\ step = \begin{array}{c} R \\ \diagdown \\ O: \rightarrow Pb(OAc)_4 \rightleftharpoons ROPb(OAc)_3 + HOAc \\ \diagup \\ H \end{array}$$

$$Tertiary\ step = (RO)_nPb(OAc)_{4-n} \longrightarrow \begin{cases} nRO\cdot + \cdot Pb(OAc)_{4-n} \\ \qquad\qquad or \\ nRO^+ + Pb(OAc)_2 + {}^-OAc \end{cases}$$

Figure 5

* It has always been *assumed* that lead diacetate is the only Pb(II) species remaining after alcohol oxidation. For this reason only alcoholates A and B of Figure 5 are shown as precursors to products. One aspect in favor of the mono- and bisalcoholates being the highest alcoholized lead species is the low alcohol to LTA ratios used in oxidations.

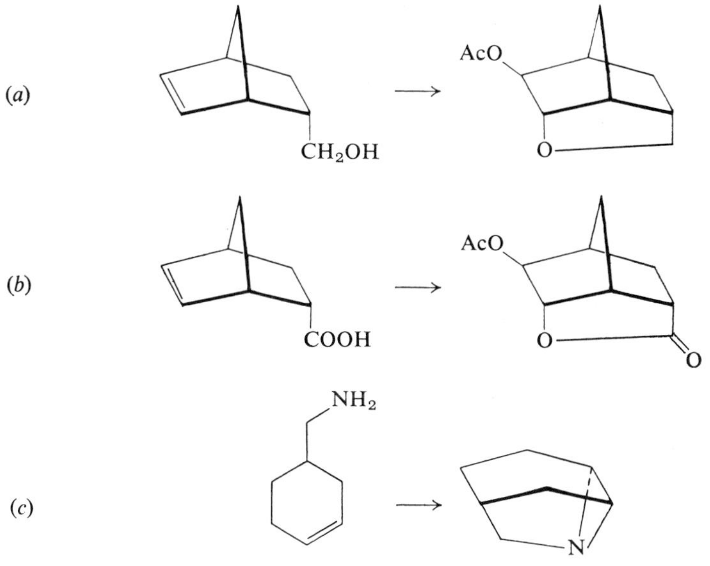

Figure 6. (13a,13c,13d)

secondary reactions, as exemplified by addition of acetate groups to a carbon–carbon double or triple bond. When a strong and a weak coordinator are contained within the same molecule, the weaker seems capable of influencing the reaction by sterically orienting the lead-containing moiety (Figure 6). Hence, a suitably situated π-bond in an alkenol orientates the lead atom in the alcoholate, allowing subsequent π-bond interaction in the decomposition (tertiary) reaction (13).

Reaction of LTA with hydroxyl groups, especially when in vicinal position to each other, came into popularity following the studies of Criegee in the 1930s (14). The cleavage reactions of vicinal diols, α-hydroxy acids, 1,2-dicarbonyl compounds, 1,2-hydroxycarbonyl compounds, and 1,2-aminohydroxy compounds to yield fragmented organic molecules in the form of carbonyl compounds have been well documented (Figure 7). As of 1959 these reactions were thought to be reasonably well understood. Within the same time period LTA was being used to oxidize other molecules such as aromatic alcohols and hydrocarbons, monohydric alcohols, and to add acetate groups to alkenes (Figures 1, 2, and 8).

The status of lead tetracarboxylates and other metal carboxylate salts as efficient oxidizers of organic compounds has profoundly increased since 1959 (15). Indeed, since that time the chemical literature demonstrates a

(a)

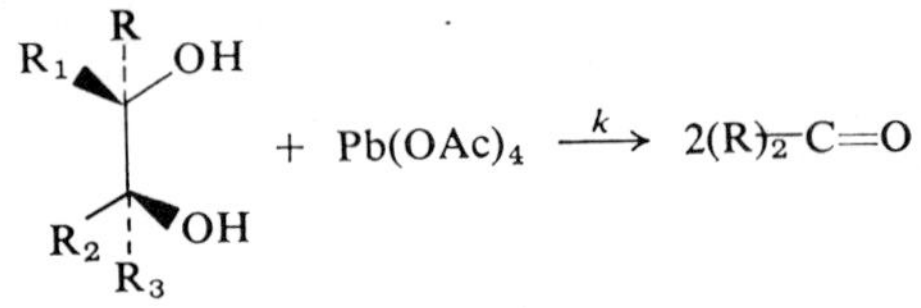

$k_{cis\text{-}erythro} > k_{trans\text{-}threo}$

(b)

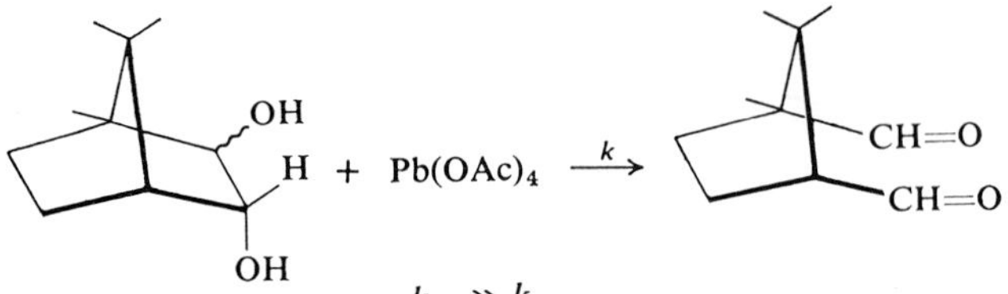

$k_{cis} \gg k_{trans}$

$k_{trans\ C_5H_5N} \gg k_{trans\text{-}C_6H_6\text{-}HOAc}$

Figure 7. (14)

(a)

(b)

(c)

(d)

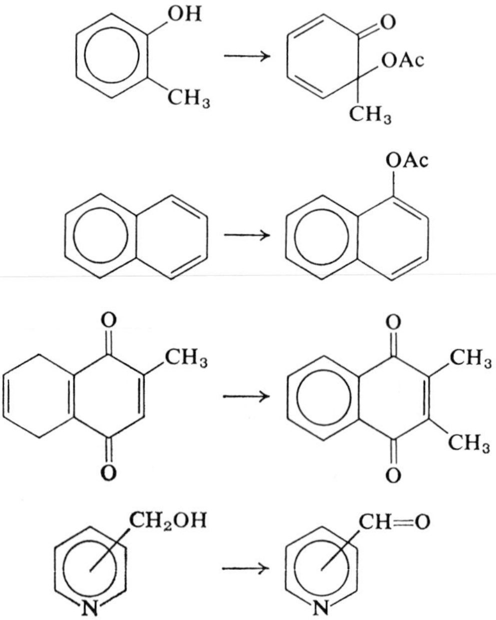

Figure 8. (1a)

continuum of reactivity for LTA (16). The impetus for the recent, critical surveys for new utility came from the discovery of cyclic, intramolecular ether formation from monohydroxyl compounds containing an *unactivated* carbon–hydrogen bond in the δ-position (17) (eq. 1). Among all the compounds used to oxidize alcohols, this reaction is peculiar, to date, to Pb(IV)

(1)

and Ce(IV) ions and to HgO–halogen and Ag₂O–halogen oxidation couples (18).* Formally, the reaction is a 1-5 dehydrogenative ring closure and is in competition with (*1*) the 1-2 dehydrogenation leading to carbonyl compound, (*2*) the fragmentation of the carbinol carbon atom from the starting alcohol, (*3*) esterification, and (*4*) dehydration.

To bring together and interpret all of the pertinent, recent data on substrate oxidation by LTA, a review must cover one functional group at a time. For this reason the present chapter deals only with alcohol oxidation by the reagent. More specifically, an attempt has been made to include only those data that relate definitive facts for discussion under the title of *selective transformations*. Excellent reviews showing the spectrum of reactivity towards various organic functional groups precede this one and should be consulted for further information (16). For this chapter the literature has been surveyed up to January 1970. The subject matter is treated so as to give the reader a feeling for the product possibilities as a function of reaction condition and alcohol structure, followed by interpretation of the data for mechanistic understanding. Finally, brief comments on other related reactions are included. It is hoped that the material presentation will stimulate new thought on an old subject—"alcohol oxidation by LTA."

---

* This is not to say, however, that closely related compounds (but not alcohols) such as ROBr, RONO, ROI, and RONO₂ do not undergo self-decomposition (with activation) in a similar manner (Figure 9). Alcohol oxidation by ceric ion is capable of generating alkyl radicals, presumably from precursory alkoxy radicals or by concerted collapse of an alcohol–ceric ion complex (19).

II. DATA AND DISCUSSION

In the introduction it was stated that no attempt has been made to include *all* data on alcohol oxidation. To do less, however, implies that a reasonable data selection method must be adhered to. The authors have attempted, therefore, to include transformations which: (*1*) give an important product in higher yield than obtainable by other methods, (*2*) show selective peculiarities of the oxidant, and (*3*) are definitive for interpreting the mechanisms involved.

For convenience of survey the data appear in sections depending on reaction conditions. In apolar and acidic-polar solvents it is possible to include thermal and photolytic reactions together. In neutral-polar or basic solvents, however, photolytic activation usually yields different product ratios than those from thermal activation. Because the alcohol classes are considered separately, for convenience of *tabular* survey, summaries of critical facts are included where pertinent.

To introduce the reader to the techniques of utilizing LTA requires comment on experimental procedures. These can be found fully explained within the references and cannot be elaborated here. The following are important among them.

(*1*) The extreme lability of the oxidant to water.

(*2*) The oxidant's low room temperature solubility in acetic acid and alkane solvents.

(*3*) Its reasonable (0.05–0.10 M) room temperature solubility in alcohols (usually considered a reactant), benzene, pyridine, methylene chloride, chloroform, dimethyl-sulfoxide, and acetonitrile.

(*4*) Its higher solubility at temperatures up to 80°C, *without rapid decomposition*, in the above solvents except alcohols.

(*5*) Its instability at room temperature in some mixed solvents, particularly pyridine–dimethylsulfoxide, and at temperatures exceeding 90°C in pure solvents.*

(*6*) Its ability to form hydrocarbon insoluble coordination complexes with ligands (complexing solvents) possessing atoms having nonbonding electrons.

Since, as will become abundantly evident, a multitude of products can be formed when alcohols are oxidized, decision is needed as to how to optimize the yield of product desired. While the alcohol structure often

---

* The LTA titer in a pyridine–dimethylsulfoxide mixture (50:50 volume) decreases as rapidly at 30° as does the titer when 1-butanol is oxidized in pyridine at 30° (7c,20). When LTA is added to pyridine, a deep red color persists. A brown precipitate is indicative of impure pyridine. The color diminishes along with the LTA titer over several days. The color fades within minutes at 90° and the LTA titer becomes zero. Continued heating to 100° causes a redarkening of the solution. One can expect LTA–solvent reactions to become prevalent at higher temperatures (21).

majorily dictates the reaction path taken, a few external conditions listed below, can be profitably varied.

(7) Cyclic ether formation is best accomplished by using an oxidant containing minor amounts of acetic acid of crystallization.

(8) Noncomplexing hydrocarbon solvents allow maximal thermal production of cyclic ether.

(9) Complexing solvents enhance carbonyl formation.

(10) Photolytic activation allows maximal cyclic ether production.

(11) Using a large excess of oxidant does not increase the cyclic ether yield greatly, even though starting alcohol often appears as a product.

(12) Reaction mixtures having undissolved LTA effect transformations similar to homogeneous solutions.

(13) Thermal cyclic ether formation is not dependent on the presence of calcium carbonate, a rather inefficient neutralizer of generated acetic acid.

A. Apolar Media

1. Data

Table 1 shows most of the monohydric alcohols oxidized by an LTA containing medium with the steroid examples being classified by hydroxyl position. Those polyols included are either nonvicinal or *trans*-vicinal and do not undergo the usual glycol cleavage reaction.

TABLE 1

Alcohols Oxidized by Lead Tetraacetate Containing Media:
Acidic or Neutral Solvents

	Ref.
A. Saturated aliphatic alcohols	
1. Primary	
a. Methanol	106
b. Ethanol	79
c. 1-Propanol	30a,79
d. 1-Butanol	30a,117
e. 1-Pentanol	30a,69,117
f. 2-Methyl-1-pentanol	30b,69
g. 2,2-Dimethyl-1-pentanol	69
h. 4,4-Dimethyl-1-pentanol	53
i. 1-Hexanol	30a,117
j. 4,4-Dimethyl-1-hexanol	53

Continued

Table 1 *(Continued)*

	Ref.
k. 1-Heptanol	117,118
l. 1-Octanol	30b,117,118
m. 6,7-Dibromo-3,7-dimethyl-1-octanol (dibromocitronellol)	13b
n. 4,8-Dimethyl-1-nonanol	50a
o. 1-Dodecanol	117
p. 2-(3'-Hydroxy-6'-methoxy-2',4',5'-trimethylbenzyl)propane-1,3-diol	62
q. 1,6-Hexanediol	55
r. 1,7-Heptanediol	55
s. 1,8-Octanediol	55
t. 1,9-Nonanediol	55
u. 1,10-Decanediol	55
v. 1,11-Undecanediol	55
2. Secondary	
a. 2-Propanol	79
b. 2-Butanol	79
c. 3,3-Dimethyl-2-butanol	81
d. 2-Pentanol	117
e. 2-Hexanol	69,117
f. 3-Methyl-2-hexanol	69
g. 3,3-Dimethyl-2-hexanol	69
h. 2-Heptanol	117
i. 2-Octanol	30b,117,118
j. 2-Decanol	117
k. 3-Heptanol	117
l. 3-Octanol	117
m. 3-Nonanol	117
n. 4-Heptanol	117
o. 4-Octanol	117
p. 4-Nonanol	117
q. 5-Nonanol	117,118
3. Tertiary	
a. 2-Methyl-2-hexanol	69
b. 2,3-Dimethyl-2-hexanol	69
B. Unsaturated aliphatic alcohols	
1. 2-Buten-1-ol	107
2. 3-Buten-1-ol	58a

3.	4-Penten-1-ol	59a
4.	5-Hexen-1-ol	59a
5.	3,7-Dimethyl-6-octen-1-ol (citronellol)	13b

C. Saturated alicyclic alcohols
  1. Monocyclic

a.	Cyclopropylcarbinol	70g
b.	Cyclopropylmethylcarbinol	70g
c.	Cyclobutanol	46a
d.	Cyclobutylcarbinol	70g
e.	Cyclopentanol	46a
f.	2-Cyclopentylethanol	50c
g.	3-Cyclopentyl-1-propanol	50c
h.	Cyclohexanol	46a
i.	2-*n*-Butylcyclohexanol	20,43
j.	2-Isopropyl-5-methylcyclohexanol ((−)-menthol)	33c
k.	2-Isopropyl-5-methylcyclohexanol ((+)-isomenthol)	33c
l.	2-Isopropyl-5-methylcyclohexanol ((+)-neomenthol)	33c
m.	Cyclohexylcarbinol	54
n.	1-Methylcyclohexylcarbinol	54
o.	Dimethyl-(3-isopropyl-4-ethyl-4-methylcyclohexyl) carbinol (tetrahydroelemol)	112
p.	2-Cyclohexyl-1-ethanol	50c,50d
q.	3-Cyclohexyl-1-propanol	50c,50d
r.	Cycloheptanol	66,46a,116
s.	Cyclooctanol	66,46a,116
t.	1-Methyl-1-cyclooctanol	66
u.	*trans*-4-Phenyl-1-cyclooctanol	59b
v.	*cis*-5-Phenyl-1-cyclooctanol	59b
w.	*trans*-5-Phenyl-1-cyclooctanol	59b
x.	Cyclodecanol	43,46a
y.	Cyclododecanol	46a
z.	Cyclohexadecanol	46a

  2. Bicyclic

a.	Borneol	20,28b
b.	Isoborneol	28b
c.	Bicyclo(2.2.1)heptane-2-methanol	72
d.	Bicyclo(2.2.2)octane-2-methanol	72
e.	Bicyclo(2.2.2)octane-2,3-bismethanol	72
f.	Bicyclo(3.2.1)octane-3-ol	46b
g.	2-[4-(1,7-Dimethylbicyclo(4.4.0)decanyl)]propan-1-ol	74b
h.	dimethyl-5-(2,8-dimethylbicyclo(5.3.0)decanyl)carbinol	74b

Continued

TABLE 1 (*Continued*)

	Ref.
i. *endo*-Camphanol	74c
j. *exo*-Camphanol	74c
k. *endo*-3-Hydroxy-bicyclo(3.3.1)nonane	53b
3. Tricyclic	
a. Longifolol	74a,74b
b. Isolongifolol	74a,74b
c. Phorbol	76d,111
d. 20-Hydroxynorlabdan	113
e. Cedrol	110
f. 9β-Hydroxy-trimethylperhydrophenanthrene derivative	50e
g. 8β-Hydroxymethyl-dimethylperhydrophenanthrene derivative	50e
h. 2β-Hydroxymanoyl oxide (iodine added)	123
D. Unsaturated alicyclic and bicyclic alcohols	
1. 1.-Allylcyclohexan-1-ol	60
2. 1-(3-Hydroxypropyl)-1-cyclohexene	50d
3. 1,1-Pentamethylene-2-(2-hydroxyethyl)ethylene	50d
4. Bicyclo(2.2.1)hept-5-ene-2-methanol	13a
5. 2-Cycloocten-1-ol	59a
6. 3-Cycloocten-1-ol	59a
7. 4-Cycloocten-1-ol	59a
8. 1-Phenyl-4-cycloocten-1-ol	59b
9. 7,8-Benzo-*trans*-2,3-dimethyl-6-hydroxy-bicyclo(2.2.2)octene	72
10. 7,8-Benzo-bicyclo(2.2.2)octene-*trans*-2,3-bismethanol	72
11. Coriamyrtin	109
12. Gaiol	108
E. Saturated hydroxy ethers (cyclic and noncyclic)	
1. β-Hydroxy ethers	
a. 2-Methoxyethanol	107
b. 2-Ethoxyethanol	107
c. 2-Pentoxyethanol	38a
d. 2-Propoxy-1-propanol	38a
e. Tetrahydrofuran-2-methanol	54
f. Tetrahydropyran-2-methanol	54
2. γ-Hydroxy ethers	
a. 3-Butoxy-1-propanol	38a

Continued

TABLE 1 (*Continued*)

	Ref.
21. 1-Phenyl-2-pentanol	2c
22. 5-Phenyl-1-pentanol	2c,2d
23. D-9	72
24. D-10	72

H. Steroidal alcohols [by position of OH group (major product)]
1. 1α-(1–10 fragmentation) — 28c
2. 1β-(1–10 fragmentation) — 28c
3. 2α-($2\alpha,9\alpha$ ether) — 138
4. 2β-($2\beta,19$ ether + ketone) — 124
   -($2\beta,19$ hemiacetal)[a] — 121,126
5. 3α-($3\alpha,9\alpha$ ether) — 36,122
   -($3\alpha,9\alpha$ ether-11-keto) — 36
   -($3\alpha,9\alpha$ ether-11,12-olefin) — 36
   -($3\alpha,9\alpha$ ether-9,11-olefin) — 36
6. 3β-(3-keto-A-nor) (CO_2 loss and position 5 epimer) — 33e
   -(3-4 fragmentation) — 56
   -($3\beta,19$ ether) — 121
   -($3\beta,19$ ether)[a] — 123
7. 4α-($4\alpha,9\alpha$ ether) — 129
8. 4β-($4\beta,19$ hemiacetal) — 121
   -($4\beta,19$ ether) — 33a,121
9. 5α-(5–10 fragmentation) — 28d
   -(5α-hydroxy-6α-formate)[a] — 78e
10. 5β-(5–10 fragmentation) — 28d
   -(4-en-5,9-ol ether) (B-nor) — 44
11. 6β-($6\beta,19$ ether) — 78a
   -($6\beta,19$ ether)[a] — 133
   -($6\beta,19$ ether + hemiacetal) (5α-chloro) — 127,130
   -($6\beta,19$ ether) (5α-bromo) — 78c,131
   -($6\beta,19$ ether) (5α-bromo-19-methylene) — 50f
   -($6\beta,19$ ether) (5α-methoxyl)[a] — 78c
   -($6\beta,3\beta$ ether) (C_{10} inversion) — 33d
   -(5-6 fragmentation) — 33d
12. 11α-($11\alpha,1\alpha$ ether) — 68
   -($11\alpha,19$ fragmentation) — 120
   -($11\alpha,1\beta$ ether) — 68
   -($11\alpha,1$ enol ether)[a] — 68,140
13. 11β-($11\beta,18$ ether) — 126,139
   -($11\beta,18:18,20\beta$ diether) — 126

114

-(11α,1α ether)	33a
-(11β,19 ether + hemiacetal)[a]	33a,134,139
-(11β,18 ether-19-iodo)[a]	139
14. 17α-(13–17 fragmentation)	56,70f
15. 17β-(13–17 fragmentation)	56,70f
16. 19-(5,10-ene-6-ol acetate)	50g,56,125
-(10-acetate + aromatic A ring)	50g,56
-(8,19 ether) (7-acetoxy or 14,15-olefin)	50g
17. 20α-(17–20 fragmentation)	34a
-(18–20 ether)	135
-(18–20 hemiacetal)[a]	136
18. 20β-(18,20 ether)	126
19. 23-(17,23 ether)	128
-(iodo, 17,23 ether)[a]	128
20. 24-(20,24 ether)	50b
-(iodo 20, 24 ether)[a]	50b

[a] Iodine added, oxidant = acetylhypoiodite.

2. Discussion

The majority of the published literature lists LTA–alcohol reactions, both thermally (60–90°C) and photolytically (15–30°C) induced, in the solvents benzene, hexane, cyclohexane, heptane, chloroform, and acetic acid. For the most part a given alcohol gives nearly the same product ratios in any of these, the major differences being lower cyclic ether yields and higher amounts of acetate ester formed in acetic acid media.* It is implied in the present discussion that those products arising from solvent–oxidant reaction are of too low percentage to merit consideration. Hence the following lists of transformations ignore side reactions not involving an alcohol reactant. Examples of alcohol, *or a fragment thereof*, becoming involved with solvent molecules or acetate anions are sparingly presented.

The data displayed in Table 1 for alcohol oxidation by LTA in apolar media has been accumulated since 1960. Its consideration reveals that the oxidant shows definite trends in its ability to form certain products in excess of others. The products that are unusual, and that therefore separate LTA from most other oxidants, are the *tetrahydrofuran* ethers. These are formed by an intramolecular process, and often in major proportions, when

---

* Oxidation rates are also quite different in acetic acid than in other solvents. This is supposedly due to suppression of the primary and secondary alcoholysis reactions (not rate determining) between alcohol and LTA (see Figure 5). However, examples of strong acid catalysis in glycol cleavage are known (22).

complicating factors are absent. That activation is needed to effect the trans-formations is evident from the essential lack of depletion of the oxidant when mixed with alcohol and benzene at 15–30°C. This medium is stable for many hours (7c).

The utility of LTA in forming tetrahydrofuran ethers is shown by two other reactions which are now being used for similar purposes. These are the alkyl nitrite and hypohalite decompositions. The more dated N-haloamine decompositions are also mechanistically related (Figure 9).

The fact that LTA is able to form the intramolecular cyclization product within one medium while multiple media are required of the other methods is intriguing. Examination of a series of alcohol products by the referenced research workers has now given the chemist a reasonable understanding of the processes that can be expected to occur. *The total mechanism of ring closure, however, has eluded specific detection.*

It is obvious from the data that many products are obtained from each alcohol. They are classified as products of esterification, dehydration, and oxidation for purposes of discussion. Only acetate esters of the starting alcohols are considered in the first class, for other acetates are the secondary products of the reactions of the oxidant with the initial products. The products of dehydrogenative ring closure, carbonyl formation and frag-mentation fall within the oxidation class.

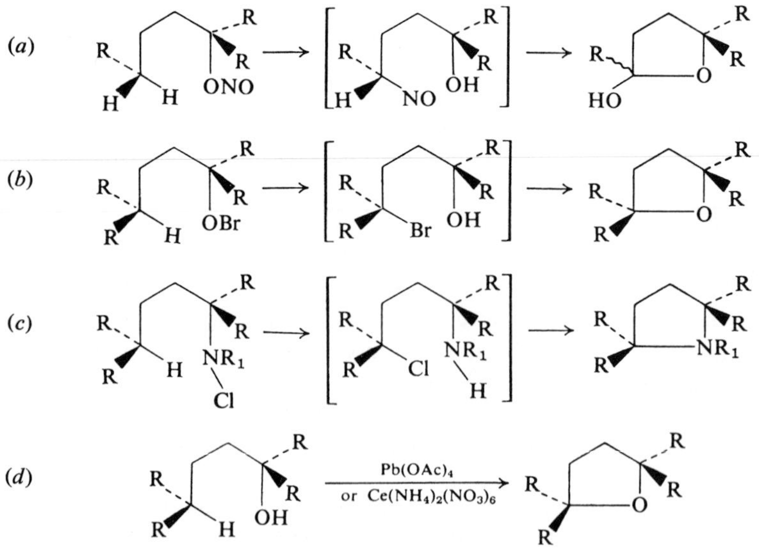

Figure 9. (1a,18)

(a) Acetate esters are formed in most LTA oxidation reactions. They are major products, in benzene medium of alcohols A.1.c, A.1.h, A.1.j, A.2.a, C.1.a, C.1.b, C.1.c, C.1.e, C.1.f, C.1.g, C.1.h, C.1.p, C.3.b, E.1.e, G.10, G.13, and G.15, most of which either cannot enter into tetrahydrofuran formation or, if they can, require intramolecular reaction at a methyl group.

One possible route leading to the acetates is direct esterification of the starting alcohol with acetic acid produced from LTA in the course of the reaction (26).* This process probably takes place to a certain extent when oxidation is carried out in benzene or in excess of the starting alcohol (no other solvent) (26b). However, esterification with acetic acid is not the only reaction producing acetates. This is demonstrated by the fact that acetates are isolated, although in lesser yield, from reactions using pyridine in, or as, the solvent (26b). In these cases no *free* acetic acid is present and the redox reactions proceed rapidly to completion. The isolation of a small amount of acetic anhydride upon LTA oxidation of 1-propanol and 2-propanol, in the presence of pyridine (26b), suggests that the formation of the corresponding acetates may be, in part, interpreted as a base-catalyzed (when pyridine is present) esterification involving acetic anhydride (i.e., the acetyl cation) and the starting alcohol or the intermediate alkoxy Pb(IV) acetate.

Acetic anhydride might arise from the reactions shown in eqs. 2 and 3.

$$2Pb(OAc)_4 \longrightarrow \begin{array}{c} AcO \quad O \quad OAc \\ \diagdown \diagup \diagdown \diagup \\ Pb \quad Pb \\ \diagup \diagdown \diagup \diagdown \\ AcO \quad O \quad OAc \end{array} + 2Ac_2O \qquad (2)$$

$$OAc)_4 \longrightarrow \begin{array}{c} Pb(OAc)_3 \\ \diagup \\ O \\ \diagdown \\ Pb(OAc)_3 \end{array} + Ac_2O \qquad (3)$$

Equation 3 is possibly more correct because organolead compounds with similar structures have been isolated and identified. Thus, treatment of diphenyllead diacetate with acetone–water gives an organodilead oxane as shown in eq. 4, large yields of which are obtained if diazomethane is added to capture the generated acetic acid (24).

$$2(C_6H_5)_2Pb(OAc)_2 + H_2O \longrightarrow \begin{array}{c} Pb(C_6H_5)_2OAc \\ \diagup \\ O \\ \diagdown \\ Pb(C_6H_5)_2OAc \end{array} + 2HOAc \qquad (4)$$

---

* Typically, alcohol–acetic acid mixtures do not produce ester products unless an "acidifying" catalyst is added. LTA has demonstrated such a catalyst effect in the dehydrative rearrangement of phenyl tritylcarbinol (23).

Definite information regarding esterification comes from the oxidation of (+)-2-hexanol and (−)-3-octanol (25a). The corresponding acetates were formed with *complete retention of configuration*. This observation indicates that acetylation of hydroxyl groups in the LTA reaction does not involve cleavage of the carbon–oxygen bond in the starting alcohol, either in an S_N2 or S_Ni type substitution on the carbinol atom, or with the formation of carbon radicals or cations followed by combination with an acetate containing species. It does proceed either by esterification of the starting alcohol or the alkoxylead(IV) acetate intermediate with acetic acid or acetic anhydride, and/or by nucleophilic attack of the hydroxyl oxygen atom of the alcohol on one of the carboxylate carbon atoms of a Pb(IV) acetate species (25).

$$
\begin{array}{c}
\overset{\displaystyle O}{\overset{\displaystyle \|}{CH_3C}}\!-\!O\!-\!X \\
\end{array}
$$

(5)

$$
X = \;-\!H,\;\; -\!Pb(OAc)_3,\;\; \text{or} \;\; -\!\overset{\displaystyle O}{\overset{\displaystyle \|}{C}}CH_3
$$

The presence of a small amount of rearranged acetate among the oxidation products of 1-propanol, 1-phenyl-2-pentanol, 2-methyl-2-hexanol, and 2-hexanol (giving 2-propylacetate, 1-phenyl-1-pentyl-acetate, and 2-butyl acetate from a fragmentation intermediate respectively) suggests that the primary coordination of alcohol to LTA (Figure 5), or the polarization tendency in alkoxylead triacetate,* stimulates some primary carbonium ion character on the α- and/or β-carbon. Favorably, a 1,2-hydride shift (concerted or not) followed by addition of acetate ion to the more stable secondary carbonium ion could yield the observed product. However, a recent demonstration of a 1,2-*hydrogen atom shift* in somewhat related systems does not allow the discarding of such a process in the present case (27). Evidence will be given that some carbon–oxygen homolytic cleavage occurs in LTA reactions and in the case of 1-propanol the subsequent transformations would be as shown in eq. 6. As in the carbonium ion mech-

$$
CH_3CH_2CH_2\!\cdot\; \rightleftarrows \; CH_3\!-\!\overset{H}{\overset{\cdot}{C}H}\!-\!CH_2 \; \rightleftarrows \; CH_3\!-\!\dot{C}H\!-\!CH_3
$$

(6)

---

* Polarization of bonds in Pb(IV) compounds containing two, three, or four different groups is open to question because dipole moment studies are lacking. It is to be noted that the 2-pentyl radical, formed by oxidative fragmentation of 3-methyl-2-hexanol and 2,3-dimethyl-2-hexanol, does not give rearranged acetate product.

anism, the hydrogen atom shift may be concerted with capture of an acetoxyl moity (3)

$$
\begin{bmatrix}
\begin{array}{c}
\text{CH}_3\text{CH} \overset{\text{H}}{-\!\!\!-\!\!\!-} \text{CH}_2 \\
\text{O} \qquad\qquad \text{Pb(OAc)}_2 \\
\text{C} -\!\!\!- \text{O} \\
\text{CH}_3
\end{array}
\end{bmatrix}^{\cdot\,or\,\dagger}
$$

3

(b). A second reaction class observed when alcohols are treated with LTA is dehydration. To date such *low yield* products have been identified in only a few experiments. The trend seems capable of generality, however, since other product analyses demonstrate materials derivable from intermediate olefins (Figure 10). In chronology, Mosher reported Example a, Figure 10, and explained it as a LTA-induced acid-catalyzed dehydration since the reaction solvent was acetic acid. Unlike this example are the others wherein reaction was carried out in the supposedly acid-free medium of refluxing benzene with added calcium carbonate. Indications that these latter dehydrations may occur via a route other than simple β-elimination reaction (eq. 7) are due to concurrent isolation of hydrocarbons derivable only from the carbonoid portion of the starting alcohol (26b). Thus, from the oxida-

$$
\begin{array}{ccc}
R\!\!-\!\!\overset{\displaystyle R}{\underset{\displaystyle H}{\diagup}}\!\!\!\!\!\overset{O}{\diagdown}\!\!\!\!\!\!\!\!\!\!\!\!\!\!\!\!\underset{\displaystyle Ac}{\overset{O}{\diagup}}\!\!\!\!\!\!\!\!\!\!\!\!\!Pb(OAc)_2 & \longrightarrow & \overset{\displaystyle R}{\underset{\displaystyle R}{\diagup\!\!\!\!\diagdown}} \;\; + \; HOAc \; + \; \overset{O}{\overset{\|}{Pb}}(OAc)_2
\end{array} \qquad (7)
$$

tion of 1-propanol, the products hexane and 1-propylbenzene imply that the n-propyl radical resides in the medium long enough to react with its duplicate or with solvent (26b).* If this is the case, then the n-propyl radical could also disproportionate to give the observed alkene. The disproportionation, which requires an alignment of the two radicals in a specific orientation, is very unlikely in sterol oxidation reactions. Hence, the minor alkene and alkane products exemplified in Example f, Figure 10, probably arise from intermediate carbonium ions and hydrogen abstraction from solvent respectively.

* n-Propylbenzene could also arise by a more ionic, Friedel-Crafts-type reaction, wherein the 1-propyl cation is never free because of the lack of formation of isopropylbenzene (26b). The relative rates of oxidation of carbon radicals (2d) and the lack of formation of cumene in the oxidation of phenylisobutyric acid (1b) suggests that the n-propyl radical reacts with solvent before being oxidized.

(a) $\quad C_6H_5CHOHC(C_6H_5)_3 \longrightarrow (C_6H_5)_2C{=}C(C_6H_5)_2$

(b)

$CH_3CH_2CH_2OH \xrightarrow{\;7\%\;}$
$\xrightarrow{\;4\%\;} CH_3CH{=}CH_2$
$CH_3CHOHCH_3$

(c)

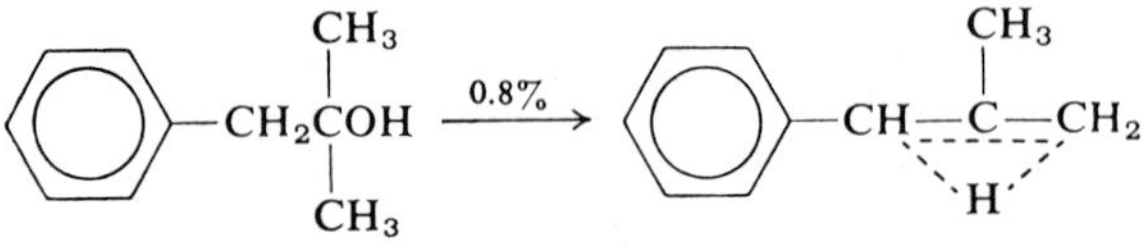

$\xrightarrow{\;0.8\%\;}$

(d)

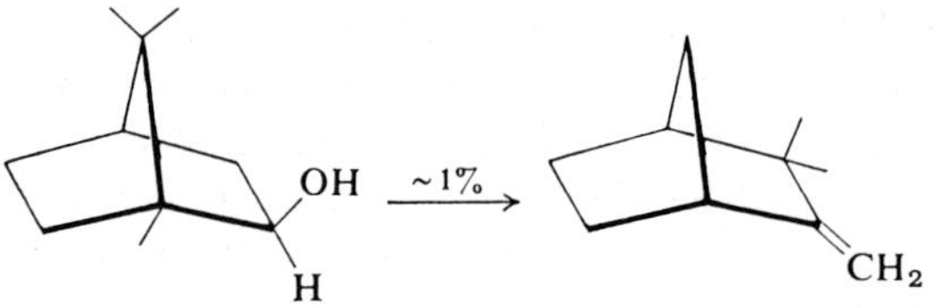

$\xrightarrow{\;\sim1\%\;}$

(e)

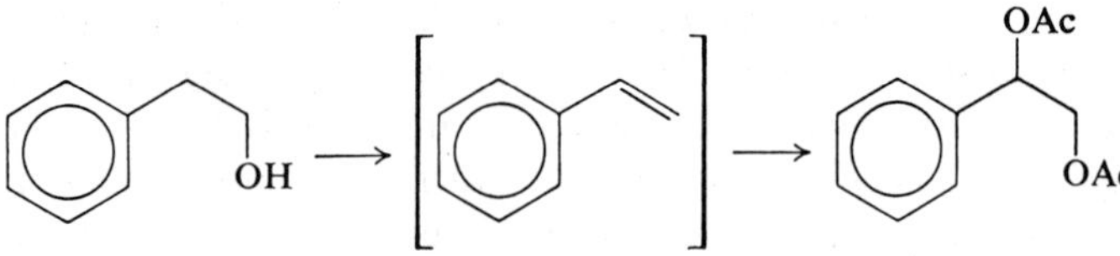

(f)

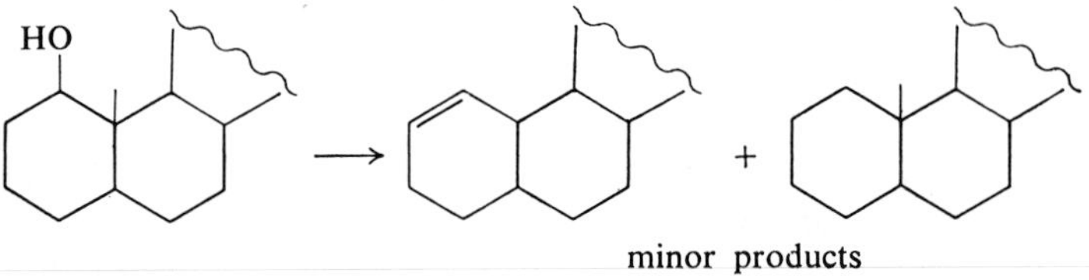

minor products

(g)

minor product

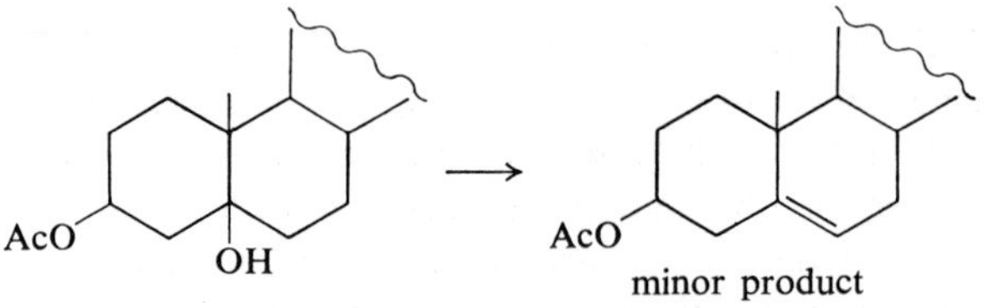

Figure 10. (23,26,28). The diacetate of (e) may alternatively be formed by benzylic acetoxylation of the acetate ester of the starting alcohol.

The small amount of camphene detected in the product of reaction of isoborneol (Figure 10d) with LTA in benzene does not clarify the mechanism picture of dehydration (28b). It is tempting to suggest a Wagner-Meerwein carbonium route, yet the pyrolyses of 2-azo-camphene and 4-(β-azoethyl)-1,3,3-trimethyl-1-cyclopentene, via obvious free radical intermediates, verify the capability of the 2-bornyl *radical* to undergo the same rearrangements (Figure 11).*

(*c*). Not the least important of the reaction classes, oxidation is discussed last because of its more complicated aspects. The three primary processes that occur are intramolecular cyclization, aldehyde or ketone formation, and fragmentation (Figure 12). Detailed information is now available on these reactions and allows discussion of particular intermediates. Figure 5 shows the primary and secondary reactions taking place between alcohol and LTA molecules. The subsequent decomposition of the alkoxylead(IV) acetate (*A* and/or *B*, Figure 5) can occur by two routes, heterolytic or homolytic. Evidence favors a *homolytic* path for both thermal and photolytic activation in apolar media. The lack of oxygen scavenging during alcohol oxidation implies that no real "free" radicals are present in the medium

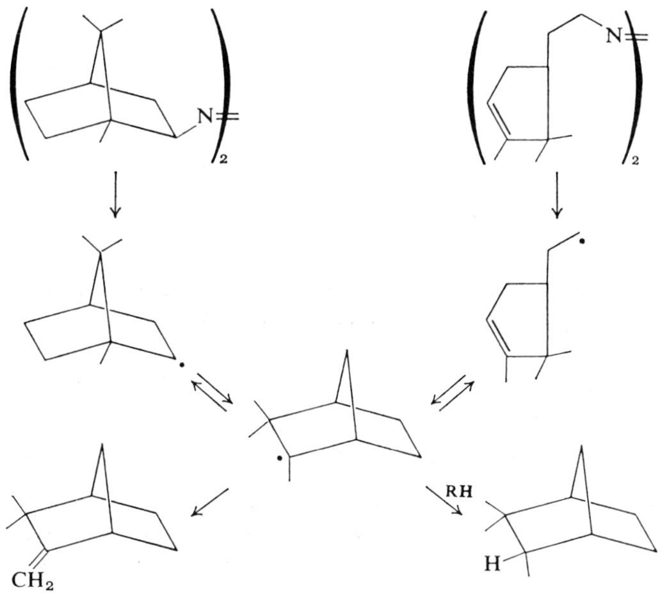

Figure 11. (29)

* The oxidation of isoborneol was carried out at 80°. The azo compounds do not decompose until activated by much higher temperatures. It is not known whether the 2-bornyl radical, formed at 80° in the absence of an oxidant, will undergo conversion to camphene.

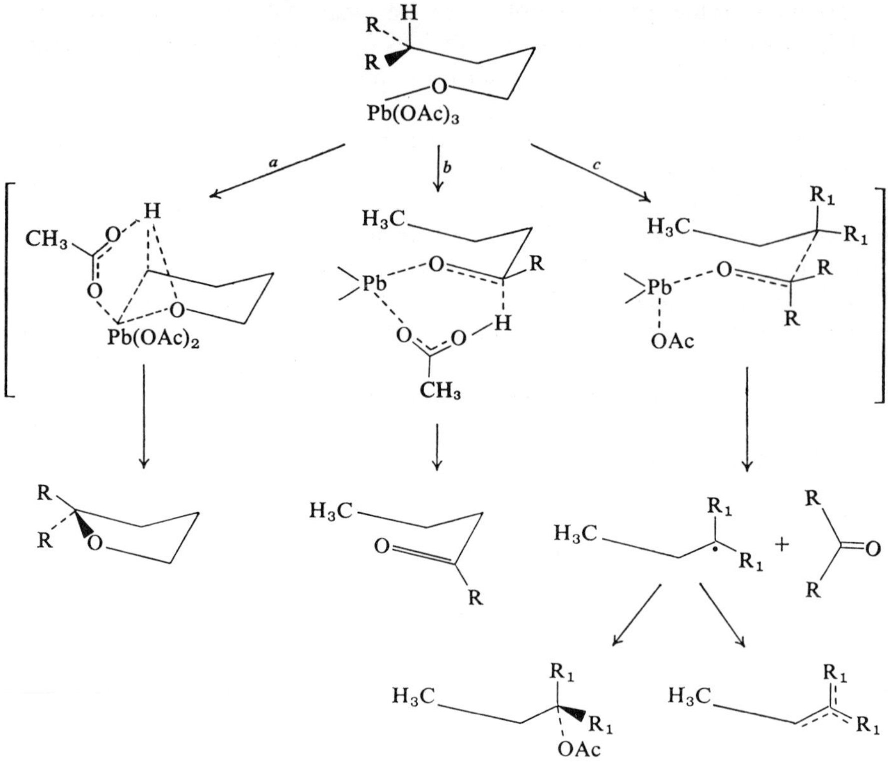

R = H, alkyl, or aryl

Figure 12. Bracketed intermediates are drawn to show the total transformations observed and are not meant to define mechanisms.

and/or the reaction chain length is very short. A listing of some critical experiments follows.

(*1*) Reactions run in apolar media, whether induced by light or heat, produce nearly the same yields of products. (30).

(*2*) Oxidations in apolar and acetic acid media carried out below 50°C, in the absence of light, are exceedingly slow. Addition of a basic substrate (e.g., pyridine) to the apolar media speeds reaction to produce major amounts of aldehyde or ketone from primary or secondary alcohols (31).

(*3*) Oxidation of mono-*p*-substituted triphenylcarbinols at 80° in a variety of benzene media gives (phenyl: substituted-phenyl) carbon to oxygen migration aptitudes suggestive of an oxygen radical intermediate. Oxidation of the same carbinols in acetonitrile gives aptitudes of ion or concerted migration (32).

(*4*) Oxidation of 4-β-hydroxy-17-β-propionyloxy-5-α-androstane and 3-β-20β-diacetoxy-11-β-hydroxy-5-α-pregnane gives cyclic ethers which have some epimerized alcohol carbon–oxygen bond (33a,33b). Likewise, oxidation of menthol, (+)-isomenthol, and (+)-neomenthol gives bicyclic ethers which have some epimerized alcohol carbon–

oxygen bonds (33c). Further, an example of reversible fragmentation with a necessary 1,3-hydrogen migration at an intermediary stage is evidenced in the oxidation of the 6-β-hydroxy steroid (eq. 8) (33d).

$$(8)$$

The 40% conversion shown by eq. 9 (33e) points out the capabilities of LTA to center its oxidizing power at one point in a molecule. Since the

$$(9)$$

3-α-carbethoxy derivative does not undergo the reaction, and because the 3-α-carboxylic acid group probably reacts faster with LTA (to give a mixed tetracarboxylate having an intramolecular hydroxyl group) than does the 3-β-hydroxyl group, and because it is difficult to rationalize epimerization at the 5-position using the radical decarboxylation product, the reaction path of eq. 10 may prevail,

$$(10)$$

+ Pb(OAc)$_2$ + $^-$OAc

(5) Oxidation of (20R)- and (20S)-3-ethylenedioxy-20-hydroxy-δ^5-pregnene gives a low yield of product containing a new methylene group (34a).

(6) Oxidation of 4,4,4-triphenyl-1-butanol gives some 4,4-diphenyl-2,3-benzodihydropyran in the presence of oxygen. Dimer products derivable only from a fragment with free radical character are obtained (eq.11) in other experiments (35).

$$\text{(11)}$$

(7) Products of reaction of an alkoxy group with solvent molecules have been isolated in several cases. One, formation of an O-cyclohexyl ether when a steroidal alcohol is oxidized in cyclohexane, must be a result of initial alkoxy radical formation (36). These reactions can be formulated as either:

$$RO \cdot \ [\text{from } ROPb(OAc)_3] + C_6H_{12} \longrightarrow ROH + \cdot C_6H_{11}$$
$$\cdot C_6H_{11} + RO \cdot \ (\text{in some form}) \longrightarrow C_6H_{11}OR$$

or

$$LTA \longrightarrow \cdot CH_3 + CO_2 + \cdot Pb(OAc)_3$$
$$\cdot CH_3 + C_6H_{12} \longrightarrow CH_4 + \cdot C_6H_{11}$$
$$\cdot C_6H_{11} + RO \cdot \ (\text{in some form}) \longrightarrow C_6H_{11}OR$$

or

$$RO \cdot [\text{from } ROPb (OAc)_3] + C_6H_{12} \longrightarrow ROH + \cdot C_6H_{11}$$
$$\cdot C_6H_{11} + LTA \longrightarrow {}^+C_6H_{11} + \cdot Pb(OAc)_3 + {}^-OAc$$
$${}^+C_6H_{11} + ROH \longrightarrow C_6H_{11}OR + H^+$$

Of these three routes, the second seems least likely due to the lack of gas evolution in most reactions. Walling has qualitatively discussed intermolecular reactions of LTA generated alkoxy radicals with hydrocarbons (37).

The formation of isopropylphenyl ether when isopropyl alcohol is oxidized by LTA in benzene (26b) is most likely a result of alkoxy radical attack (addition to) on benzene, followed by loss of a hydrogen atom or proton (from a cyclohexadienyl carbonium ion) (38). The formation of biphenyl in the same reaction most likely results from a similar attack of phenyl radical on benzene. The resistance of phenyl radicals to be oxidized to phenyl cations is exemplified by the stable phenyllead triacetate (as opposed to the unstable cyclohexyllead triacetate) (39).

Results (1), (2), and (3) strongly suggest that the oxidation mechanism prevailing is mostly a function of solvent.* Alkoxy radical formation is indicated in apolar media. If the alkoxylead(IV) triacetate intermediate decomposed *in apolar media* via heterocyclic Pb—O bond cleavage to give alkoxonium ion, lead diacetate, and acetate ion, considerably more aldehyde or ketone should be produced by α-proton removal than cyclic ether by δ-proton removal. Heterolysis of the Pb—O and α C—H bonds is the only reasonable explanation for the efficiencies of the low temperature reactions in the more polar, basic media.

Result (*3*) shows another interesting application of migration aptitudes of aryl groups in determining the nature of the alkoxy oxygen atom after Pb—O bond cleavage. It is well known that both alkoxy radicals and cations are electron deficient and demonstrate "acceptor" properties (41). Even with this similarity however, substituent effects on free radical migrations (to oxygen in RO·) are opposite those in ionic counterparts (42). Thus, the *p*-nitrophenyl group in this experiment would migrate poorly, in comparison to phenyl, to an oxygen cation but more efficiently to an oxygen radical. The fact that the aptitude favors *p*-nitrophenyl in a variety of media indicates the oxygen radical character is more pronounced than the cation character.

Evidence that substituent groups are doing more than shifting electron density comes from experiments using substituted benzenes as additives. The yield lowering, migration aptitude changing capability of added nitrobenzene, for instance, in the experiments of Starnes (32) and Norman (32) has been rationalized as due to inhibition of a free radical propogation step when the reactions are run in benzene. The same argument implies that any percentage of free radical reaction taking place in the acetonitrile medium should be lowered by addition of nitrobenzene. On this point results seem to be ambiguous and do not allow a clear-cut definition. Additionally, there is evidence that such groups as acetyl, benzoyl, nitro, cyano, halo, and methoxyl enter into coordination with Pb(IV), the net result of which is yet to be clarified. To date it is only known that these tend to enhance reaction rates (20, 43).

Result (*4*) demonstrates conclusively that alkoxy radicals are the first intermediate products of alcohol oxidation by LTA in apolar media. The essential features are fragmentation reactions cleaving a bond between two carbon atoms, at least one of which is asymmetric (33). Epimerization of the original C—O bond occurs when the reverse of fragmentation takes place. The newly formed alkoxy radical can then enter into its own set of subsequent reactions (Figure 13). If the alkoxy–lead bond cleaved to give an alkoxonium ion, followed by fragmentation of the α-carbon–β-carbon bond, the resulting

---

* LTA decomposes to CO_2 and ·CH_3 products by heat and light activation in cyclohexane, benzene, and carbon tetrachloride (40).

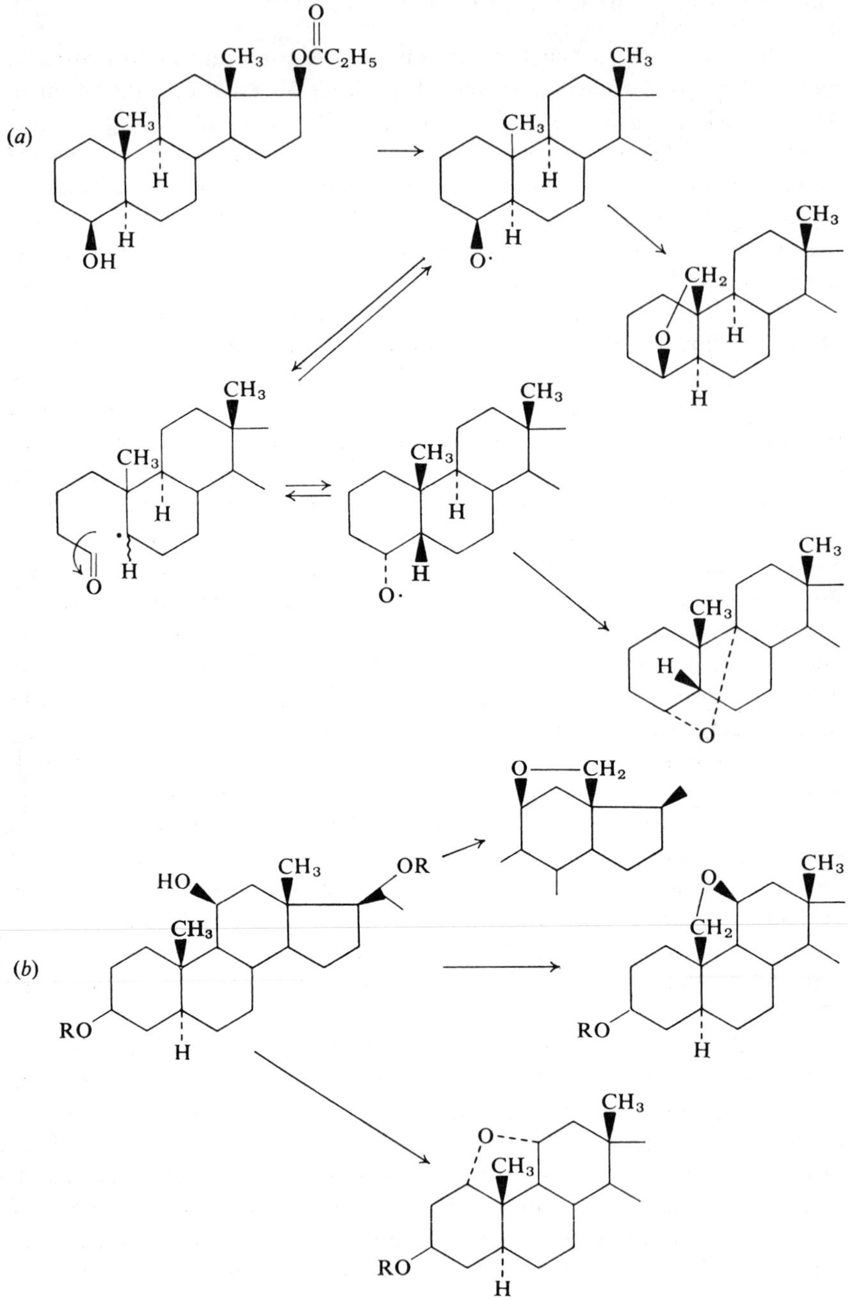

R = Ac

Figure 13

(*c*)

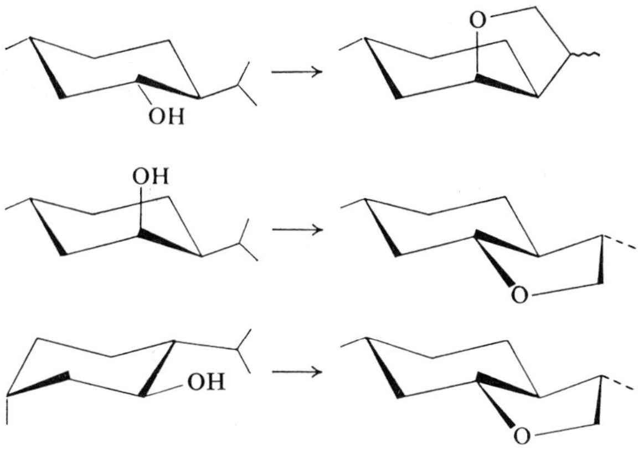

Figure 13—*Continued*

carbonium ion would hardly be expected to reattach itself to the rather electron deficient carbonyl carbon atom (Figure 14). In fact, the natural polarization in the carbonyl group, formed by the fragmentation process, would direct the reverse reaction to give *O*-alkylation if carbonium ions were involved. Consideration must be given here to a very interesting LTA

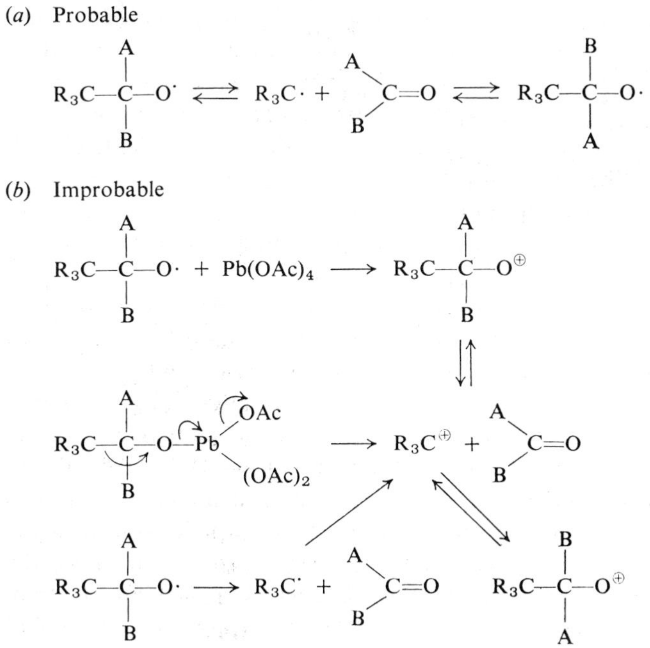

(*a*) Probable

$$R_3C-\underset{B}{\overset{A}{C}}-O\cdot \rightleftharpoons R_3C\cdot + \underset{B}{\overset{A}{C}}=O \rightleftharpoons R_3C-\underset{A}{\overset{B}{C}}-O\cdot$$

(*b*) Improbable

$$R_3C-\underset{B}{\overset{A}{C}}-O\cdot + Pb(OAc)_4 \longrightarrow R_3C-\underset{B}{\overset{A}{C}}-O^{\oplus}$$

$$R_3C-\underset{B}{\overset{A}{C}}-O-Pb\overset{OAc}{\underset{(OAc)_2}{\bigg\langle}} \longrightarrow R_3C^{\oplus} + \underset{B}{\overset{A}{C}}=O$$

$$R_3C-\underset{B}{\overset{A}{C}}-O\cdot \longrightarrow R_3C\cdot + \underset{B}{\overset{A}{C}}=O \qquad R_3C-\underset{A}{\overset{B}{C}}-O^{\oplus}$$

Figure 14

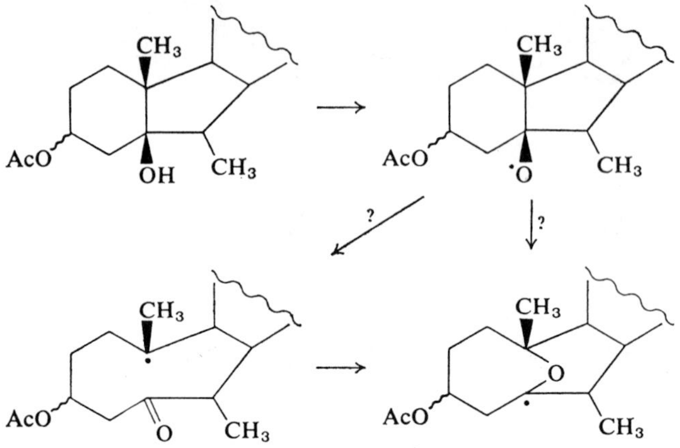

Figure 15. (44)

oxidation reported by Rosenthal (44). In the conversion of a 5-β-hydroxy-B-norsteroid to a mixture of enol ether and hemiketal acetate (Figure 15), an explanation has been advanced in support of a novel 1,2-radical migration from carbon to oxygen.

Further evidence substantiating alkoxy radical character is found in results (5), (6), and (7). The formation of a methylene group (5) from a carbinol molecular fragment in LTA reactions must be by carbon radical capture of a hydrogen atom and not by capture of an extremely improbable hydride ion by an intermediate carbonium ion. Likewise, capture of oxygen by the 4,4,4-triphenylbutanol fragmentation product, i.e., 3,3,3-triphenyl-propyl, must be due to the primary propyl radical (Example 6) (Figure 16). The peroxide radical thus formed undergoes oxygen–oxygen bond cleavage to give an alkoxy radical of optimal chain length for intramolecular cyclization to the benzopyran (35).

The conclusive establishment of an alkoxy radical as the first decomposition intermediate in apolar media allows one to consider next the known factors that dictate its subsequent transformations.

The major oxidation reaction of the alkoxy radical is often cyclization to a tetrahydrofuran (THF) derivative. These products are only realized, however, if the starting alcohol has a suitably situated δ-C—H bond and other factors are favorable. If the alcohol molecule does not conform to this temporarily vague demand, the alternative oxidation reactions of aldehyde or ketone formation, or fragmentation, may become dominant. Since the results of numerous investigations point out the generality of THF formation, we will direct our thought to those examples that (1) point out alcohol

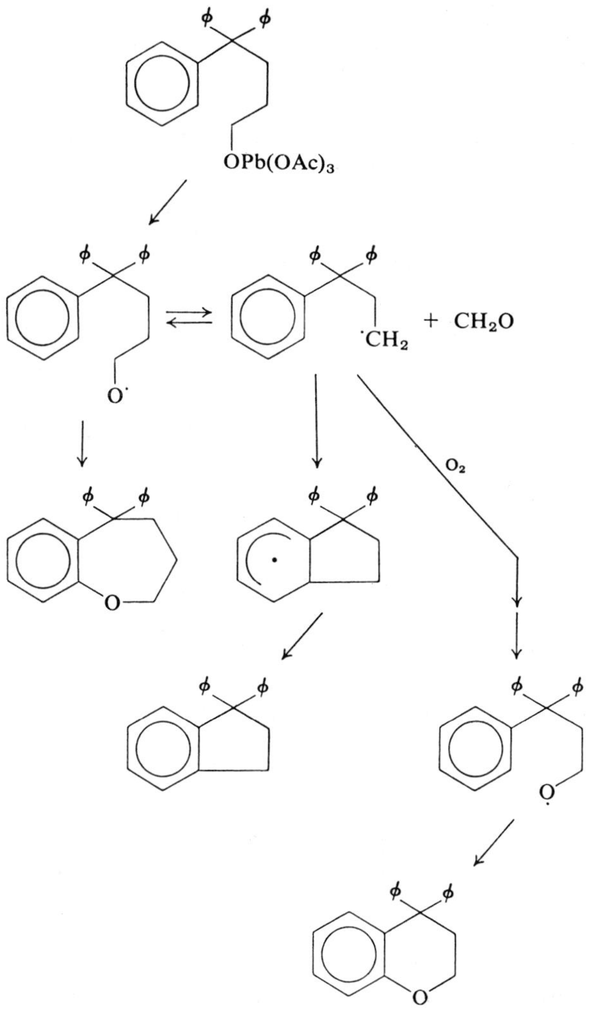

Figure 16. (35). In the case at hand, if the 3,3,3-triphenylpropyl radical had been oxidized to the corresponding carbonium ion (45), or if concerted heterolytic fragmentation produced the ion directly, a subsequent Friedel-Crafts reaction would have produced more of the indane derivative, proton loss would give 3,3,3-triphenyl-1-propene, or hydride and phenide migration would yield 1,1,2-triphenyl-1-propene.

structural features that enhance or hinder this ether forming reaction, and (2) give a mechanism definition to the three oxidation reaction sequences.

Apart from those *tabulated* alcohols that can form the tetrahydrofuran derivatives on the basis of possession of a δ-C—H bond, are methanol, ethanol, propanol, and 2-butanol on the one hand; and 3-*n*-butoxy-1-propanol, 3-benzyloxy-1-propanol, 3-phenyl-1-propanol, 4,4-dimethyl-1-pentanol, 4,4-dimethyl-1-hexanol, 3,3,3-triphenyl-1-propanol, and 4,4,4-triphenyl-1-butanol on the other. Intuitively the first group could be oxidized to their respective carbonyl compound or to cyclic epoxide or oxetan ethers. The data show that in no case is intramolecular cyclization to give an ether ring smaller than five members competitive with carbonyl formation (46). The second group, no molecules of which possess a δ-C—H bond, cannot undergo cyclization to THF derivative either. They do prefer reaction with ε-C—H bonds, over the γ counterpart, to yield tetrahydropyran (THP) products. Simultaneously, the total yield of cyclization is usually diminished (47).

Having just introduced examples of oxidative cyclization into the discussion, it is necessary to consider the possible reaction paths leading to such products. In apolar media the alkoxy radical produced from the alkoxylead(IV) triacetate intermediate finds stabilization by intramolecular hydrogen transfer. The total process is exothermic by approximately 8.5 kcal/mole (48). This reaction, which is in competition with intermolecular hydrogen abstraction (an entropy factor for which is unfavorable relative to the intramolecular process) and the other two oxidative processes (carbonyl formation and fragmentation), is highly favored when the transition state is a cycle of six atoms.* The same type of intermediate has been established for homolytic

$$[R_2C^{\delta+}\text{---}H\text{---}^{\delta-}\cdot OR]$$

$$(12)$$

decomposition of alkyl hypohalites and nitrites, as well as dialkyl-*N*-chloro-ammonium ions. In all cases, deviation from the six-membered transition state causes the competing reactions to excell (18a). Additionally, the latter are pronounced in cases where steric factors decrease the stability of the desired six-membered hydrogen transfer structure.

The δ-carbon radical, depending on its structural environment, is next oxidized more or less rapidly to a carbonium ion. Evidence for this comes from three experimental results. First, oxidation of optically active 4,8-

---

* Note should be made of substituent constant correlation in hydrogen atom abstraction by alkoxy radicals. The results show the importance of σ+ constants in describing the reactions and suggest a dipolar transition state (41,49).

dimethyl-1-nonanol yields an optically inactive mixture of THF ethers (Figure 17 (50a). In this example the δ, tertiary carbon radical is susceptible to rapid oxidation by Pb(IV) and/or Pb(III) (45). Once it is converted to the hydroxy carbonium ion the intermediate possesses no optical activity and can easily cyclize to the observed product by loss of a proton from oxygen. At this point in the discussion, however, one cannot conclude that the δ-carbon radical *must* be oxidized to the ion prior to cyclic ether formation. The carbon radical could undergo rapid inversion and thus may provide inactive products if it is the ether precursor.

Two other products isolated from this reaction are shown by structures **4** and **5**, the formation of which is best described via δ-hydrogen abstraction

4   5

by the alkoxy radical followed by carbonium ion formation, and then proton loss in two different directions to give two isomeric olefins. These then cyclize in what is best described as an oxyplumbation reaction.

Additional demonstrations of nonstereospecificity, but ring strain control in the cyclization reactions, are provided in the oxidation of 2-cyclohexylethanol, 3-cyclohexylpropanol, 2-cyclopentylethanol, and 3-cyclopentylpropanol (50c, 50d). A myriad of products are accountable from intermediates formed after intramolecular δ- and ϵ-hydrogen abstraction by the alkoxy radicals. As expected, *cis*-fused 7-*oxa*-bicyclo(4.3.0)nonane and 2-*oxa*-bicyclo(3.3.0)-octane predominate the *trans* isomers. Equal amounts of *cis*- and *trans*-2-*oxa*-bicyclo(4.4.0)decane are formed.

In Figure 17 a Pb(III) species is used in Step *C* to oxidize the δ-carbon radical. This may or may not be the case. If not, a Pb(IV) species, most likely alkoxylead triacetate, would serve as effectively. Proponents of this latter possibility, and therefore a chain reaction sequence with Step *C* being the critical propagation step, must subscribe to very short chain lengths because of the lack of oxygen inhibition in these reactions.

I $ROPb(OAc)_3 \longrightarrow RO\cdot + \cdot Pb(OAc)_3$

           $RO\cdot \longrightarrow \cdot ROH + \cdot Pb(OAc)_3$

P $\cdot ROH + ROPb(OAc)_3 \longrightarrow {}^+ROH + RO\cdot + Pb(OAc)_2 + {}^-OAc$
           $\cdot ROH + Pb(OAc)_4 \longrightarrow {}^+ROH + \cdot Pb(OAc)_3 + OAc^-$

T $\cdot ROH + \cdot Pb(OAc)_3 \longrightarrow {}^+ROH + Pb(OAc)_2 + OAc^-$

where I = initiation; P = propagation; T = termination; RO· = e.g.,
$CH_3CH_2CH_2CH_2CH_2O·$; ·ROH = e.g., $CH_3\overset{.}{C}HCH_2CH_2CH_2OH$; and
^+ROH = e.g., $CH_3\overset{+}{C}HCH_2CH_2CH_2OH$.

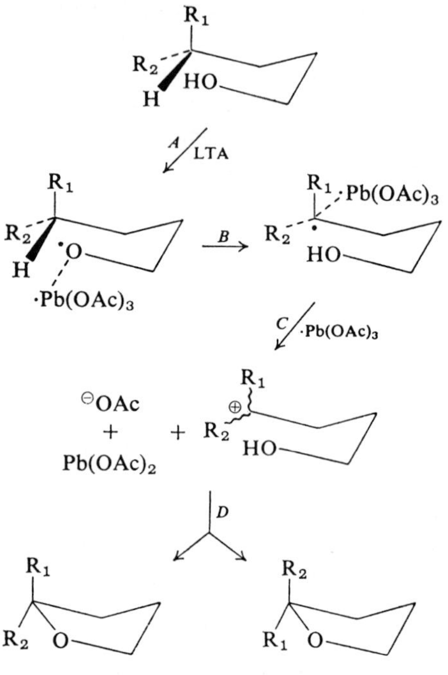

Figure 17. (50)

Another noticeable part of Figure 17 is that the species are always
paired as radicals or ions.* Fragmentary evidence supports this view; i.e.,
oxidation of a diol, wherein the two hydroxyl groups end up equidistant

---

* If the lead triacetate fragment associated with the δ-carbon radical to form a
carbon–lead bond, similar to the formation of δ-chloro alcohols in alkyl hypochlorite
decompositions, it would be expected to immediately dissociate into a carbonium ion,
lead diacetate, and acetate ion (39). It is not known whether a δ-hydroxy carbonium ion
formed via such a route would preferentially react with an acetate ion or with the intra-
molecularly situated hydroxyl group. A suggestion of preference comes from the lack
of formation (isolation) of δ-acetoxy alcohols in these alcohol oxidation reactions. The
processes of "oxidative substitution" and "oxidative elimination" apparently do not
compete with each other, while in oxidative decarboxylations they do (2d).

It is not likely that δ-acetoxy alcohols are *solvolysis* precursors to the observed
cyclic ether products because of facile photolytic ether formation in benzene at room
temperature.

An interesting comparison is the instability of alkyllead triacetates and the stability
of alkyllead trihydroxides (51).

from the δ-carbon atom (presumably a carbonium ion), yields only one cyclic ether, that derived from the hydroxyl group entering into initial reaction (see Figures 5 and 12a) with LTA. This constraint imposes either an intimate radical pair or concerted mechanism into the method of decomposition of the alkoxylead triacetate intermediate. An opposing view has been suggested (35).

It is important to notice that Steps B, C, and D of Figure 17 cannot be replaced by one conjugate of all parts. That is, if the alkoxylead triacetate decomposed to give only a nonclassical radical directly,* followed by oxidation to a nonclassical ion, the stereochemistry at the δ-carbon would have been retained in the THF product (eq. 13). However, such a nonclassical bridged transition state may be operative in rigid systems in which both reacting centers (C—OH and δ-C—H) are geometrically fixed, for example in various steroid alcohols (16c, 50g).

$$\text{(13)}$$

optically active   optically active

A second experiment also points to δ-carbonium ion character in a reaction intermediate. Oxidation of 3-cyclohexyl-1-propanol gives a major yield of 1-(3-hydroxypropyl)-1-cyclohexene (50d). In this example, after intramolecular hydrogen transfer and oxidation of the δ-carbon to a carbonium ion, the loss of a proton from carbon to give a stable, endocyclic, trisubstituted double bond apparently competes favorably with loss of a proton from oxygen to give a cyclic ether (eq. 14).†

A secondary reaction, probably an oxyplumbation type oxidation of the product ene-ol, has been shown capable of producing the low yield ene-ether and acetoxy-ether products observed in several cases (50a,50c–e).

$$\text{(14)}$$

* This implies insertion into a carbon–hydrogen bond by the oxygen radical (52).

† Interpretation of these results by a free radical path would necessitate a mechanism whereby a δ-carbon radical undergoes hydrogen atom loss, possibly by disproportionation, to give the alkene. The rapidity of tertiary carbon radical oxidation makes this sequence unlikely (2d).

This is most certainly the case in the 60% conversion of eq. 15 (50f).

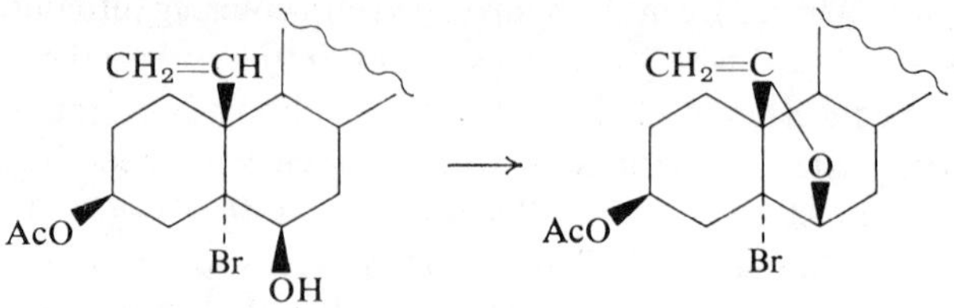

$$(15)$$

Finally, the formation of 2-ethyl-2-methyltetrahydrofuran in the oxidation of 4,4-dimethyl-1-pentanol is indirect evidence for intermediate carbonium ion character (53a). The product is best formulated as arising via a 1,2-methide shift [rather than methyl radical (42)] to give a more stable tertiary carbonium ion, followed by ring closure, The methide shift is in competition with ring closure to give 3,3-dimethyltetrahydropyran (Figure 18).* The formation of THP ethers occurs in lower yield than THF ethers, presumably because of the higher energy transition state for intramolecular hydrogen transfer (6).

An exception is with structurally ideal *endo*-3-hydroxy-bicyclo(3.3.1)nonane derivatives, which produce 80% of the 2-oxaadamantane upon LTA oxidation (53b,53c).

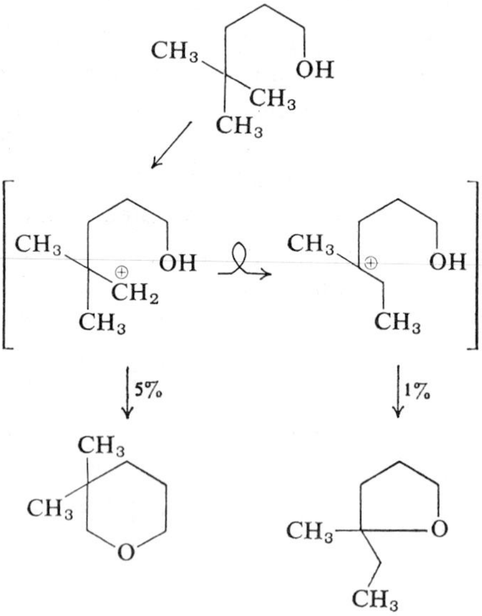

Figure 18. (53a)

* Since the precursor to the supposed carbonium ion intermediate is the free radical, and since Cu(II) is known to be a more efficient oxidant of such radicals (9b), it would be desirable to know the effect of added cupric ion on the relative proportions of cyclic products.

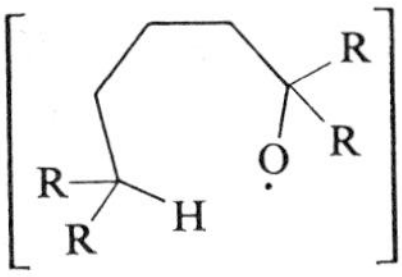

6

Considerable data indicate that structural features play important roles in modifying THP and THF yields. Replacement of an aliphatic unit next to a δ- or ε-carbon by an alkoxy, keto, vinyl, or phenyl group enhances the respective ether yields, sometimes considerably. The largest increases are observed with alkoxy substituents (Figure 19). As shown in the steroid examples, the "substituents" have a tremendous yield enhancing effect (36).

Exceptions to allylic activation giving rise to higher yields of cyclic ether are known. Thus, 2-butenol yields crotonaldehyde instead of dihydro-furan, and fragmentary loss of formaldehyde dominates THP and THF

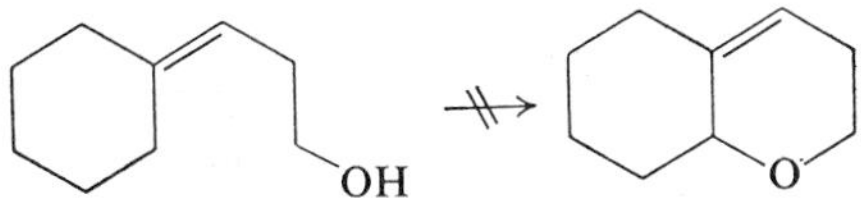

(16)

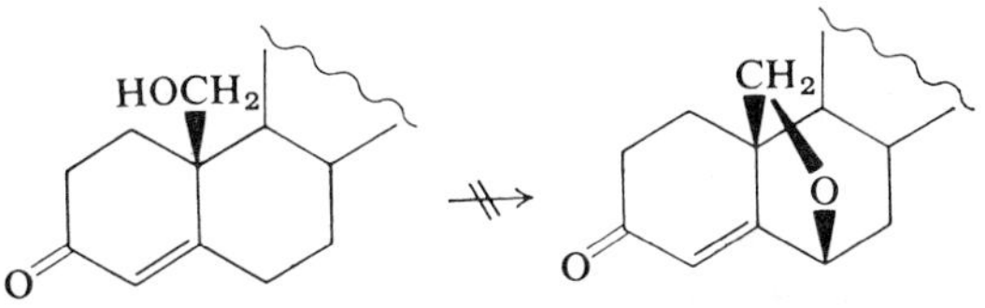

(17)

formation in eqs. 16 and 17 (50d,56). Citronellol oxidation gives none of the product resulting from allylic activation of a C—H bond (eq. 18) (13b).

Besides the yield altering effect, these same substituents enhance the reaction rates. This can be due to low energy stabilization of the δ-carbon radical and/or subsequent carbonium ion intermediate.

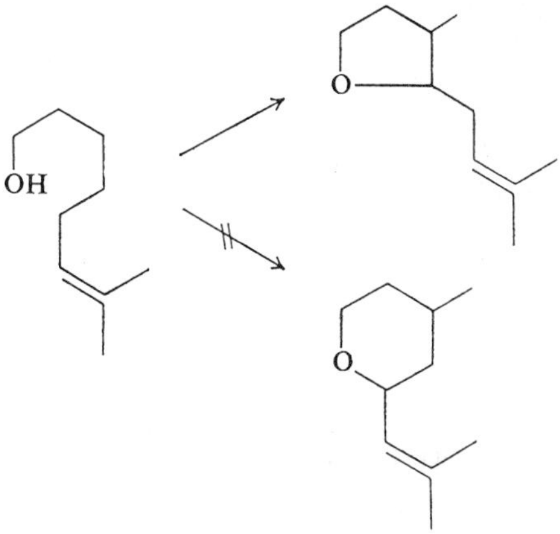

(18)

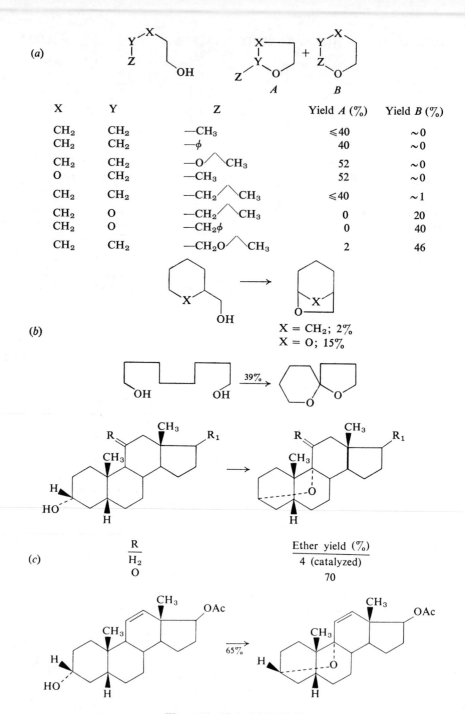

X	Y	Z	Yield A (%)	Yield B (%)
CH₂	CH₂	—CH₃	⩽40	~0
CH₂	CH₂	—φ	40	~0
CH₂	CH₂	—O⁀CH₃	52	~0
O	CH₂	—CH₃	52	~0
CH₂	CH₂	—CH₂⁀CH₃	⩽40	~1
CH₂	O	—CH₂⁀CH₃	0	20
CH₂	O	—CH₂φ	0	40
CH₂	CH₂	—CH₂O⁀CH₃	2	46

(b)

X = CH₂; 2%
X = O; 15%

39%

(c)

R	Ether yield (%)
H₂	4 (catalyzed)
O	70

65%

Figure 19. (2b,2c,36,38,54,55)

136

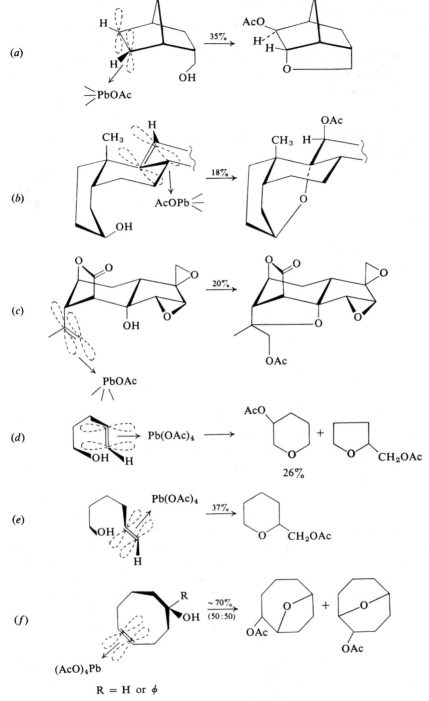

(a)

(b)

(c)

(d)

(e)

(f)

$R = H$ or ϕ

Figure 20. (13a,36,57,58a,59)

137

When the δ-carbon atom of an alcohol is sp^2 hybridized, it participates in cyclization reactions different from the above. If the carbon atom is part of a nonconjugated double bond, evidence suggests that a π-complex between the carbon–carbon double bond and a tetravalent lead containing species is produced in the alkoxylead triacetate intermediate (13). The π-complex then often dominates the subsequent reactions of the intermediate by the routes shown in Figure 20 (13a,36,57,58a,59).

Attempts to incorporate π-bond interaction into the oxidative cycliza-tion of 2- and 3-penten-1-ol type molecules have failed. Thus, 3-butenol (58), 2- and 3-cyclohexen-1-ol (59), and 1-allyl-cyclohexan-1-ol (60) give none of the acetoxy-ether products.

Aside from alcohol oxidation, but within the topic of π-bond participa-tion, is the recent use of LTA to effect 1,4-substituted-cis-decalin formation from cis,trans-1,5-cyclodecadiene (1c), and dihydrofuran formation as in eq. 19 (61).

(19)

If the δ or ε sp^2 hybrid carbon atom of an alcohol is a member of an aromatic ring, it must possess a hydrogen atom or it will not enter into a reaction producing cyclic ether (Figure 21). It must be pointed out that the formation of benzodihydropyrans in greater yield than benzodihydrofurans is in direct conflict to preferential THF formation from purely aliphatic alcohols. Seemingly the imposition of the benzo system, in directing the bond angles emanating from it, is such that the most favorable transition state for intramolecular hydrogen transfer is seven-membered instead of six. This mechanism, however, may not be operative in these examples. It is possible that these reactions take place via benzenoid complexed LTA, and since the alkyl or alkoxy fragments attached to the benzene ring in the aryl-alkanols tend to increase electron density *around that ring*, complexation should not involve only solvent.* Unlikely, however, is a reaction sequence whereby LTA acts as a Lewis acid inducing attack of oxygen on the aromatic ring

---

* This assumes that the solvent is benzene, for which Pb(IV) coordination complexes have been spectrally suggested. Specifically, the species $PbCl_4 \cdot 3C_6H_6$ has been considered (12a). It will be interesting to see if substituents on the aromatic ring in 2-phenylethanol influence the yield of benzodihydrofuran. If the alkoxide radical depends on its electro-negativity for cyclization, electron withdrawing groups should hinder the process.

(eq. 20).* If this was prevalent, product ratios of aliphatic alcohol oxidation in benzene solvent would be quite different than those observed.

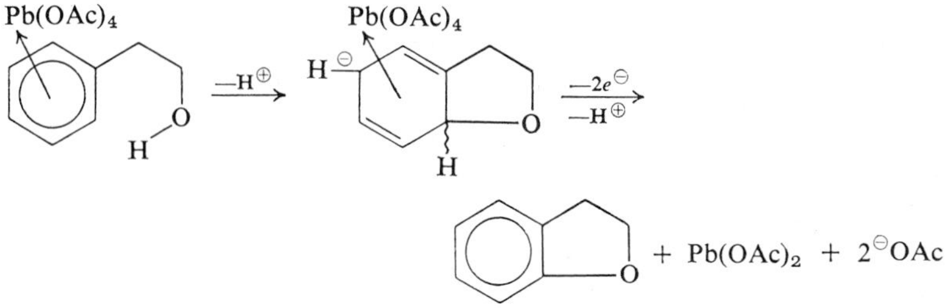

A more favorable process is alkoxy radical formation followed by intra-molecular attack on the aromatic ring to give an intermediate cyclohexadienyl radical (2b,38). Subsequently, a second one-electron transfer and proton

* An example of this process may be found in the oxidation of the complex phenol (62):

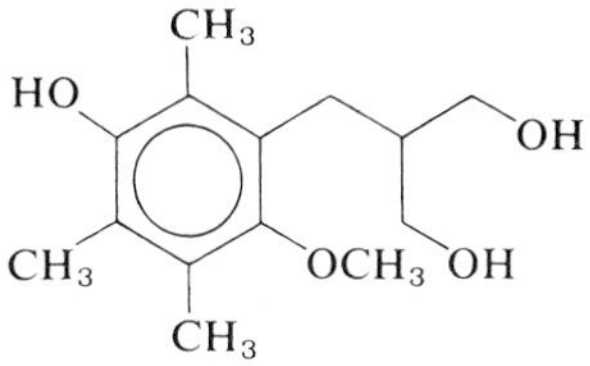

In this example the phenolic hydroxylic group would be expected to react with LTA first (due to the more acidic hydrogen atom), followed by the possible sequence:

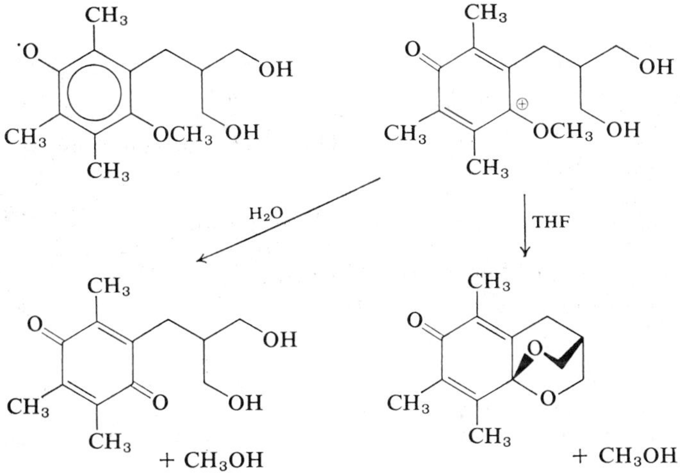

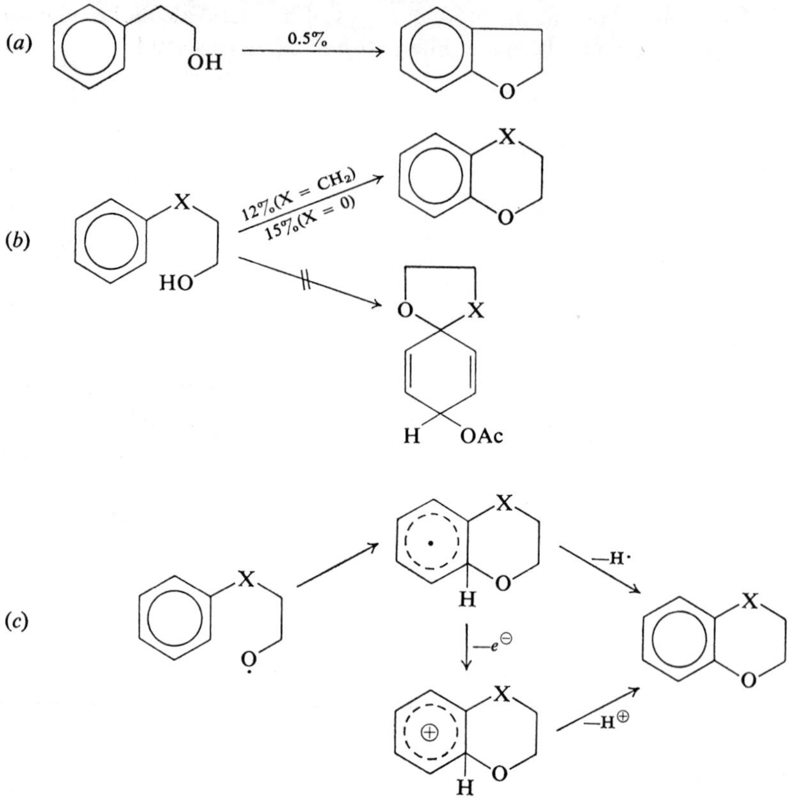

Figure 21. (2b,2c,38)

loss nets the observed products (Figure 21c).* In the oxidation of 2-phenyleth-anols (homobenzylic alcohols) the yield of benzodihydrofuran is very low also because of the favorable, competing fragmentation process which affords an intermediate benzyl radical (2b).

Alcohols with the oxygen *attached to* an sp^2 hybrid carbon atom are either enolic or phenolic and their reactions are not part of the present chapter. For the most part they are transformed into α-acetoxy ketones. An exception is the case of methyl podocarpate where oxidation in acetic acid leads to the 7-hydroxy derivative (eq. 21) (63).

---

* No data are available on the relative rates of intramolecular alkoxy radical abstraction of hydrogen from sp^3 hybrid carbon, as opposed to attack of the radical on an intramolecular aromatic ring. Cyclohexadienyl radicals are oxidized rapidly by Pb(IV) (2d).

$$(21)$$

We now consider alcohol conformational effects on cyclic ether yields. This subject has been carefully considered in a review by Heusler and Kalvoda (16c). Careful work from the laboratories of Mihailović, Jeger, and Heusler has established an optimum nonbonded distance between the hydroxyl and δ-carbon groups for cyclization to be competitive with oxidative carbonyl formation and fragmentation. Other molecular structure factors being equal, this distance of 2.5–2.7 Å allows prediction of the types of products that might be formed upon LTA oxidation.

In the formation of five-membered cyclic ethers, secondary hydrogen atoms on δ-methylene carbons are considerably more reactive than primary hydrogen atoms on δ-methyl carbons. This order of ease of 1,5-hydrogen transfer (i.e., secondary over primary) in the alkoxy radical is consistent with the order of stability of alkyl radicals and simple carbonium ions (64). In the mechanism proposed for ether formation the intermediate secondary hydroxy alkyl carbon radicals formed by 1,5-hydrogen abstraction from the alkoxy radicals, and the corresponding secondary carbonium ions, are stabilized by additional hyperconjugation and positive inductive effects of the alkyl group, while in the analogous primary intermediate species, derived from 1-butanol, 2-pentanol, 4-heptanol, etc., these stabilization factors are absent. In addition the $+I$ inductive effect of the alkyl group will increase electron density on the δ-carbon atom in the alkoxy radical and thus facilitate hydrogen abstraction by the electrophilic oxygen radical.

If the order of reaction times in the series of alcohols

$$\overset{\delta}{R}-CH_2-CH_2-CH_2-\overset{\alpha}{C}HOH$$
$$\underset{R''}{|}$$

with R = variable and R″ = constant are compared, for R″ = H the order is   1-pentanol   (R = CH_3) > 1-hexanol   (R = C_2H_5) > 1-heptanol (R = n-C_3H_7) < 1-octanol   (R = n-C_4H_9) < 1-dodecanol   (R = n-C_8H_{17}); for R″ = CH_3 the order is 2-hexanol (R = CH_3) > 2-heptanol (R = C_2H_5) > 2-octanol   (R = n-C_3H_7) < 2-decanol   (R = n-C_5H_{11}); for   R″ = C_2H_5 the order is 3-heptanol (R = CH_3) > 3-octanol (R = C_2H_5) > 3-nonanol

$(R = n\text{-}C_3H_7)$; and for $R'' = n\text{-}C_3H_7$ the order is 4-octanol $(R = CH_3) >$ 4-nonanol $(R = C_2H_5)$. It is evident that all the δ-methylene groups are not equivalent and that the reaction times decrease when the length of the alkyl group R attached to the δ-carbon atom increases up to $R = n\text{-}C_3H_7$; further lengthening of R again slightly increases the reaction time. Up to $R = n\text{-}C_3H_7$ the positive $(+I)$ inductive effect of R will enhance the reactivity of the δ-carbon atom in the alkoxy radical and increase the stability of the carbonium ion in the same way as discussed above. However, beginning with $R = n\text{-}C_4H_9$, the alkyl rest R is long enough to hinder, when in a coiled conformation, the abstraction of hydrogen from the δ-carbon atom in the alkoxy radical and therefore the duration of the reaction will again increase.

On the other hand, in the series of alcohols

$$R\overset{\delta}{-}CH_2-CH_2-CH_2-\overset{\alpha}{C}HOH,$$
$$\underset{R''}{|}$$

with R = constant and R'' = variable, the order of reaction times for R = H is 1-butanol $(R'' = H)$ < 2-pentanol $(R'' = CH_3)$; for R = CH_3 it is 1-pentanol $(R'' = H)$ < 2-hexanol $(R'' = CH_3)$ < 3-heptanol $(R'' = C_2H_5)$ < 4-octanol $(R'' = n\text{-}C_3H_7)$ > 5-nonanol $(R'' = n\text{-}C_4H_9)$; for R = C_2H_5 it is 1-hexanol $(R'' = H)$ < 2-heptanol $(R'' = CH_3)$ < 3-octanol $(R'' = C_2H_5)$ < 4-nonanol $(R'' = n\text{-}C_3H_7)$; and for $R = n\text{-}C_3H_7$ it is 1-heptanol $(R'' = H)$ < 2-octanol $(R'' = CH_3)$ < 3-nonanol $(R'' = C_2H_5)$. These orders show that by increasing the alkyl rest R'' attached to the carbinol carbon atom the reaction times increase. It appears that the important factor in this case is the positive $(+I)$ inductive effect of the electron releasing alkyl group R'', which by increasing electron density on the radical oxygen in the alkoxy radical will diminish its "electrophilic" properties as a hydrogen abstracting agent (in the order corresponding to the length of R'', i.e., methyl < ethyl < n-propyl) and thus slow down the reaction. This effect is nonexistent when $R'' = H$ and therefore the reaction of primary alcohols with LTA to produce five-membered cyclic ethers is relatively fast.

The steric effect of the rest R'' is probably only of minor importance, since alkyl groups from methyl to n-propyl are not large enough to hinder appreciably the formation of the alkoxylead triacetate and the attack of oxygen on the δ-hydrogens in the alkoxy radical.

According to these qualitative findings on the rates of the LTA reaction with primary and secondary aliphatic alcohols, the ease of tetrahydrofuran formation can be roughly predicted, i.e., that when the hydroxyl group is moved along an unbranched carbon chain the reaction times will increase in the order 1-alkanol < 2 alkanol < 3-alkanol < 4-alkanol, because R decreases and R″ increases in the same order. This is particularly useful in preparing a five membered cyclic ether which can be obtained from two alcohols; the alcohol with a larger group R and a shorter group R″ will be preferred as substrate for the reaction (e.g., 1-heptanol over 4-heptanol for preparing 2-*n*-propyl-tetrahydrofuran; 2-heptanol over 3-heptanol for obtaining 2-ethyl-5-methyltetrahydrofuran; 2-octanol over 4-octanol for the synthesis of 2-methyl-5-*n*-propyl-tetrahydrofuran).

Cyclobutanol and cyclopentanol, when treated with LTA, do not yield intramolecular ethers since the initially produced four- and five-membered cycloalkoxy radicals cannot undergo, across the shorter atom chain, a homolytic 1,5-hydrogen transfer from carbon to oxygen via the respective six-membered cyclic transition state (46a). The structure of cyclohexanol permits 1,5-hydrogen transfer in the corresponding alkoxy radical, but because of the energetically unfavorable boat conformation that the cyclohexane ring must assume in such a process the yield of 1,4-epoxycyclohexane 7 is only 1% (46a). This finding is similar to that reported for the photolysis of cyclohexyl nitrite (18a).

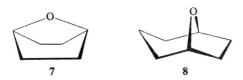

In going from cyclohexanol to cycloheptanol and to cyclooctanol ring flexibility increases and the possible conformations in the six-membered cyclic transition states and corresponding intramolecular ether products result in an appreciable increase in yield which amounts to 15% for 1,4-epoxycycloheptane (**8**) and to over 35% for 1,4-epoxycyclooctane (46a). Molecular structure features improving cyclic ether yields are shown in eq. 22 (46b). Apparently the *exo* alcohol (R = OH) does not epimerize by

$$\text{(22)}$$

LTA induced reversible fragmentation, a process evident in other examples (33a,33b).

Although cyclooctanol can also form a 1,5-ether, the yield of this cyclization product is only about 0.8%, approximately the amount of six-membered tetrahydropyran ethers obtained in the LTA oxidation of un-branched secondary aliphatic alcohols (46a). This result indicates that in the case of cyclooctanol, similarly to acylic alcohols, the main prerequisite for intramolecular hydrogen transfer from carbon to oxygen in the corre-sponding alkoxy radical is attack at the 4-position of the ring leading to a 1,4-ether.* Proximity effects, which might be expected to increase the ease (and yield) of the THP-type 1,5-ether formation, do not seem to be operative in the case of cyclooctyloxy radicals generated from cyclooctanol and LTA. However, Cope et al., have reported that the tertiary alcohol 1-methyl-cyclooctanol affords 18% of the 1,5-ether and only 6% of the 1,4-ether upon oxidation with LTA (eq. 23) (66). A pronounced proximity effect, involving attack on hydrogen at the 5-position, has also been observed in the photolysis of 1-methylcyclooctyl hypochlorite (67). These results suggest that the 1-methyl group appreciably changes the geometry of the cyclo-octane ring in the reacting, tertiary 1-methylcyclooctyloxy radical, and thus decreases the activation energy of the seven-membered cyclic transition state necessary for 1,5-ether formation via ϵ-hydrogen abstraction.

$$(23)$$

R = H	0.8%		36%
R = CH$_3$	18%		6%

Other geometry and activation effects are shown in the oxidation of 4- and 5-phenyl-cyclooctanols (eqs. 24) (59b).

---

* Photolysis of cyclohexyl and cyclooctyl nitrites gives about 21% of homolytic hydrogen abstraction from the δ-carbon. Further, cyclooctyloxy radical from this same source does not measurably attack the ϵ-C—H position (65).

$$(24)$$

Cyclodecanol also undergoes intramolecular ether formation to a considerable extent (27%), but here *several* ether compounds are isolable, the major product (12%) being *trans*-expoxycyclodecane accompanied by 1% yield of the *cis*-isomer (eq. 25) (43). This appears to be the first case of oxirane ring formation from alcohol and LTA. Transannular effects are no

$$(25)$$

doubt operative. Interestingly, when the hydrogen atom on the carbinol carbon of cyclodecanol is replaced by deuterium, the yield of ketone is considerably decreased whereas that of epoxide remains unchanged. Based on other deuterium isotope effects in LTA reactions (7c,43), this is an expected result since competing carbonyl forming reactions would be less favorable.

Intramolecular cyclic ether formation from cyclohexadecanol and LTA parallels the cyclization reaction of secondary aliphatic alcohols, such as 2-octanol, whereby only one oxide product is formed in 45% yield (46a).

Thus far in the discussion on oxidative cyclization we have considered factors that structurally influence the yields. One more example of subtle molecular features will be presented and then the alternatives to cyclization, carbonyl formation and fragmentation, will be presented with appropriate examples. It is important to keep in mind the fact that these are competing

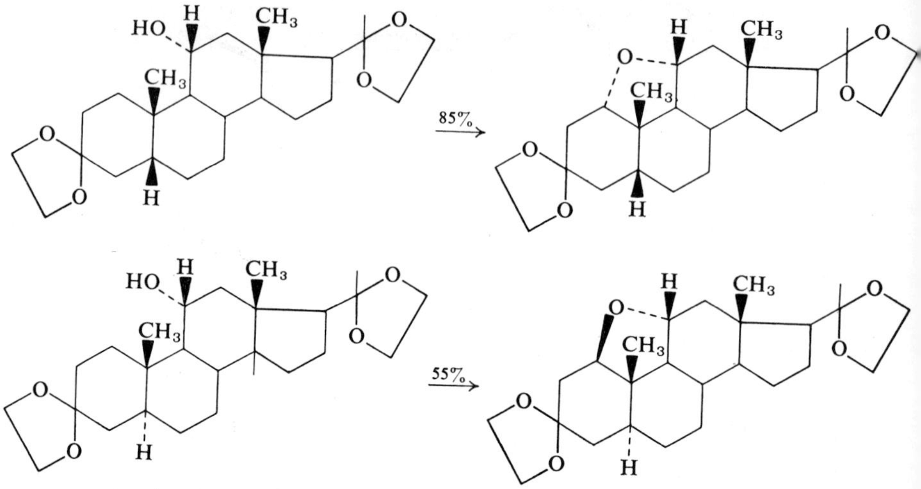

Figure 22. (68)

reactions and all are observable together even though they are discussed separately here.

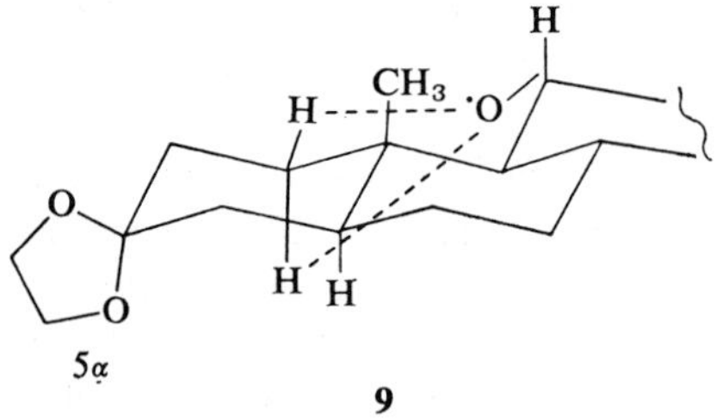

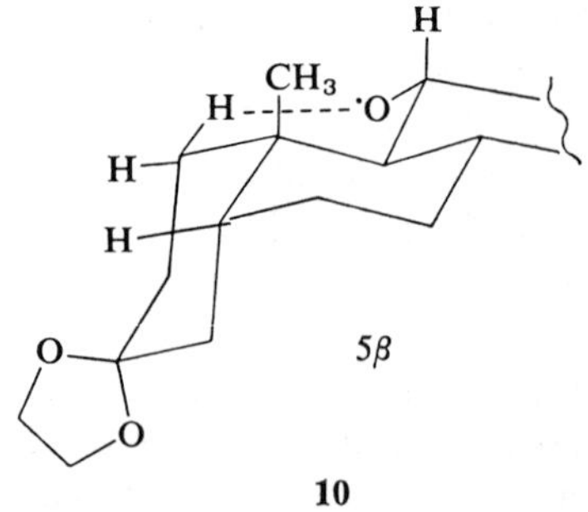

The steroidal alcohols have been widely studied as reducing agents for LTA, both because of the excitement of activating carbon centers difficult to reach by other means and because the stereochemistries of transformations are easier to follow. It is appropriate to consider one example in detail. The oxidation of 3,20-diethylenedioxy-11-α-hydroxy-5-β-pregnane gives 85% of the corresponding 1-α,11-α ether. The 5-α-epimeric compound oxidizes to the 1β,11-β ether in 55% yield (Figure 22). In order to suggest reasons for this change in yield and ether configuration, the authors (68) considered the conformation aspects of the A-ring and the positioning of the two hydrogen atoms on the C_1 carbon atom. In the 5-β-compound (10) only one C_1 hydrogen atom is available for reaction with the developing 11-αrad-oxy ical. In the 5-α example (9) the *trans A/B* ring fusion allows two suitable conformations giving good proximity of both the C_1 α- and C_1 β-hydrogen atoms to the 11-α-oxy radical. It is apparent that the lack of formation of the 1-α,11-α-ether is due to nonbonded interactions in the required, unstable boat form of the A-ring (11).

11

Evidence given previously certifies the fragmentation route of oxidation to be initially free radical and reversible in character. Notable yields of fragmentation products are obtained when the alcohol being oxidized is branched at the α- and/or β-carbon atom. This sequence, being one of the three competing oxidation patterns, seems to be in preponderance when high stabilization can be given to the carbon radical resulting from the fragmenting of the α–β-carbon bond (Figure 23) (70). This stabilization factor can be electronic or steric in nature. As β branching converts the carbon at that position from —CH_2— to —$CH(CH_3)$— to —$C(CH_3)_2$—, the bond to the α-carbon is more easily homolyzed because of the incipient greater stability of the β-carbon radical. Steric factors do play a part, however as exemplified by the facile elimination of formaldehyde from the 4,4,4-triphenylbutoxy radical (35.). It is interestering that a similar molecule, 4,4-dimethylpentanol, reportedly gives no products derived from the possible fragmentation product, 3,3-dimethylbutyl radical (47).

The oxidation of 4,4,4-triphenylbutanol deserves further comment. A notable product is 1,1-diphenylindane when oxygen is purged from the reaction medium. Its formation most likely comes about by oxidation of the intermediate cyclohexadienyl radical rather than the primary methylene radical (Figure 16). Support for this route comes from the suggestion that primary carbon radicals are not rapidly oxidized by Pb(IV) but that cyclohexadienyl radicals are (2d,35).

The oxidation of isoborneol (28b) and related benzobicycles (72) to give a preponderance of ring opened product relative to very little of the

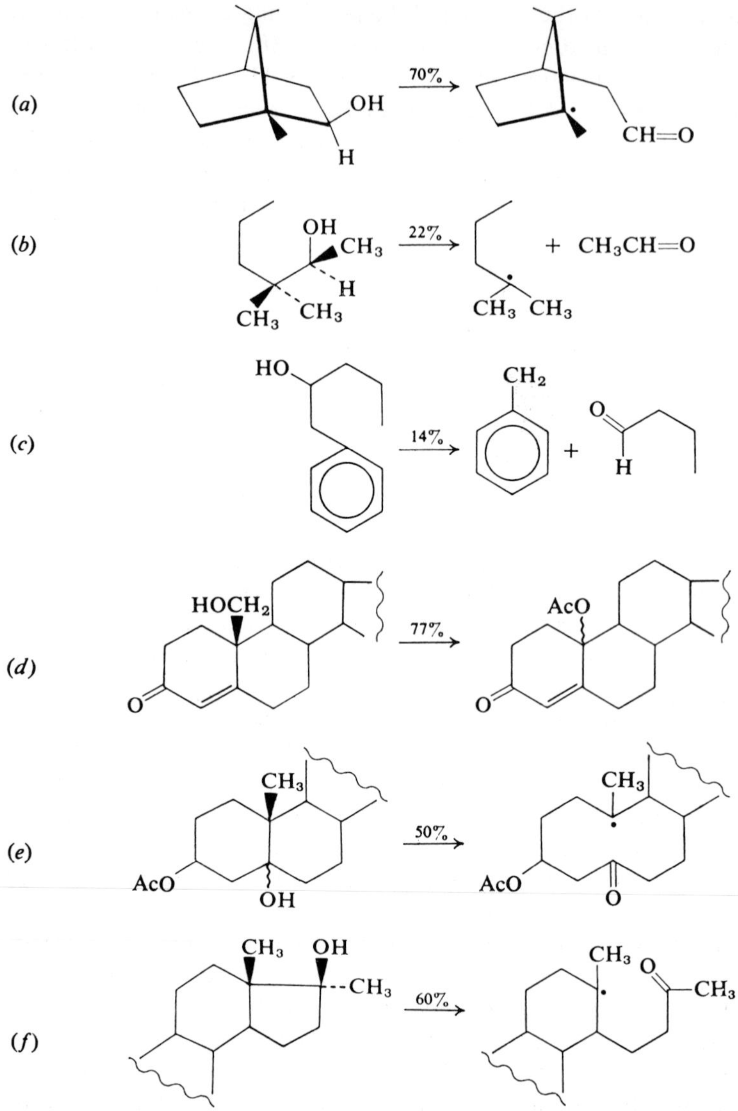

Figure 23. (70). For Example *a*, the oxidation of fenchyl alcohol, under the same conditions as those used for the LTA oxidation of isoborneol, results majorily in fragmentation *on both sides* of the carbinol carbon atom (20). The stabilities of the resulting carbon radicals are nearly equal in this case. Ceric ion oxidation of bicyclo-alkanols has also been reported to yield fragmentation products (71). Additionally, ceric ion will oxidatively fragment norcamphor while Pb(IV) will not. Contrary to these results, chromium trioxide converts the borneols to camphor in 96% yield (73).

148

carbonyl product can be rationalized as release of the bond strain inherent in such ring systems. The lack of oxidative cyclization has been explained as due to the δ-carbon–hydrogen bond being too distant from the secondary hydroxyl group. The same bicyclic materials having a primary hydroxyl group are found to convert easily to the desired ethers. While the functional group distances may be altered in favor of cyclization in these latter compounds, a more likely explanation of lack of success in obtaining ether in the former compounds is found in the stabilities of the carbon free-radicals resulting from fragmentation (Figure 24).

Molecular conformation plays a rather strict role in allowing fragmentation to compete with the other oxidative processes. Figure 13c has demonstrated that menthol oxidation yields cyclization products of epimerized alcohol groupings. More careful study of the yields of cyclization vs. fragmentation products demonstrates particular conformational requirements for facile fragmentation (33).

Further, conformational requirements are shown in Figure 25, wherein nonbonded interactions have been invoked to explain the product differences. Possibly the most discrete report of molecular structure influences in LTA reactions comes from the work of Ourisson (74a). The use of an apparently very special set of molecules, longifolol and isolongifolol, has allowed interpretation of the mechanisms involved in fragmentation reactions, as well as in an isolated example of cyclopropane formation during oxidative carbonyl formation (Figure 26). Of most importance here is the fact that the *act of fragmentation*, a known free radical process, is followed by oxidation of the carbon radical fragment to a carbonium ion. This ion then enters

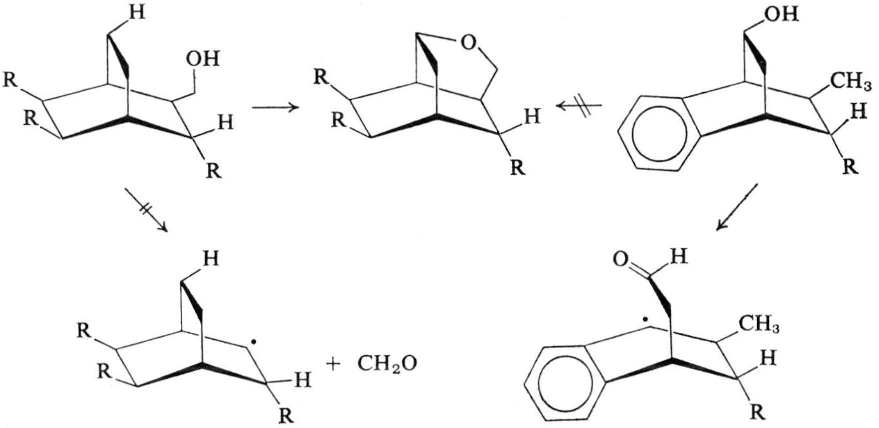

Figure 24. (72)

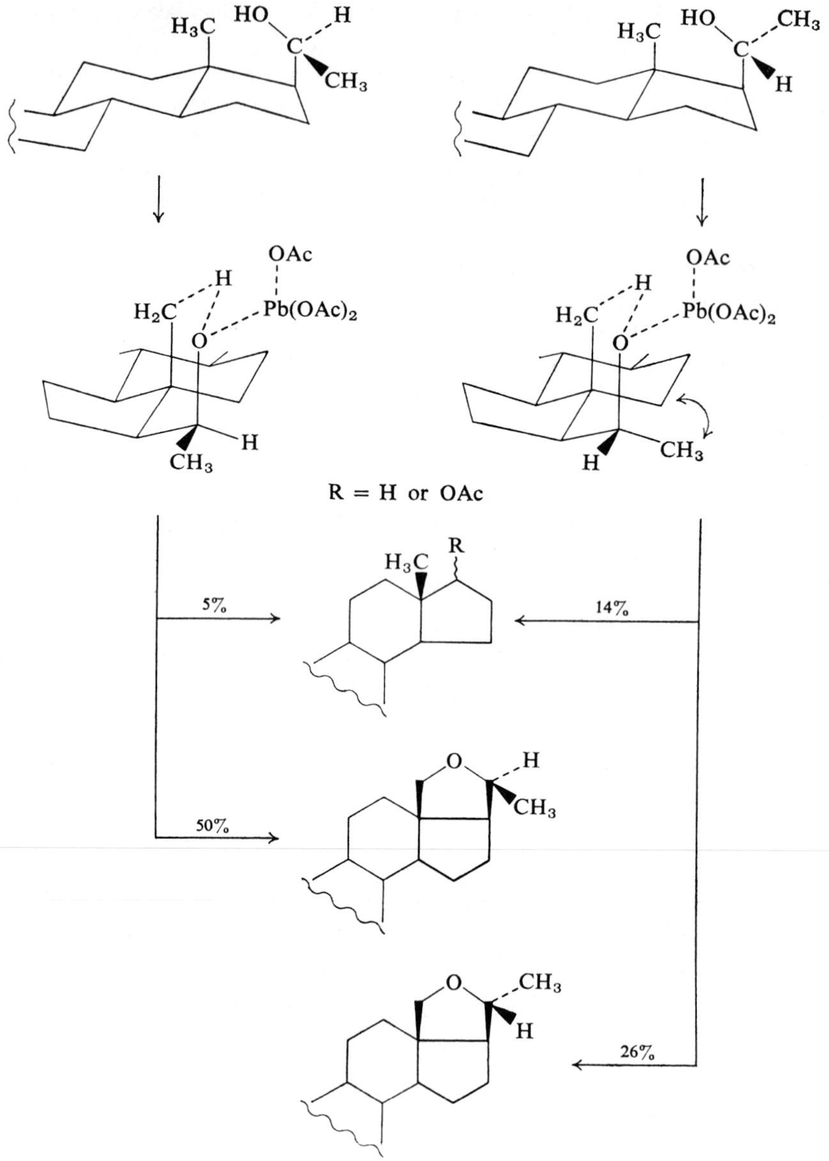

R = H or OAc

Figure 25. (34a)

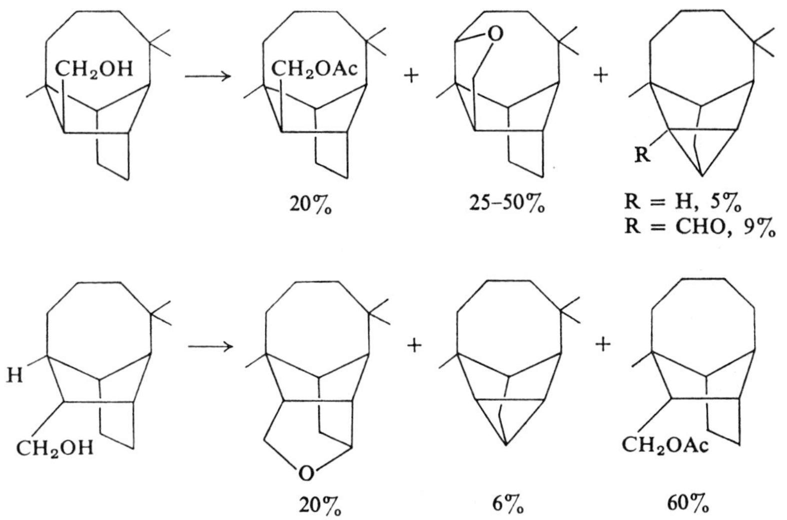

Figure 26. (74a,74b)

into the usual reactions of α-proton elimination and/or capture of an anion (2d); or, unusually in this case, the loss of a β-proton (to yield a cyclopropane derivative). In separate experiments that generated the radical and carbonium ion fragments independently, it was observed that only the cation route afforded the cyclopropane product.

These experiments lend support to the suggestion that oxidative cyclization and fragmentation, while having a common, free radical precursor, continue a similar path of carbonium ion formation *but at different positions in the original molecule* (30b,70g).*

Another example of formation of cyclopropane containing fragmentation hydrocarbon is the LTA oxidation of 1-methylcyclohexane-1-methanol (Figure 27). A small amount of 1-methyl-bicyclo(3.1.0)hexane(**17**) is formed along with other, "normal" fragmentation products (54). Since in this case the steric requirements in the intermediate 1-methylcyclohexyl-1-cation fragment for 1,3-bond formation leading to the fused cyclopropane compound are not particularly favorable, and, as known so far, common cyclohexyl cations do not convert to bicyclo(3.1.0)hexanes, it is believed that **17**

---

* Other evidence supporting carbonium ion formation after fragmentation is found in Ref. 1b. The formation of α-methyl styrene without cumene suggests the lack of disproportionation of radical species. Compare oxidation by chromium trioxide (75).

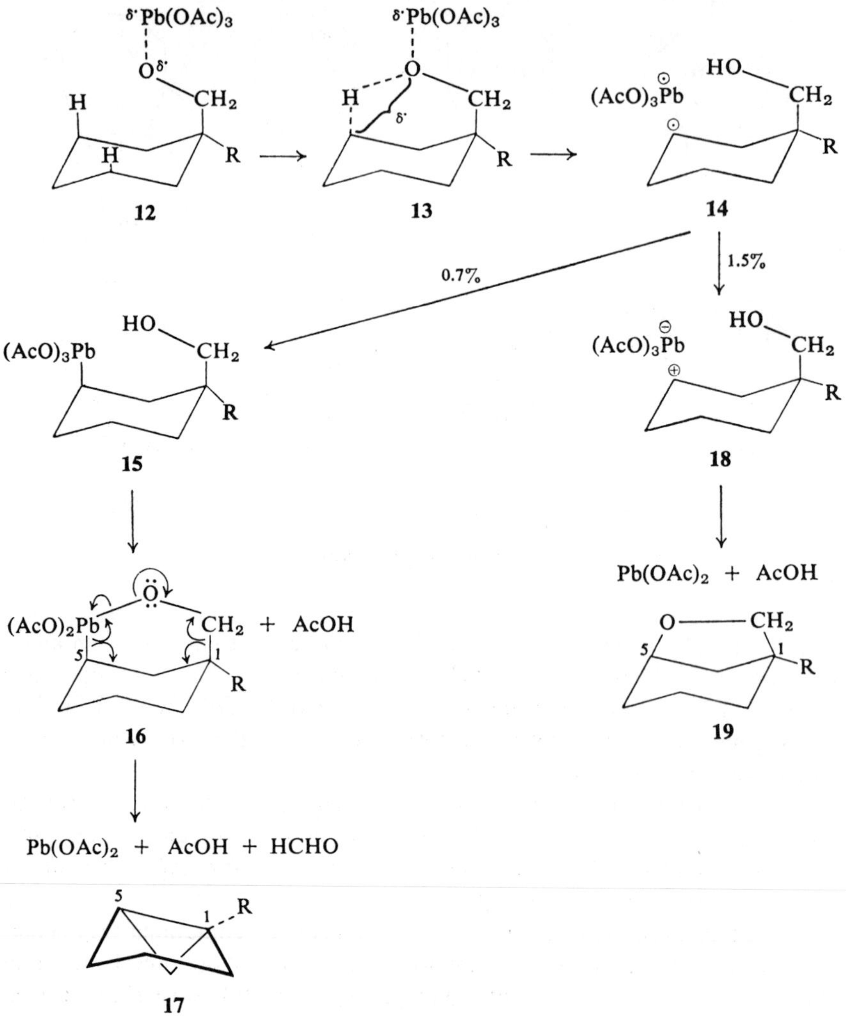

Figure 27. (54)

could be formed by a pathway involving the novel 1,1-diacetoxy-3-methyl-1-plumba-2-*oxa*-bicyclo(3.3.1)nonane (16).*

The subject of cyclopropane structures as a part of a molecule *entering*

* The process by which 16 might be formed and differently decomposed has been presented (30a).

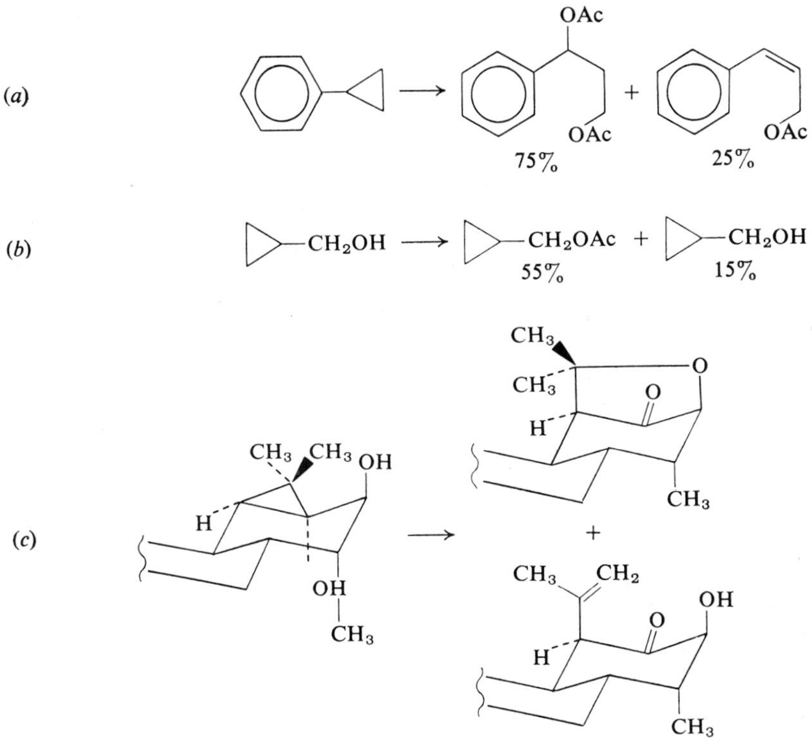

Figure 28. (76)

into reaction with LTA has been considered by several investigators (76). When no other functional group is present, or when a hydroxyl group is attached directly to the cyclopropane ring, cleavage occurs to yield non-cyclic products (Figure 28). When the structure is of the cyclopropylcarbinol type there is little or no fragmentation to give an intermediate cyclopropyl radical. Presumably oxidation of 2-cyclopropyl-1-ethanol would lead to fragmentation of formaldehyde with inherent stability of the cyclopropyl-carbinyl fragment being the driving force. This same characteristic is responsible for the fragmentation observed when cyclobutylcarbinol is oxidized (eq. 26) (70g, compare 2d).

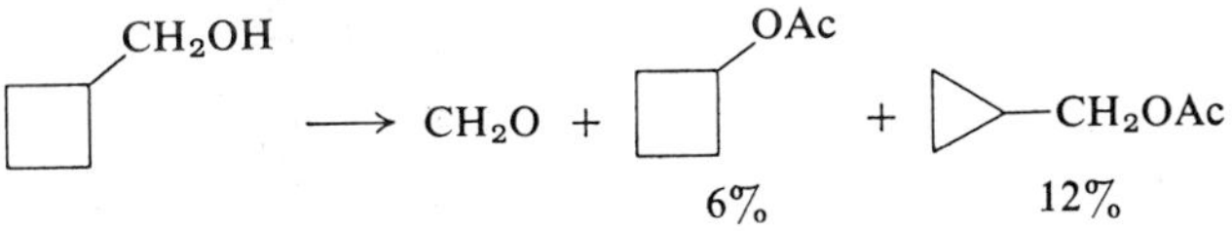

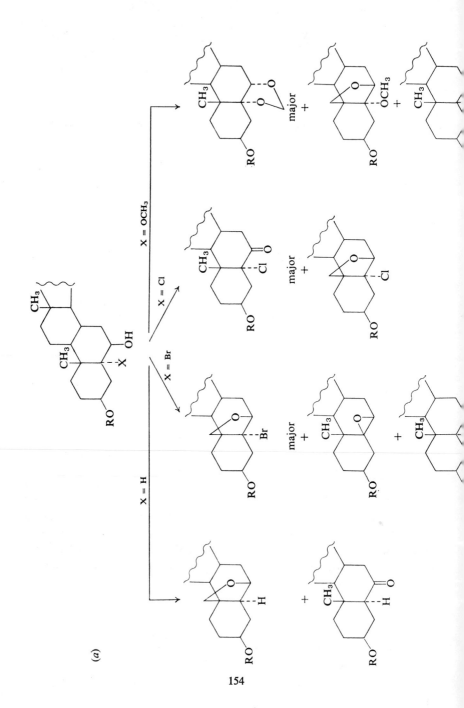

(a)

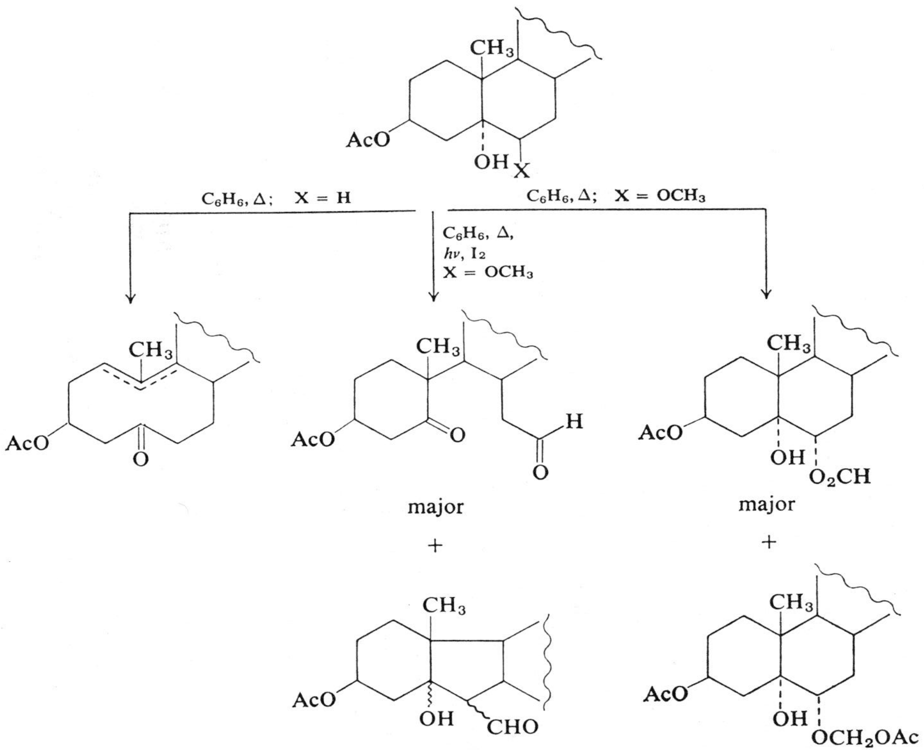

Figure 29. (78). In Example *a* (X = H) considerable discrepancy exists for product yields between laboratories (78a,78b). This is not unusual in the studies of oxidation with LTA and may be due to reaction times, method of addition of oxidant to alcohol (or vice versa), and/or the purity of the oxidant. Various worldwide samples of Pb_3O_4 used for the preparation of LTA do not transform into the more powerful oxidant, lead tetrakis-trifluoracetate (20).

155

It is known that cyclobutyl radicals are not capable of giving cyclo-propylcarbinyl radicals, hence radical oxidation to a carbonium in LTA reaction is necessary for product explanation (77).

That LTA oxidations are influenced by neighboring groups is exemplified by comparing a series of 5-α-hydroxy and 6-β-hydroxy steroid reactions wherein the 6-β and 5α (respectively) substituent is varied (Figure 29).

Direct oxidation of saturated primary and secondary alcohols to the corresponding aldehydes and ketones generally occurs in low yield when the LTA reaction is performed in apolar or acidic media. Even when the alternative routes of cyclization or fragmentation are not favored, carbonyl compound formation does not increase accordingly and rarely exceeds 15–20% (54,70g,79).* In the case of methanol, ethanol, and propanol as well as others, there is further conversion upon prolonged heating with excess starting alcohol and LTA to carbonyl compound derivatives such as acetals, acids, esters, α-acetoxy and/or α-alkoxy aldehydes (or ketones). Thus, in the case of 1-propanol, free propionaldehyde is not present in the reaction mixture after completion of the LTA reaction (no LTA titer) (79). However, if the reaction is interrupted after 90 minutes about 4% yield of free aldehyde may be isolated.

The ease of carbonyl compound formation from secondary cycloalkanols and LTA in refluxing benzene parallels the trend observed in the rate constants for various S_N1 (and radical) reactions involving a change of hybridization of the reacting carbon center from sp^3 to sp^2, either in a transition state or in the final product (46a). As expected, the highest yield of ketone was obtained from cyclodecanol, since in this case the conversion of the carbinol C-atom to a keto-carbonyl C-atom is associated with considerable relief of bond angle strain and decrease in transannular nonbonded interaction involving the α-proton.

To establish the precedent of molecular structures showing favoritism for oxidative carbonyl formation, consideration should be given to the conversion of 3β,20β-diacetoxy-5α-pregnan-6β-ol to the -6-one in 62% yield (78a,78b).

Conclusions as to the mechanism of oxidative carbonyl formation can be derived from the experiments of Partch, Mihailović, and Heusler (30). The facts suggest that the above ascertained alkoxy radical intermediate, formed by homolytic decomposition of an alkoxylead triacetate species, is not favorably disposed to disproportionation to carbonyl compound plus

---

* Rather than disproportionation, the process of carbonyl formation may be one of hydrogen migration from the α-carbon to the oxygen radical with subsequent oxidation of the α-carbon radical to the carbonium ion. Proton loss would then net carbonyl product.

alcohol. While this would explain the lingering presence of the starting alcohol in all LTA reaction products, regardless of the quantity of oxidant used, the sequence is as unfavorable as oxidation of the alkoxy radical to an oxonium ion, followed by loss of an α-proton (45a,80). Thus, in apolar or acidic media the low yields of carbonyl compound can be justified by assuming they are formed via a heterolytic process (80). If a fraction of the LTA was converted to a bisalkoxylead diacetate **20**, even when the molar ratio of alcohol to LTA is 1:1, intramolecular collapse to generate 1 mole of carbonyl product, 1 mole of alcohol, and 1 mole of lead diacetate seems particularly facile (81).

$$\boxed{\text{HOAc}} \;=\; \text{coordinated HOAC}$$

20

To be more precise regarding the mechanism of carbonyl formation, a discussion of solvent effects will be most helpful. With this in mind we now terminate the section on oxidation in apolar and acidic media and commence a discussion of oxidation in neutral-polar and basic media.

B. Polar-Neutral and Basic Media

1. Introduction

In an attempt to solidify thought as to the processes involved when LTA is used as an alcohol oxidant, as well as to find conditions for optimum yields of certain products, investigators have utilized other solvents than the above-mentioned apolar and acidic solvents. These now include alcohols, pyridine, acetonitrile, dimethylsulfoxide (DMSO), and nitrobenzene; sometimes pure and sometimes mixed with benzene.* In addition, benzene solutions of pyridine-N-oxide have been utilized (7c).

---

* An unusual solvent effect of hexafluorobenzene is that it will not allow pentanol oxidation by LTA even after lengthy reflux periods. LTA is recovered. Addition of benzene or pyridine does not alter these results (20). Hexafluorobenzene does not inhibit hydrocarbon oxidation by lead tetrakis-trifluoroacetate (82, Appendix).

Comments regarding the structure of LTA, as a solid or in benzene solution, appear earlier in this chapter. It is pointed out that LTA undergoes a structural change when placed in solvents of high coordinating power.* As shown in Figure 4, solutions of LTA in pyridine or DMSO demonstrate ir absorption in regions ordinarily compared with ester-type carbonyl groups. This is a drastic change from the spectrum in benzene (83). To accommodate the spectral and conductance data it must be assumed that the acetate groups of LTA rearrange from bidentate to monodentate, that they remain in very close contact with the lead atom, and that the lead atom captures at least two ligands, L, to form an apparently favorable hexacoordinate system (21) (6b,11).† This same structure is predictable if LTA is

21

placed in a large excess of alcohol, and indeed this medium does show absorption at 1700 and 1250 cm^{-1}.

Opposed to the spectrum of LTA *dissolved in pyridine* is that of the adduct LTA, C_6H_5N (7c,20). This compound is isolable from dry benzene containing dissolved LTA by adding from 10 to 40 mole ratios *excess* of pyridine. Its C=O ir spectrum (mull) is very similar to that of uncomplexed LTA. Because of its insolubility and the instability of the Pb(IV) ion, it is difficult to suggest a correct structure without more physical data. Temporarily, evidence favors a nine-coordinate lead atom with a single pyridine ligand coordinated to lead through one of the faces of the LTA square antiprism. Presumably Pb(IV) carboxylate coordination complexes with other ligands are obtainable by the above procedure.

Solutions of LTA in coordinating media are ordinarily colored. An exception is the solution in acetic acid. Attempts to demonstrate by es reso-

---

* Comparison should be made to the structural predictions for $R_nSn(OAc)_{4-n}$ and $R_nPb(OAc)_{4-n}$ compounds (8). Nmr data for the chemical shift of the CH_3 groups in LTA are as follows (ppm, solvent): 1.58, C_6H_6; 2.06, C_5H_5N; 1.92, DMSO. Singlet CH_3 absorption appears from $+30°$ to $-45°$ (7c).

† Bisdecarboxylation reactions (84a) studied by Professor Grob, University of Basel, Basel, Switzerland, show catalysis up to LTA: pyridine ratios of 1:2.

nance that the color producing mechanism is conversion of the ligand to a radical-cation failed (20).

When alcohols are oxidized in coordinating media several different phenomena are observed. The reaction rates are dependent on the type of coordinating ligand, pyridine being the ligand having the best rate enhancing effect.* Additionally, the products formed are dependent on the type of activation energy applied to the reacting medium. Thermal activation (room temperature or above) causes oxidative carbonyl formation while photolysis enhances oxidative cyclization and/or fragmentation. These latter two processes are reminiscent of those taking place in apolar and acidic media.

2. Data

Table 2 shows most of the monohydric alcohols oxidized by an LTA–pyridine medium.

TABLE 2

Alcohols Oxidized by Lead Tetraacetate–Pyridine Media:
Thermal and Photolytic Activation

	Ref.
1. 2-Chloroethanol	31a
2. 1-Propanol	79
3. 2-Propanol	79
4. 1,3-Propanediol	116
5. 1-Butanol	31a
6. 1-Pentanol	31a,31b
7. 2-Methyl-1-pentanol	30b
8. 4,4-Dimethyl-1-pentanol	31b
9. 1-Hexanol	31b
10. 2,5-Hexanediol	31a,116
11. 1,3,6-Hexanetriol	116
12. 1-Heptanol	31b
13. 4-Heptanol	31b
14. 1-Octanol	30b
15. 2-Octanol	30b
16. Allyl alcohol	31a

Continued

* This same phenomenon is observed for the oxidative reactivity of tin tetranitrate and cupric acetate. Pyridine greatly enhances the reactivity of Sn(IV) towards organic compounds while it alters the product of interaction of Cu(II) with diazoketones (85). Pb(IV) phosphate salts are stabilized by coordinated triphenylphosphine (86).

TABLE 2 (*Continued*)

	Ref.
17. 5-Hexen-1-ol	59a
18. Cyclohexanol	31b
19. 1-Methylcyclohexylcarbinol	54
20. 2-Propylcyclohexanol	31a
21. Tetrahydropyran-2-methanol	31a
22. Isoborneol	31a
23. Isophorol	31a
24. Benzyl alcohol	2d,31a
25. α,α-Dimethylbenzyl alcohol	2d
26. *p*-Nitrobenzyl alcohol	31a
27. Benzhydrol	2d,31a
28. 1-Phenylethanol	2d
29. 2-Phenylethanol	2d
30. 1,1-Dimethyl-2-phenylethanol	2d
31. 3-Phenyl-1-propanol	2c,2d
32. Cinammyl alcohol	31a
33. Furfuryl alcohol	31a
34. Benzoin	31a
35. Pyridylmethanol	119
36. Steroidal (by OH position)	
a. 4β	1b,30c
b. 6β	30c,92
c. 11β	30c

3. Discussion

a. Thermal Activation. The interaction of alcohols with LTA to form a lead alcoholate and acetic acid (Figure 5) is favored in a basic medium capable of accepting the acid formed. Undoubtedly this is a kinetically important difference between the reactions run in basic and nonbasic solvents (see footnote on page 101). That it is not the only role played by the base, however, is evident from the fact that the base effectively inhibits oxidative cyclization, an initially free radical process which can take place in apolar and acidic media. This suggests that the base is capable of influencing the decomposition of the alkoxylead triacetate formed in the primary reaction.

Facts point directly to a structural factor *within the alkoxylead triacetate species* as being the defining force for mode of decomposition. Ordinarily alcohols are oxidized by using an excess of LTA. This not only has the effect of giving maximal yields of alcohol oxidation product at the expense of the cheaper oxidant, but also disfavors equilibrium formation of polyalcoholates of lead such as $(RO)_nPb(OAc)_{4-n}$ where $n = 2$, 3, or 4. The net result, assuming that lead prefers the hexacoordinate state (11), is that the alkoxylead triacetate formed has a different structure than the presumably octacoordinate LTA. The slower oxidation rate of alcohols to cyclic ethers in acetic acid is due to competitive coordination of alcohol and acid with the lead atom; and the abundance of acid protons to associate with the (basic) acetate groups inhibits acetate-catalyzed cleavage of the α-carbon–hydrogen bond in the alcohol. Thus oxidative carbonyl formation is also reduced.* Regardless of the specific structure of $ROPb(OAc)_3$ in apolar and acidic media, its decomposition via homolytic fission requires less energy than heterolytic fission. In these media activation by heat or light is required to secure a reasonable rate of reaction, demonstrating that even homolytic fission is not automatic. Once separated from each other, the alkoxy radical and lead triacetate undergo transformations other than oxidative carbonyl formation (16c).

The situation is reversed in basic media. No additional energy above room temperature is needed to effect alcohol oxidation. Of course, the more heat applied, the faster the rate of alcohol oxidation. However, higher reaction temperatures lead to excessive self-decomposition of LTA (21b). The conclusion that the base (uncomplexed) catalyzes α-proton removal from the alkoxylead triacetate intermediate by a type of β-elimination reaction (eq. 27) has been suggested wrong by the experiments shown in

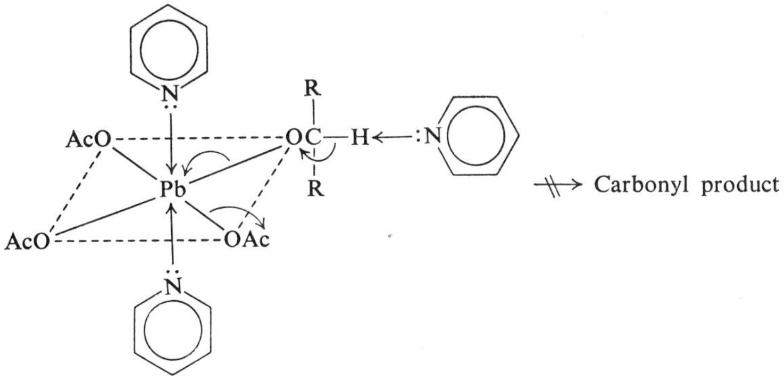

$$\longrightarrow\!\!\!\not\longrightarrow \text{Carbonyl product} \qquad (27)$$

Figure 30. A possible explanation for catalysis and product dictation is α-proton removal by a *coordinated base* molecule (eq. 28). This process

* Variations in yields of ester products when alcohols are oxidized in benzene vs. hexane solvents may have a partial explanation in differences in lead atom coordination. The extent of ester formation may be directed by the ease of acetoxyl escape from lead, which in turn may be influenced by coordination (2c).

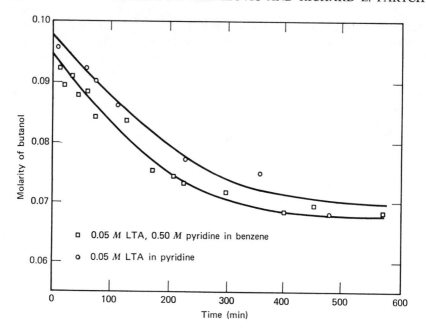

Figure 30. Butanol oxidation at different pyridine concentrations (30°C).

would be facilitated if pyridine coordination to lead was by π-cloud inter-action. However, in very few cases does pyridine not coordinate via the nitro-

$$\longrightarrow \text{Carbonyl product} \qquad (28)$$

gen atom (87). Deuterium substitution demonstrates that α-proton removal is rate determining (Figure 31).

Grob's experiments on pyridine-catalyzed, LTA bisdecarboxylation* (see footnote on page 158) and Kochi's demonstation (2d) of probable acetonitrile-catalyzed cupric ion induced decarboxylation *when the aceto-*

nitrile concentration is high enough suggest that reactions using just 2 moles of pyridine per mole of LTA will not give clear-cut oxidative carbonyl formation (2d). The equilibrium constant for formation of $LTA \cdot C_5H_5N$ and $LTA \cdot (C_5H_5N)_x$ is probably very small compared to formation of $ROPb(OAc)_3$.

The effects of small amounts of pyridine on oxidation rates and product yields have been studied (30,32b). The studies show variable effects of added pyridine in mole ratios of 2:1 to 10:1 to LTA but do not contain sufficient kinetic data for interpretation of concentration effects. $LTA \cdot C_5H_5N$, like LTA, serves as a heterogeneous oxidant in heated benzene to convert alcohols to cyclic ethers (20). This is probably the most conclusive evidence pointing to the requirement of more than one basic ligand for oxidative carbonyl formation and the necessity for ligand rearrangement to other than octa-coordinate lead.

While solvent dielectric constants do not affect the rate of carbonyl formation, other solvent properties do (7c). Thus, DMSO has been shown to coordinate strongly with LTA yet it has no catalyzing effect on alcohol oxidation. Since DMSO-coordinated LTA should undergo alcoholysis as rapidly as LTA in other, nonbasic coordinating media, there is the suggestion that in the hexacoordinate complex, $ROPb(OAc)_3 \cdot 2DMSO$, the fixed DMSO

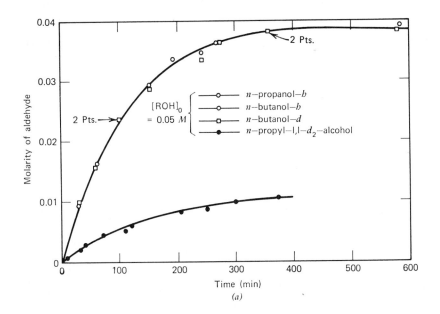

Figure 31a

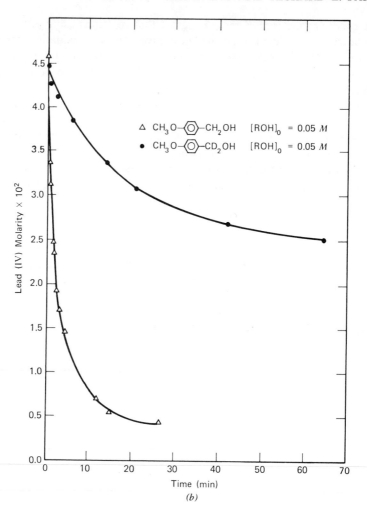

Figure 31. (7c,20,43). Alcoholic oxidation in pyridine, $[LTA]_0 = 0.05$ M (30°C).

ligands are not effective in stimulating the removal of the α-proton on the alkoxy group (88).

Figure 32 qualitatively shows the alcohol oxidation rate enhancing effects for a number of ligands. It is difficult to demonstrate a direct relation between rate and basicity as opposed to one between rate and ligand strength (85,88). The slower rate of the 2,6-lutidine-catalyzed reaction suggests that steric factors hindering coordination are more regulating than basicity. DMSO apparently has stronger coordination tendencies with Pb(IV) than

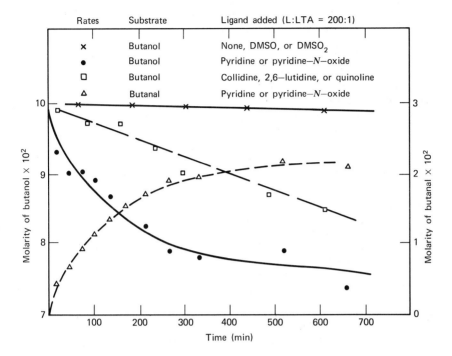

Rates	Substrate	Ligand added (L:LTA = 200:1)
×	Butanol	None, DMSO, or DMSO$_2$
●	Butanol	Pyridine or pyridine–N–oxide
□	Butanol	Collidine, 2,6–lutidine, or quinoline
△	Butanal	Pyridine or pyridine–N–oxide

Figure 32. (7c). Butanol oxidation in benzene.

does pyridine since pyridine added to a solution containing ROPb(OAc)$_3$·2-DMSO shows no alcohol oxidation rate enhancement.

The above discussion pertains primarily to reactions using *equal molar ratios* of alcohol to LTA. That alcohol concentration affects not only the reaction rate but also product distribution is shown in Figure 31*b* and Table 3. Since solvent media much more polar than alcohol–benzene mixtures do not stimulate alcohol oxidation as much as excess alcohol, the solvent polarity must be a minor factor in enhancing reaction rates. The excess alcohol is more likely to facilitate bisalcoholate (*B*, Figure 5) formation, which is structurally suited for intramolecular redox production of alcohol, carbonyl, and lead diacetate products.

The data in Figure 33 demonstrate that ligand interactions with LTA are complex. Oxidation of benzyl alcohols in pyridine proceeds at a rapid rate compared to oxidation of aliphatic alcohols. Substituent effects on reaction rates are impossible to correlate with any of the σ constants. More important is the observation that nitrobenzene, anisole, and benzonitrile, added to the benzyl alcohol oxidation reaction, causes the rate to increase

TABLE 3[a]

Alcohol Concentration Effects on Oxidation

Amount of 2-octanol (moles)[b]	Amount of benzene (ml)	Reaction time (min)	Yields of reaction products (%)		
			2-Octanone	Cyclic ethers[c]	2-Octyl acetate
0.7	None[d]	45–60[e]	18	20	29[f]
0.1	50	90	6	32	15
0.1	100	135–150[e]	3	40	9
0.1	200	165–180[e]	1.5	44	6

[a] Ref. 31b.

[b] Reactions in boiling benzene were carried out with an equimolar proportion of 2-octanol and lead tetraacetate.

[c] Total yield of 2-methyl-5-n-propyl-tetrahydrofuran (cis- and trans-isomers) and 2-ethyl-6-methyltetrahydropyran (cis- and trans-isomers).

[d] This reaction was performed at ∼80° in excess of starting alcohol (0.7 mole of 2-octanol and 0.1 mole of lead tetraacetate).

[e] Reaction times (until disappearance of tetravalent lead) varied slightly from run to run.

[f] In this case most of the ester is probably formed by direct esterification of the starting alcohol (large excess) with acetic acid produced in the course of the reaction.

to approximately that of p-nitrobenzyl-, p-methoxybenzyl-, and p-cyano-benzyl-alcohol oxidation, respectively (20). Other than suggesting that the substituent groups are coordinating (in pyridine medium) with the alkoxylead triacetate and that the mixed ligand field accelerates Pb(IV) reduction, no explanation has been given (89). Deuterium isotope effects show the rate-determining step to be α-proton removal (Figure 31).

Nitro and methoxy groups have been utilized as alcohol substituents in certain LTA reactions to help decipher the prevailing reactions (32,90). Nitrobenzene has also been used as a solvent. It would seem that in the light of the kinetic data of Figure 33 a more careful interpretation of results is needed.

b. Photolytic Activation. Photolytic activation of alcohol–pyridine–LTA mixtures creates competition between the room temperature oxidative carbonyl forming reaction and the oxidative cyclization reaction. The latter is usually dominant (1b,30). The small amount of carbonyl product is probably formed by the same heterocyclic alkoxylead triacetate decomposition that takes place with no light activation. However, the tetrahydrofuran cyclization product must arise by homolytic decomposition generating an

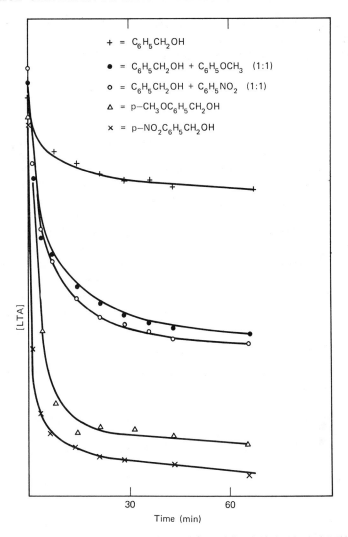

Figure 33. (20,43). Ligand interactions with LTA in pyridine (30°C). Alcohol:LTA = 2:1.

alkoxy radical and lead triacetate. It is interesting that even though Pb(IV) species in pyridine (and other coordinating media) are of different coordination than in benzene, photolytic activation converts them to the same products observed from oxidation in benzene.* Because LTA has a much longer half-

* Pyridine has a drastic effect on the reaction of alkoxy radicals generated from alkyl nitrites (91a). Solvent effects also have an effect on the photodecomposition of organolead compounds (91b).

life in boiling or cold-photolyzed benzene or pyridine than does alkoxylead triacetate, and CO_2 is formed only in minute quantities, homolytic alkoxylead bond cleavage must be favored over acetoxylead bond cleavage (7b).

III. APPENDIX

A. Effect of Carboxylate Structure

A discussion of alcohol oxidation by LTA, a singular example of a lead tetracarboxylate, leaves open the question of the effect of changing R in $Pb(OOCR)_4$. Early work on glycol cleavage demonstrated that carboxylate functionality had little influence on product yield or rate of formation.

Contrary to this, Heusler has shown that increased alkyl substitution in the carboxylate moity diminishes thermal and photolytic oxidative cyclization of monohydroxy compounds (Table 4) (92). The yield of carbonyl product is increased at the same time, but the over-all conversion is reduced to less than 20% in some cases. These changes suggest that there is an increasing tendency for the lead tetracarboxylate to undergo self-decomposition

TABLE 4
6β-Hydroxy Steroid Oxidation by Lead Tetracarboxylates

Pb^{+4}-acylate	Moles $6\beta OH:Pb^{+4}$	Time (hrs)	6β,19-Ether	6 Ketone	6β-ol
In Benzene					
Acetate	1:4.4	30	90.2	6.5	—
Benzoate	1:3.1	15	80.6	15.0	2.4
Propionate	1:3.92	48	25.8	29.0	45.0
Butyrate	1:3.97	48	21.3	26.5	49.0
Isobutyrate	1:3.4	48	3.5	13.8	79.4
Cyclohexanoate	1:2.5	22	1.5	7.9	90.6
Pivalate	1:3.0	52	0.9	65.2	24.8
Trichloroacetate	1:5	2	5.7	20.3	69.5
In Cyclohexane					
Propionate	1:3.9	48	64.1	27.6	2.2
Butyrate	1:3.9	24	50.1	26.8	18.7
Isobutyrate	1:3.9	48	10.3	11.5	29.8
Cyclohexanoate	1:2.5	30	9.8	7.0	78.8
Trichloroacetate	1:2.5	2	8.0	20.2	72.0

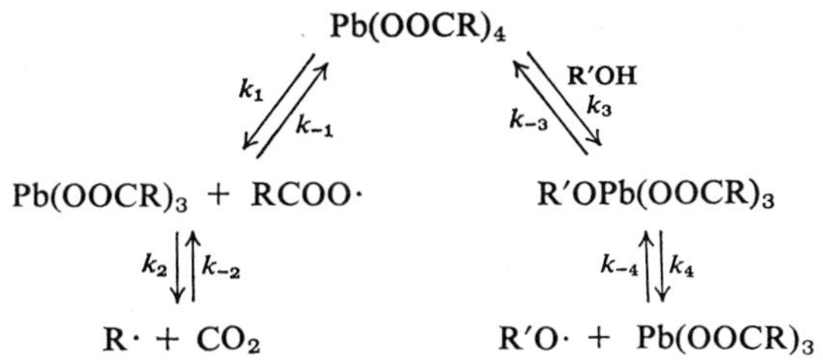

Pb(OOCR)4
k1
k−1
R′OH
k3
k−3
Pb(OOCR)3 + RCOO·
R′OPb(OOCR)3
k2
k−2
k−4
k4
R· + CO2
R′O· + Pb(OOCR)3

for R = —CH_3 or —C_6H_5; $k_1 < k_3 \lll k_2$
for R = —$CH(CH_3)_2$; $k_3 < k_1 \lll k_2$

Figure 34. (7b)

as steric compression increases and as the stability of the decarboxylated radicals increase. The competing equilibria are shown in Figure 34. There may also be a hindered rate of alkoxylead triacetate formation because of steric hindrance to primary coordination (alcoholation) of alcohol to the lead atom (Figure 5). Experimental evidence of this phenomenon is given in Figure 35.

Substitution of other than alkyl or aryl groups has not been extensive. Hydroxyl substitution on an α- or β carbon in RCOO— leads to loss of CO_2

0.04
0.03
0.02
0.01
0
[Aldehyde], mole/liter
100
200
300
400
500
600
700
Time (min)
Pb(OOCCH3)4
Pb(OOCC6H5)4

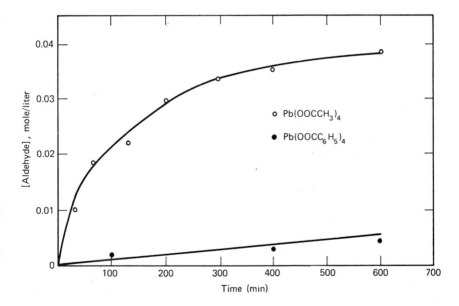

Figure 35. (20). Rate of butanal formation by butanol oxidation in pyridine (30°C).

with the formation of an aldehyde or ketone fragment (14,33c). The same process (a type of glycol cleavage) occurs with 1,1-dicarboxylic acids (eqs. 29 and 30) (93a–f).

$$\underset{\text{RCHCOOH}}{\overset{\text{OH}}{|}} \xrightarrow{\text{LTA}} \underset{(\text{RCHCOO})_n\text{Pb(OAc)}_{4-n}}{\overset{\text{OH}}{|}} \longrightarrow \begin{cases} \text{RCHO} \\ \text{CO}_2 \\ \text{Pb(OAc)}_2 \\ \text{HOAc} \end{cases} \qquad (29)$$

$$R_2C(COOH)_2 \xrightarrow{\text{LTA}} R_2CO + Pb(OAc)_2 + 2CO_2 + HOAc \qquad (30)$$

Transannular interaction of a hydroxyl group with a carbonium ion, the latter resulting from decarboxylation, results in cyclic ether formation (eq. 31) (93g).

(31)

The use of lead tetrakis-trifluoroacetate as an oxidant demonstrates an unusual reactivity for lead tetracarboxylates (82,94). This oxidant (LTTFA) reacts quickly with solvents normally used for alcohol oxidation by LTA. The representative products are shown in Figure 36. Data are not yet available on product formation when LTTFA is mixed with an alcohol, neat, or in the relatively inert solvent hexafluorobenzene.

Comparison of reaction of LTA, of LTA·BF$_3$ (etherate), and of LTTFA with aromatic compounds suggests that the LTTFA reaction may be an ionic Pb(IV) reaction catalyzed by trifluoroacetic acid (96). The product distributions in the reaction of benzotrifluoride with LTTFA, as well as the reaction of thallic trifluoroacetate with aromatics to give ArTl(OOCCF$_3$)$_2$ compounds (which cleave with nucleophiles) (97), suggest an aromatic electrophilic substitution reaction pathway for the trifluoroacetoxylation.

Reaction of LTTFA with saturated hydrocarbons is not as fast as with

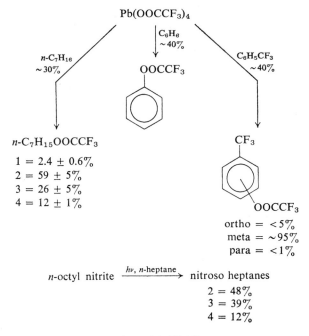

$$Pb(OOCCF_3)_4$$

n-C$_7$H$_{16}$
~30%

C$_6$H$_6$
~40%

OOCCF$_3$

C$_6$H$_5$CF$_3$
~40%

n-C$_7$H$_{15}$OOCCF$_3$

1 = 2.4 ± 0.6%
2 = 59 ± 5%
3 = 26 ± 5%
4 = 12 ± 1%

CF$_3$

OOCCF$_3$

ortho = <5%
meta = ~95%
para = <1%

n-octyl nitrite $\xrightarrow{h\nu,\ n\text{-heptane}}$ nitroso heptanes

2 = 48%
3 = 39%
4 = 12%

Figure 36. (82,95)

aromatics and the trifluoroacetoxyheptane isomer distribution suggests that the reaction proceeds by a free radical path. If LTTFA was acting like a "super acid" hydride abstractor from heptane, extensive rearrangement would be expected (98).

The esr signal generated from photolyzed LTTFA in a frozen matrix of hexafluorobenzene shows only the spectrum of ·CF$_3$ radicals (83c).

B. Lead Tetraacetate Oxidation of Thiols and Amines

1. Thiols

In general LTA reacts with thiols to convert them to disulfides (1a). The process is most likely initiated by the formation of RSPb(OAc)$_3$ which then decomposes to RS· and ·Pb(OAc)$_3$ radicals. Dimerization is more energetically favorable than hydrogen abstraction by almost 7 kcal mole, hence tetrahydrothiophene is not a product. Solvent effects on these reactions have been studied (99) and stable Pb(IV) compounds having Pb—S bonds have been prepared (100).

2. Amines*

Primary amines containing a methylene group in the α-position have been converted to nitriles in 50% yield by LTA oxidation in benzene (101). Conversely, primary amines possessing suitably situated intramolecular double bonds have been converted to tricyclic aziridines (13d). Intermediates

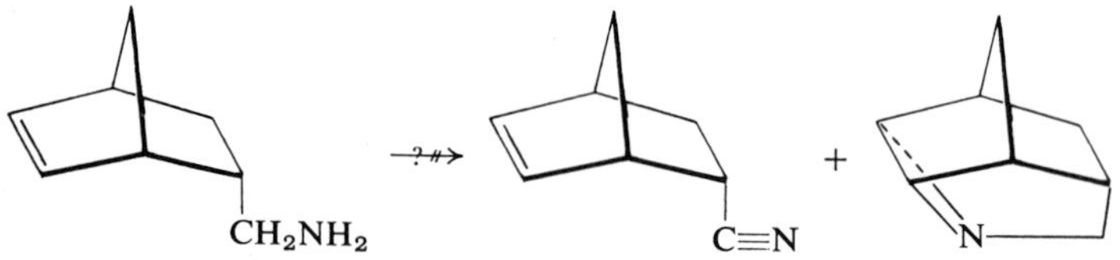

(32)

precursory to these products are not established with certainty. Two are shown in Figure 37. Interestingly, oxidation of *endo*-bicyclo(2.2.1)hept-5-

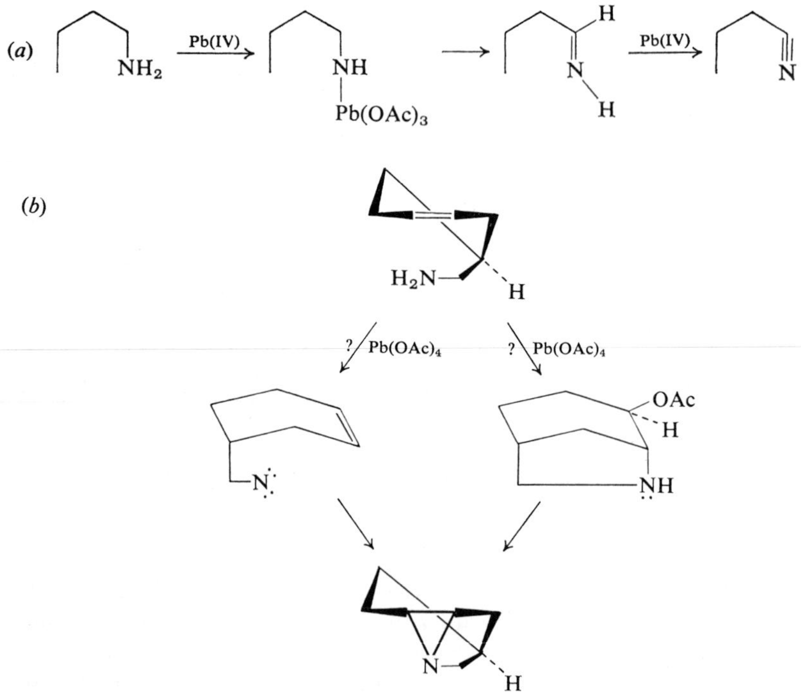

Figure 37. (13d,101)

---

* Oxidation of amides, hydrazines, and oximes will not be considered (see 1a,14).

ene-2-methylamine does not cleanly produce either the nitrile or the aziridine (eq. 32) (20).*

Primary aromatic amines are known to undergo oxidation to more or less complex products depending on the number and position of nitrogen atoms (eqs. 33–35). Nitrene intermediates are preferred (102).

$$(33)$$

$$(34)$$

$$(35)$$

Secondary and tertiary araliphatic amines follow reactions involving 1,2-carbon to nitrogen rearrangement. The product imines are fragmented to secondary amines after hydrolysis (103).

Heterocyclic amines of the pyridine type are known to form coordination complexes with LTA at room temperature but are alkylated at higher temperatures (7c,21).

Unreported reactions of amines are shown in Figure 38. Cyclizations of secondary amines in strong acid solution via the N-chloro intermediate are well known (18c). It is important to consider that this is feasible because the iminium radical intermediate possesses electronegativity approaching that of the radical, and both proceed to cyclize via δ-hydrogen abstraction (see footnote on page 130). Neutral amine radicals do not cyclize (18f). It could be that the transformation of eq. 36 is energetically unfavorable

$$(36)$$

* This was unexpected because the methanol derivative is very easy to ringclose (13a).

(a) 　　　$R'NH_2 + RCH{=}CHR \xrightarrow{Pb(IV)}$ R'NHCH—CHOAc

(b)

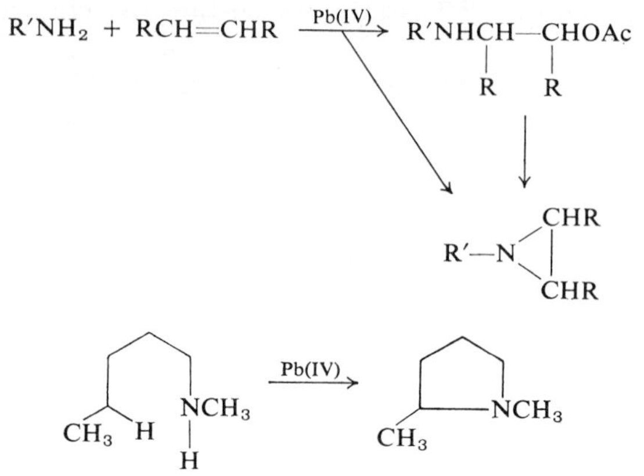

Figure 38

compared to eq. 37.

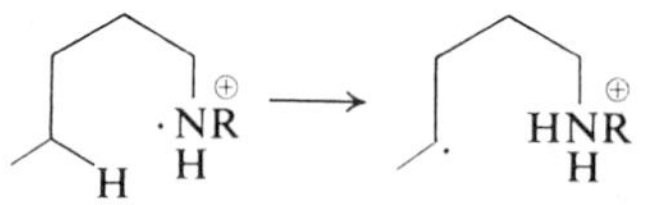

　　　　　　　　　　　(37)

　　　The products of a last example of an amine derivative entering into reaction with LTA is given in Table 5 (20). The stimulus for the study came from the well-established generation of N-aminonitrene intermediates in the oxidation of N-amino heterocycles (102c). Some decompose with loss

TABLE 5

LTA Oxidation of O-Benzylhydroxylamine (S)

Solvent	Temp. (°C)	LTA:S[a]	Time (min)[b]	Major product
C_6H_6	20	1:1	10	Benzaldoxime
C_6H_6	20	1.5:1	15	Phenylaldazine-bis-N-oxide
CH_2Cl_2	20	1:1	10	Benzaldoxime
CH_2Cl_2	0	1:1	100	α-Nitrosotoluene dimer
$CH_2Cl_2 + C_6H_{10}$[c]	0	1:1	100	α-Nitrosotoluene dimer
HOAc	70	1.5:1	30	Benzyl acetate

[a] Mole ratios.
[b] Approximate time until negative Pb(IV) titer.
[c] 50:50 mixture; C_6H_{10} = cyclohexene.

of gaseous by-products (104). The structure of an alkoxyamine (*O*-sub-stituted hydroxylamine) is such that oxidation, if similar to the *N*-amino heterocycles, could produce the hitherto unknown *N*-alkoxynitrene (105), which could produce dimer (with or without the elements of nitric oxide), rearrangement, or insertion products. Table 5 shows that, at least with the *O*-benzyl derivative, oxidation in acidic media gives fragmentation products while oxidation in aprotic media give rearrangement products. Preliminary attempts to demonstrate the presence of a benzyloxynitrene intermediate have been unsuccessful (20). Using methoxyamine, tetramethylethylene, and LTA, Brois (141) has prepared *N*-methoxy tetramethylaziridine for stereo-chemical studies. Neither this work nor that of Carey (142) on the oxidation of O-diphenylmethylhydroxylamine precludes a multistep insertion process not involving a nitrene. It will be interesting to follow new developments in this area.

References

1. (a) R. Criegee, in K. B. Wiberg, Ed., *Oxidation in Organic Chemistry*, Part A, Academic Press, New York, 1965, pp. 277–366. (b) K. Heusler, *Chimia*, **21**, 557 (1967). (c) J. G. Traynham, G. R. Franzen, G. A. Knesel, and D. J. Northington, Jr., *J. Org. Chem.*, **32**, 3285 (1967).
2. All examples for Figure 2 are from the following sources. (a) K. B. Wiberg, Ed., *Oxidation in Organic Chemistry*, Academic Press, New York, 1965. (b) M. Lj. Mihailović, L. Živković, Z. Maksimović, D. Jeremić, Ž. Čeković, and R. Matić, *Tetrahedron*, **23**, 3095 (1967). (c) S. Moon and P. R. Clifford, *J. Org. Chem.*, **32**, 4017 (1967). (d) J. K. Kochi and J. D. Bacha, *J. Org. Chem.*, **33**, 2746 (1968); J. D. Bacha and J. K. Kochi, *Tetrahedron*, **24**, 2215 (1968). (e) E. I. Heiba, R. M. Dessau, and W. J. Koehl, Jr., *J. Amer. Chem. Soc.*, **90**, 1082 (1968). (f) D. R. Bryant, J. E. McKeon, and B. C. Ream, *J. Org. Chem.*, **33**, 4123 (1968). (g) J. B. Bush, Jr., and H. Finkbeiner, *J. Amer. Chem. Soc.*, **90**, 5903 (1968). (h) S. I. Goldberg and R. L. Matteson, *J. Org. Chem.*, **33**, 2926 (1968).
3. See discussions on pp. 10–11, 69–75, 188–194, and 244–247 in Ref. 1a.
4. See Appendix II in W. M. Latimer and J. H. Hildebrand, *Reference Book of Inorganic Chemistry*, 3rd ed., Macmillan, New York, 1964.
5. (a) H. J. Kabbe, *Justus Liebigs Ann. Chem.*, **656**, 204 (1962). (b) J. D. Bacha and J. K. Kochi, *J. Org. Chem.*, **33**, 83 (1968).
6. (a) G. Rudakoff, *Z. Naturforsch.*, **17b**, 623 (1962). (b) B. Kamenar, *Acta Cryst.*, **16A**, 34 (1963). (c) R. O. C. Norman and M. Poustie, *J. Chem. Soc., B*, **1968**, 781.
7. (a) F. R. Preuss and I. Janshen, *Arch. Pharm.*, **295**, 284 (1962). (b) K. Heusler, H. Labhart, and H. Loeliger, *Tetrahedron Lett.*, **1965**, 2847. (c) R. E. Partch and J. Monthony, *Tetrahedron Lett.*, **1967**, 4427. (d) V. Dvorak, I. Nemec, and J. Zyka, *Microchem. J.*, **11**, 153 (1966). (e) K. Heusler and H. Loeliger, *Helv. Chim. Acta*, **52**, 1495 (1969).
8. (a) Ref. 6b. (b) P. B. Simmons and W. A. G. Graham, *J. Organometal. Chem.*, **8**, 479 (1967); *ibid.*, **10**, 457 (1967). (c) M. J. Janssen, J. G. A. Luijten, and G. J. M. Van der Kerk, *Rec. Trav. Chim. Pays-Bas.*, **82**, 90 (1963). (d) B. F. E. Ford, B. V.

Liengme, and J. R. Sams, *Chem. Commun.*, **1968**, 1333. (e) R. Okawara and H. Soto, *J. Inorg. Nucl. Chem.*, **16**, 204 (1961).
9. (a) E. A. Evans, J. L. Huston, and T. H. Norris, *J. Amer. Chem. Soc.*, **74**, 4985 (1952). (b) J. K. Kochi, R. A. Sheldon, and S. S. Lande, *Tetrahedron*, **25**, 1197 (1969).
10. (a) A. W. Davidson, W. C. Lanning, and Sr. M. M. Zeller, *J. Amer. Chem. Soc.*, **64**, 1523 (1942). (b) R. E. Partch in Ref. 7c, and Figure 30.
11. (a) Refs. 7a and 7c. (b) R. C. Olberg and M. Stammler, *J. Inorg. Nucl. Chem.*, **26**, 565 (1964). (c) R. Frydrych, *Chem. Ber.*, **100**, 3588 (1967); *ibid.*, **99**, 3930 (1966). (d) W. Beck, E. Schuierer, P. Poellmann, and W. P. Fehlhammer, *Z. Naturforsch.*, **21b**, 811 (1966). (e) F. Huber, H. Horn, and H. J. Haupt, *Z. Naturforsch.*, **22b**, 918 (1967); F. Huber and H. J. Haupt, *ibid.*, **21b**, 808 (1966); F. Huber and M. Enders, *ibid.*, **20b**, 601 (1965). (f) See K. B. Wiberg and H. Schäfer, *J. Amer. Chem. Soc.*, **91**, 927 (1969), for a discussion of preoxidation equilibria between alcohols and chromic acid. (g) M. Gazikalović, Z. Pavlović, and T. Marković, *Arch. Tehnol.*, **5**, 51 (1967); cf. *Chem. Abstr.*, **70**, 51292z (1969).
12. (a) Evidence for PbCl$_4$·(C$_6$H$_6$)$_3$ has been reported by J. Szychlinski, E. Latowska, and W. Moska, *Rocz. Chem.*, **38**, 1427 (1964). (b) K. Ichikawa and S. Uemura, *J. Org. Chem.*, **32**, 493 (1967). (c) Y. Yukawa and M. Sakai, *Nippon Kagaku Zasshi*, **87**, 81 (1966).
13. (a) R. M. Moriarty and K. Kapadia, *Tetrahedron Lett.*, **1964**, 1165. (b) C. F. Seidel, D. Felix, A. Eschenmoser, K. Biemann, E. Palluy, and M. Stoll, *Helv. Chim. Acta*, **44**, 598 (1961). (c) R. M. Moriarty, H. G. Walsh, and H. Gopal, *Tetrahedron Lett.*, **1966**, 4363; R. M. Moriarty, H. Gopal, and H. G. Walsh, *ibid.*, **1966**, 4369. (d) W. Nagata, S. Hirai, K. Kawata, and T. Aoki, *J. Amer. Chem. Soc.*, **89**, 5045 (1967). (e) D. Evans, Petroleum Research Fund Report, American Chemical Society, August 1968, p. 120.
14. (a) W. A. Waters, *Mechanisms of Oxidation of Organic Compounds*, Methuen, London, 1964. (b) C. A. Bunton, in K. B. Wiberg, Ed., *Oxidation in Organic Chemistry*, Part A, Academic Press, New York, 1965, pp. 398–405. (c) R. Criegee, *Angew. Chem.*, **70**, 173 (1958). (d) A. S. Perlin, in R. L. Augustine, *Oxidation*, Vol. 1, Dekker, New York, 1969, pp. 190–212. (e) Y. Omote and Y. Fujinuma, *Shinku Kagaku*, **16**, 39 (1968); cf. *Chem. Abstr.*, **70**, 56862v (1969).
15. Consult Ref. 1a and the following. (a) Y. Yukawa and M. Sakai, *Nippon Kagaku Zasshi*, **87**, 81, 84 (1966). (b) H. Baumann, N. C. Franklin, H. Möhrle, and U. Scheidegger, *Tetrahedron*, **24**, 589 (1968), and references therein. (c) J. Halpern and H. B. Tinker, *J. Amer. Chem. Soc.*, **89**, 6427 (1967). (d) F. G. Bordwell and M. L. Douglass, *J. Amer. Chem. Soc.*, **88**, 993 (1966). (e) R. J. Ouellette, D. L. Shaw, and A. South, Jr., *J. Amer. Chem. Soc.*, **86**, 2744 (1964). (f) A. Lattes and J. J. Périé, *Tetrahedron Lett.*, **1967**, 5165.
16. Almost all types of functional groups enter into oxidation reactions with LTA. Useful reviews for some of these reactions are from the following. (a) J. K. Kochi, *Rec. Chem. Progr.*, **27**, 207 (1966). (b) J. Zyka, *Pure Appl. Chem.*, **13**, 569 (1966). (c) K. Heusler and J. Kalvoda, *Angew. Chem.*, **76**, 518 (1964); *ibid.*, Int. Ed. Eng., **3**, 525 (1964). (d) M. Lj. Mihailović and Z. Čeković, *Synthesis*, **1970**, 209. (e) Ref. 1a.
17. This cyclization reaction effected by LTA was first discovered in the middle 1950s with aliphatic alcohols by Mićović and Mihailović. (a) The first published literature describing ring closure to cyclic ether in a steroid alcohol appeared in 1959: G. Cainelli, M. Lj. Mihailović, D. Arigoni, and O. Jeger, *Helv. Chim. Acta*, **42**, 1124

(1959). (b) Tetrahydrofuran formation from aliphatic alcohols and LTA was first reported in 1961: V. M. Mićović, R. I. Mamuzić, D. Jeremić, and M. Lj. Mihailović, *Glas de l'Académie Serbe des Sciences et des Arts, CCXLIX, Classe des Sciences mathématiques et naturelles,* No. **22**, 161 (1961).

18. Major references are the following. (a) M. Akhtar, *Advan. Photochem.*, **2**, 263–303 (1964). (b) C. Walling, *Pure Appl. Chem.*, **15**, 69 (1967). (c) M. E. Wolff, *Chem. Rev.*, **63**, 55 (1963). (d) R. S. Neale and R. L. Hinman, *J. Amer. Chem. Soc.*, **85**, 2666 (1963). (e) A. L. Buchachenko and O. P. Sukhanova, *Russ. Chem. Rev.*, **36**, 192 (1967). (f) R. S. Neale, *J. Org. Chem.*, **32**, 3263 (1967). (g) W. S. Trahanovsky, M. G. Young, and P. M. Nave, *Tetrahedron Lett.* **1969**, 2501. (h) M. Lj. Mihailović, Ž. Čeković, and J. Stanković, *Chem. Commun.*, **1969**, 981. (i) A. Deluzarche, A. Maillard, P. Rimmelin, F. Schue, and J. Sommer, *J. Chem. Soc.*, D, **1970**, 976.
19. D. Greatorex and T. J. Kemp, *Chem. Commun.*, **1969**, 383.
20. R. E. Partch and/or M. Lj. Mihailović, Unpublished Results.
21. (a) D. H. Hey, C. J. M. Stirling, and G. H. Williams, *J. Chem. Soc.*, **1954**, 2747. (b) D. H. Hey, C. J. M. Stirling, and G. H. Williams, *J. Chem. Soc.*, **1955**, 3963.
22. R. P. Bell, V. G. Rivlin, and W. A. Waters, *J. Chem. Soc.*, **1958**, 1696.
23. W. A. Mosher and H. A. Neidig, *J. Amer. Chem. Soc.*, **72**, 4452 (1950).
24. E. M. Panov, N. N. Zemlyanskii, and K. A. Kocheshkov, Dokl. Akad. Nauk SSSR, **143**, 603 (1962).
25. (a) M. Lj. Mihailović, R. I. Mamuzić, Lj. Žigić-Mamuzić, J. Bošnjak, and Ž. Čeković, *Tetrahedron*, **23**, 215 (1967). (b) S. Moon and J. M. Lodge, *J. Org. Chem.*, **29**, 3453 (1964). (c) K. Heusler, Footnote 30 in Ref. 25a.
26. (a) W. A. Mosher, C. L. Kehr, and L. W. Wright, *J. Org. Chem.*, **26**, 1044 (1961). (b) M. Lj. Mihailović, Z. Maksimović, D. Jeremić, Ž. Čeković, A. Milovanović, and Lj. Lorenc, *Tetrahedron*, **21**, 1395 (1965).
27. (a) J. A. Bosoms, *Dissertation Abstr.*, **27B**, 401 (1966). (b) T. Nagai, K. Nishitomi and N. Tokura, *Tetrahedron Lett.*, **1966**, 2419.
28. (a) Ref. 2b. (b) R. E. Partch, *J. Org. Chem.*, **28**, 276 (1963). (c) M. Stefanović, M. Gašić, Lj. Lorenc, and M. Lj. Mihailović, *Tetrahedron*, **20**, 2289 (1964). (d) M. Lj. Mihailović, Lj. Lorenc, M. Gašić, M. Rogić, A. Melera, and M. Stefanović, *Tetrahedron*, **22**, 2345 (1966).
29. J. A. Berson, C. J. Olsen, and J. S. Walia, *J. Amer. Chem. Soc.*, **82**, 5000 (1960).
30. (a) R. E. Partch, *J. Org. Chem.*, **30**, 2498 (1965). (b) M. Lj. Mihailović, M. Jakovljević, and Ž. Čeković, *Tetrahedron*, **25**, 2269 (1969). (c) K. Heusler, *Tetrahedron Lett.*, **1964**, 3975.
31. (a) R. E. Partch, *Tetrahedron Lett.*, **1964**, 3071. (b) M. Lj. Mihailović, J. Bošnjak, Z. Maksimović, Ž. Čeković, and Lj. Lorenc, *Tetrahedron*, **22**, 955 (1966). (c) Ref. 30c. (d) Ref. 1b.
32. (a) R. O. C. Norman and R. A. Watson, *J. Chem. Soc.*, B, **1968**, 184, 692. (b) W. H. Starnes, Jr., *J. Amer. Chem. Soc.*, **89**, 3368 (1967). (c) For migration aptitudes in the chromic acid oxidation of trityl alcohols, see R. Stewart and F. Banoo, *Can. J. Chem.*, **47**, 3207 (1969).
33. (a) K. Heusler, J. Kalvoda, G. Anner, and A. Wettstein, *Helv. Chim. Acta*, **46**, 352 (1963). (b) Ref. 1b. (c) G. Ohloff, K. H. Schulte-Elte, and B. Willhalm, *Helv. Chim. Acta*, **49**, 2135 (1966). (d) K. Heusler and J. Kalvoda, *Helv. Chim. Acta*, **46**, 2732 (1963). (e) H. R. Nace and E. M. Holt, *J. Org. Chem.*, **34**, 2692 (1969).
34. (a) G. Cainelli, B. Kamber, J. Keller, M. Lj. Mihailović, D. Arigoni, and O. Jeger, *Helv. Chim. Acta*, **44**, 518 (1961). (b) Ref. 28c.
35. W. H. Starnes, Jr., *J. Org. Chem.*, **33**, 2767 (1968).

36. H. Immer, M. Lj. Mihailović, K. Schaffner, D. Arigoni, and O. Jeger, *Helv. Chim. Acta*, **45**, 753 (1962).
37. C. Walling and J. Kjellgren, *J. Org. Chem.*, **34**, 1488 (1969).
38. (a) M. Lj. Mihailović and M. Miloradović, *Tetrahedron*, **22**, 723 (1966). (b) Ref. 2b.
39. See p. 364 of Ref. 1a.
40. (a) J. Ujhazy and E. R. Cole, *Nature*, **209**, 395 (1966). (b) Ref. 5b. (c) V. Franzen and R. Edens, *Justus Liebigs Ann. Chem.*, **735**, 47 (1970).
41. R. S. Davidson, *Quart. Rev.*, **21**, 249 (1967).
42. (a) R. Kh. Freidlina, *Advan. Free Radical Chem.*, **1**, 211–278 (1965). (b) C. Walling, in P. de Mayo, Ed., *Molecular Rearrangements*, Part 1 Wiley-Interscience, New York, 1963, pp. 407–455. (c) M. S. Kharasch, A. C. Poshkus, A. Fono, and W. Nudenberg, *J. Org. Chem.*, **16**, 1458 (1951). (d) C. Rüchardt and R. Hecht, *Chem. Ber.*, **98**, 2471 (1965).
43. M. Lj. Mihailović, V. Andrejević, D. Jeremić, A. Stojiljković, and R. E. Partch, *J. Chem. Soc.*, D, **1970**, 854.
44. (a) D. Rosenthal, C. F. Lefler, and M. E. Wall, *Tetrahedron Lett.*, **1965**, 3203; *Tetrahedron*, **23**, 3583 (1967); D. Rosenthal, *Experientia*, **23**, 686 (1967). (b) D. Helmlinger and G. Ourisson, *Tetrahedron*, **25**, 4895 (1969).
45. (a) J. K. Kochi, J. D. Bacha, and T. W. Bethea III, *J. Amer. Chem. Soc.*, **89**, 6538 (1967). (b) For thermodynamic considerations, see K. Torssell, *Ark. Kemi*, **31**, 401 (1969).
46. (a) M. Lj. Mihailović, Ž. Čeković, V. Andrejević, R. Matić, and D. Jeremić, *Tetrahedron*, **24**, 4947 (1968). A 1,2-epoxide is formed as a major ether product in the LTA oxidation of cyclodecanol (but this reaction involves additional transannular hydrogen shifts). See Ref. 43. (b) P. Brun, M. Pally, and B. Waegell, *Tetrahedron Lett.*, **1970**, 331. (c) J.-C. Richer and N. T. T. Hoa, *Can. J. Chem.*, **47**, 2479 (1969), should be consulted for comparison to chromic acid oxidation.
47. M. Lj. Mihailović, Ž. Čeković, and D. Jeremić, *Tetrahedron*, **21**, 2813 (1965).
48. P. Gray and A. Williams, *Chem. Rev.*, **59**, 239 (1959).
49. B. R. Kennedy and K. U. Ingold, *Can. J. Chem.*, **44**, 2381 (1966).
50. (a) D. Hauser, K. Schaffner, and O. Jeger, *Helv. Chim. Acta*, **47**, 1883 (1964). (b) Y. Shalon, Y. Yanuka and S. Sarel, *Tetrahedron Lett.*, **1969**, 957, 961; compare also A. S. Vaidya, S. M. Dixit, and A. S. Rao, *ibid.*, **1968**, 5173. (c) S. Moon and B. H. Waxman, *J. Org. Chem.*, **34**, 288 (1969). (d) M. Lj. Mihailović, S. Konstantinović, A. Milovanović, J. Janković, Ž. Čeković, and D. Jeremić, *Chem. Commun.*, **1969**, 236. (e) R. E. Ireland and L. N. Mander, *J. Org. Chem.*, **34**, 142 (1969). (f) Y. Watanabe, Y. Mizuhara, and M. Shiota, *Chem. Commun.*, **1969**, 984. (g) D. Hauser, K. Heusler, J. Kalvoda, K. Schaffner, and O. Jeger, *Helv. Chim. Acta*, **47**, 1961 (1964).
51. R. W. Leeper, L. Summers, and H. Gilman, *Chem. Rev.*, **54**, 101 (1954).
52. (a) D. H. R. Barton and L. R. Morgan, Jr., *J. Chem. Soc.*, **1962**, 622. (b) G. Smolinsky and B. I. Feuer, *J. Amer. Chem. Soc.*, **86**, 3085 (1964).
53. (a) M. Lj. Mihailović, Ž. Čeković, and D. Jeremić, *Tetrahedron*, **21**, 2813 (1965). (b) M. Fisch, S. Smallcombe, J. Gramain, M. McKervey, and J. Anderson, *J. Org. Chem.*, **35**, 1886 (1970). (c) W. Ayer, D. Law, and K. Piers, *Tetrahedron Lett.*, **1964**, 2959.
54. M. Lj. Mihailović, A. Milovanović, S. Konstantinović, J. Janković, Ž. Čeković, and R. E. Partch, *Tetrahedron*, **25**, 3205 (1969).
55. V. M. Mićović, S. Stojčić, M. Bralovic, S. Mladenović, D. Jeremić, and M. Stefanović, *Tetrahedron*, **25**, 985 (1969).

56. M. Amorosa, L. Caglioti, G. Cainelli, H. Immer, J. Keller, H. Wehrli, M. Lj. Mihailović, K. Schaffner, D. Arigoni, and O. Jeger, *Helv. Chim. Acta*, **45**, 2674 (1962).
57. D. R. Harvey and R. O. C. Norman, *J. Chem. Soc.*, **1964**, 4860.
58. (a) S. Moon and J. M. Lodge, *J. Org. Chem.*, **29**, 3453 (1964). (b) Compare the reaction of propargyl alcohol; p. 351 in Ref. 1a.
59. (a) S. Moon and L. Haynes, *J. Org. Chem.*, **31**, 3067 (1966). (b) A. C. Cope, M. A. McKervey, and N. M. Weinshenker, *J. Amer. Chem. Soc.*, **89**, 2932 (1967).
60. E. A. Braude and O. H. Wheeler, *J. Chem. Soc.*, **1955**, 320.
61. C. Bodea, E. Nicoara, and T. Salontai, *Justus Liebigs Ann. Chem.*, **648**, 147 (1961); C. Bodea and V. Tămaş, *ibid.*, **671**, 57 (1964).
62. N. Nakabayashi, G. Wegner, and H. G. Cassidy, *J. Org. Chem.*, **33**, 2539 (1968).
63. (a) C. R. Bennett and R. C. Cambie, *Tetrahedron*, **23**, 927 (1967). (b) LTA oxidation of phenolic compounds is presented in Ref. 1a and in M. Lj. Mihailović and Ž. Čeković, in S. Patai, Ed., *Chemistry of the Hydroxyl Group*, Wiley-Interscience, New York, 1970.
64. (a) M. Lj. Mihailović, Ž. Čeković, Z. Maksimović, D. Jeremić, Lj. Lorenc, and R. I. Mamuzić, *Tetrahedron*, **21**, 2799 (1965). (b) C. Walling, *Free Radicals in Solution*, Wiley, New York, 1957, chap. 8. (c) The possibility of methyl groups being electron withdrawing is stated in F. W. Baker, R. C. Parish, and L. M. Stock, *J. Amer. Chem. Soc.*, **89**, 5677 (1967).
65. P. Kabasakalian and E. R. Townley, *J. Org. Chem.*, **27**, 2918 (1962).
66. A. C. Cope, M. Gordon, S. Moon, and C. H. Park, *J. Amer. Chem. Soc.*, **87**, 3119 (1965).
67. (a) A. C. Cope, R. S. Bly, M. M. Martin, and R. C. Petterson, *J. Amer. Chem. Soc.*, **87**, 3111 (1965). (b) S. Wawzonek and P. J. Thelen, *J. Amer. Chem. Soc.*, **72**, 2118 (1950).
68. J. Kalvoda, G. Anner, D. Arigoni, K. Heusler, H. Immer, O. Jeger, M. Lj. Mihailović, K. Schaffner, and A. Wettstein, *Helv. Chim. Acta*, **44**, 186 (1961).
69. M. Lj. Mihailović, M. Jakovljević, V. Trifunović, R. Vukov, and Ž. Čeković, *Tetrahedron*, **24**, 6959 (1968).
70. (a) Ref. 28b. (b) Ref. 69; and F. D. Greene, M. L. Savitz, F. D. Osterholtz, H. H. Lau, W. N. Smith, and P. M. Zanet, *J. Org. Chem.*, **28**, 55 (1963). (c) Refs. 2c and 2b. (d) Ref. 56. (e) Ref. 28d. (f) G. Cainelli and S. Morrocchi, *Atti Accad. Naz. Lincei, Rend., Cl. Sci. Fis. Mat. Nat.*, **40**, 591 (1966). (g) See M. Lj. Mihailović and Ž. Čeković, *Helv. Chim. Acta*, **52**, 1146 (1969) for further discussion on fragmentation.
71. W. S. Trahanovsky, P. J. Flash, and L. M. Smith, *J. Amer. Chem. Soc.*, **91**, 5068 (1969).
72. K. Kitahonoki and A. Matsuura, *Tetrahedron Lett.*, **1964**, 2263.
73. (a) W. A. Mosher and E. O. Langerak, *J. Amer. Chem. Soc.*, **73**, 1302 (1951). (b) I. Rothberg and R. V. Russo, *J. Org. Chem.*, **32**, 2003 (1967).
74. (a) J. Lhomme and G. Ourisson, *Tetrahedron*, **24**, 3177 (1968). (b) S. G. Patnekar and S. C. Bhattacharyya, *Tetrahedron*, **23**, 919 (1967). (c) J. Lhomme and G. Ourisson, *Tetrahedron*, **24**, 3201 (1968).
75. (a) S. Landa, J. Vais, and J. Burkhard, *Z. Chem.*, **7**, 233 (1967). (b) D. I. Davies and C. Waring, *Chem. Commun.*, **1965**, 263.
76. (a) R. J. Ouellette and D. L. Shaw, *J. Amer. Chem. Soc.*, **86**, 1651 (1964). (b) A. South, Jr., R. Robins, and R. J. Ouellette, Organic Chemistry Division Abstracts, American Chemical Society Meeting, March 1966, Paper 45. (c) Ref. 70g. (d) L.

Crombie, M. L. Games, and D. J. Pointer, *J. Chem. Soc., C,* **1968**, 1347. (e) S. E. Schaafsma, H. Steinberg, and T. J. de Boer, *Rec. Trav. Chim. Pays-Bas,* **85**, 70, 73 (1966). (f) L. B. Young and W. S. Trahanovsky, *J. Org. Chem.,* **32**, 2349 (1967). (g) S. Moon, *J. Org. Chem.,* **29**, 3456 (1964).

77. (a) H. Hart and D. P. Wyman, *J. Amer. Chem. Soc.,* **81**, 4891 (1959). (b) C. Walling and P. S. Fredricks, *J. Amer. Chem. Soc.,* **84**, 3326 (1962). (c) D. B. Denney and J. W. Hanifin, Jr., *J. Org. Chem.,* **29**, 732 (1964). (d) R. Breslow, in P. de Mayo, Ed., *Molecular Rearrangements,* Part 1 Wiley-Interscience, New York, 1963, pp. 233–294. (e) A. Streitwieser, Jr., *Solvolytic Displacement Reactions,* McGraw-Hill, New York, 1962.

78. (a) A. Bowers, E. Denot, L. C. Ibáñez, M. E. Cabezas, and H. J. Ringold, *J. Org. Chem.,* **27**, 1862 (1962). (b) J. F. Bagli, P. F. Morand, and R. Gaudry, *J. Org. Chem.,* **28**, 1207 (1963). (c) P. F. Morand and M. Kaufman, *J. Org. Chem.,* **34**, 2175 (1969). (d) Ref. 28d. (e) W. H. W. Lunn, *J. Org. Chem.,* **30**, 1649 (1965).

79. M. Lj. Mihailović, Z. Maksimović, D. Jeremić, Ž. Čeković, A. Milovanović, and Lj. Lorenc, *Tetrahedron,* **21**, 1395 (1965).

80. Compare M. S. Kharasch, H. N. Friedlander, and W. H. Urry, *J. Org. Chem.,* **16**, 533 (1951).

81. Compare W. A. Mosher, C. L. Kehr, and L. W. Wright, *J. Org. Chem.,* **26**, 1044 (1961).

82. R. E. Partch, *J. Amer. Chem. Soc.,* **89**, 3662 (1967).

83. (a) UV data are found in Ref. 1b, 7b, 7d, and 7e. (b) IR data are found in Refs. 7a, 7c, and 12c. (c) ESR data are found in Ref. 7e; and H. Loeliger, *Helv. Chim. Acta,* **52**, 1516 (1969). (d) Lead tetracarboxylates showing ester-type carbonyl absorption in noncomplexing media are reported in Ref. 12c.

84. (a) C. A. Grob, M. Ohta, and A. Weiss, *Angew. Chem.,* **70**, 343 (1958). (b) E. E. van Tamelen and S. P. Pappas, *J. Amer. Chem. Soc.,* **85**, 3297 (1963). (c) E. J. Corey and J. Casanova, Jr., *J. Amer. Chem. Soc.,* **85**, 165 (1963).

85. M. Takebayashi, T. Ibata, H. Kohara, and B. H. Kim, *Bull. Chem. Soc. Jap.,* **40**, 2392 (1967).

86. F. Huber and M. S. El-Meligy, *Angew. Chem. Int. Ed. Eng.,* **7**, 946 (1968); *Chem. Ber.,* **102**, 872 (1969).

87. H. Taube and E. S. Gould, *Accounts Chem. Res.,* **2**, 321 (1969).

88. (a) Coordination of pyridine-N-oxide and DMSO to metal ions is most often through the oxygen atom. See J. V. Quagliano, J. Fujita, G. Franz, D. J. Phillips, J. A. Walmsley, and S. Y. Tyree, *J. Amer. Chem. Soc.,* **83**, 3770 (1961); R. Carlin, *ibid.,* **83**, 3773 (1961); V. G. K. Das and W. Kitching, *J. Organometal. Chem.,* **13**, 523 (1968); P. W. N. M. van Leeuwen and W. L. Groeneveld, *Rec. Trav. Chim. Pays-Bas,* **86**, 1217 (1967). (b) The base strength of pyridine-N-oxide is less than that of pyridine but greater than that of DMSO. See Ref. 88a; I. I. Grandberg, G. K. Faizova, and A. N. Kost, *Khim. Geterotsikl. Soedin.,* **1966**, 561, cf. *Chem. Abstr.,* **66**, 10453b (1967); P. Haake and R. D. Cook, *Tetrahedron Lett.,* **1968**, 427; F. A. Cotton and R. Francis, *J. Amer. Chem. Soc.,* **82**, 2986 (1960).

89. Compare mixed ligand effects on Mn(II) catalyzed decarboxylations; J. V. Rund and K. G. Claus, *J. Amer. Chem. Soc.,* **89**, 2256 (1967).

90. Compare substitutent effects on oxidation by other metal ions. See W. S. Trahanovsky, L. B. Young, and G. L. Brown, *J. Org. Chem.,* **32**, 3865 (1967); R. Robson and H. Taube, *J. Amer. Chem. Soc.,* **89**, 6487 (1967); T. G. Clarke, N. A. Hampson, J. B. Lee, J. R. Morley, and B. Scanlon, *Can. J. Chem.,* **47**, 1649 (1969). Bromine oxidation of alcohols is presented by P. Aukett and I. R. L. Barker, *Chem. Ind.* (London), **1967**, 193.

91. (a) P. Kabasakalian and E. R. Townley, *J. Amer. Chem. Soc.*, **84**, 2711 (1962).
(b) N. A. Maier and Yu. A. Oldekop, *Dokl. Akad. Nauk SSSR*, **172**, 349 (1967).
92. K. Heusler, *Helv. Chim. Acta*, **52**, 1520 (1969).
93. (a) J. J. Tufariello and W. J. Kissel, *Tetrahedron Lett.*, **1966**, 6145. (b) T. D. Walsh and H. Bradley, *J. Org. Chem.*, **33**, 1276 (1968). (c) C. M. Cimarusti and J. Wolinsky, *J. Amer. Chem. Soc.*, **90**, 113 (1968). (d) N. B. Chapman, S. Sotheeswaran, and K. J. Toyne, *Chem. Commun.*, **1965**, 214. (e) Ref. 2d. (f) W. J. Kissel, *Dissertation Abstr.*, *B*, **29**(2), 546 (1968). (g) B. Stokes and R. Partch, American Chemical Society Northeast Regional Meeting, October 1970, Paper 177.
94. (a) Comparative reactivities of radicals are ˙CF_3 > ˙CH_3 > ˙CCl_3. See Ref. 18e; M. Levy and M. Szwarc, *J. Amer. Chem. Soc.*, **76**, 5981 (1954); S. W. Charles, J. T. Pearson, and E. Whittle, *Trans. Faraday Soc.*, **59**, 1156 (1963). (b) P. Sartori and M. Weidenbruch, *Chem. Ber.*, **100**, 2049 (1967).
95. (a) R. E. Partch and T. Jenks, Division of Fluorine Chemistry, American Chemical Society Meeting, September 1969, Paper 8. (b) Ref. 91a.
96. (a) J. B. Aylward, *J. Chem. Soc.*, *B*, **1967**, 1268. (b) L. Eberson, *J. Amer. Chem. Soc.*, **89**, 4669 (1967). (c) J. R. Shelton and C. W. Uzelmeier, *J. Amer. Chem. Soc.*, **88**, 5222 (1966). (d) H. C. Brown and R. A. Wirkkala, *J. Amer. Chem. Soc.*, **88**, 1447 (1966).
97. A. McKillop, J. S. Fowler, M. J. Zelesko, J. D. Hunt, E. C. Taylor, and G. McGillivray, *Tetrahedron Lett.*, **1969**, 2423, 2427.
98. G. A. Olah and R. H. Schlosberg, *J. Amer. Chem. Soc.*, **90**, 2726 (1968), and references therein.
99. (a) L. Suchomelova and J. Zyka, *J. Electroanal. Chem.*, **5**, 57 (1963). (b) T. Mukaiyama and T. Endo, *Bull. Chem. Soc. Jap.*, **40**, 2388 (1967). (c) R. Criegee, *Angew. Chem.*, **70**, 173 (1958). (d) H. E. Barron, G. W. K. Cavill, E. R. Cole, P. T. Gilham, and D. H. Solomon, *Chem. Ind.* (London), **1964**, 76. (e) W. M. Doane, B. S. Shasha, C. R. Russell, and C. E. Rist, *J. Org. Chem.*, **30**, 3071 (1965).
100. F. H. Fink, J. A. Turner, and D. A. Payne, Jr., *J. Amer. Chem. Soc.*, **88**, 1571 (1966).
101. A. Stojiljković, V. Andrejević, and M. Lj. Mihailović, *Tetrahedron*, **23**, 721 (1967).
102. (a) J. H. Hall and E. Patterson, *J. Amer. Chem. Soc.*, **89**, 5856 (1967). (b) W. J. Irwin and D. G. Wibberley, *Chem. Commun.*, **1968**, 878. (c) C. D. Campbell and C. W. Rees, *J. Chem. Soc.*, *C*, **1969**, 742, 748, 752; C. W. Rees and R. C. Storr, *ibid.*, **1969**, 756, 760, 765; R. W. Hoffmann, G. Guhn M. Preiss, and B. Dittrich, *ibid.*, **1969**, 769; R. S. Atkinson and C. W. Rees, *ibid.*, **1969**, 772, 778; D. Nasipuri, R. Bhattacharya and C. K. Ghosh, *ibid.*, **1969**, 782; and references in these papers.
103. A. J. Sisti, *Chem. Commun.*, **1968**, 1272.
104. C. W. Rees and M. Yelland, *Chem. Commun.*, **1969**, 377.
105. J. H. Boyer and J. D. Woodyard, *J. Org. Chem.*, **33**, 3329 (1968).
106. R. Criegee, L. Kraft, and B. Rank, *Justus Liebigs Ann. Chem.*, **507**, 159 (1933).
107. R. E. Partch, T. Jenks, and J. Maroski, Organic Chemistry Division, American Chemical Society Great Lakes Regional Meeting, Chicago, 1966.
108. C. Ehret and G. Ourisson, *Bull. Soc. Chim. Fr.*, **1968**, 2629.
109. T. Okuda and T. Yoshida, *Chem. Ind.* (London), **1965**, 37.
110. K. H. Baggaley, T. Norin, and S. Sundin, *Acta Chem. Scand.*, **22**, 1709 (1968).
111. M. Gschwendt and E. Hecker, *Z. Naturforsch.*, **23b**, 1584 (1968).
112. A. D. Wagh, S. K. Paknikar, and S. C. Bhattacharyya, *Tetrahedron*, **20**, 2647 (1964).
113. U. Scheidegger, K. Schaffner, and O. Jeger, *Helv. Chim. Acta*, **45**, 400 (1962).
114. K. Ichikawa and Y. Takeuchi, *Nippon Kagaku Zasshi*, **79**, 1060 (1958).

115. H. R. Goldschmid and A. S. Perlin, *Can. J. Chem.*, **38**, 2280 (1960).
116. R. Moriarty and H. G. Walsh, *Tetrahedron Lett.*, **1965**, 465.
117. M. Lj. Mihailović, Ž. Čeković, Z. Maksimović, D. Jeremić, Lj. Lorenc, and R. I. Mamuzić, *Tetrahedron*, **21**, 2799 (1965).
118. V. M. Mićović, R. I. Mamuzić, D. Jeremić, and M. Lj. Mihailović, *Tetrahedron Lett.*, **1963**, 2091; *Tetrahedron*, **20**, 2279 (1964).
119. V. M. Mićović and M. Lj. Mihailović, *Rec. Trav. Chim. Pays-Bas*, **71**, 970 (1952).
120. H. Wehrli, M. S. Heller, K. Schaffner, and O. Jeger, *Helv. Chim. Acta*, **44**, 2162 (1961); M. S. Heller, H. Wehrli, K. Schaffner, and O. Jeger, *ibid.*, **45**, 1261 (1962).
121. K. Heusler, J. Kalvoda, P. Wieland, G. Anner, and A. Wettstein, *Helv. Chim. Acta*, **45**, 2575 (1962).
122. A. Bowers and E. Denot, *J. Amer. Chem. Soc.*, **82**, 4956 (1960).
123. E. Wenkert and B. L. Mylari, *J. Amer. Chem. Soc.*, **89**, 174 (1967).
124. P. N. Rao and J. C. Uroda, *Naturwiss.*, **50**, 548 (1963).
125. R. M. Moriarty and K. Kapadia, *Tetrahedron Lett.*, **1964**, 1165.
126. P. F. Beal and J. E. Pike, *Chem. Ind.* (London), **1960**, 1505.
127. A. Wettstein, G. Anner, K. Heusler, J. Kalvoda, and H. Ueberwasser, Swiss Patent 453,345; cf. *Chem. Abstr.*, **70**, 4440m (1969).
128. Y. Yanuka, S. Sarel, and M. Beckermann, *Tetrahedron Lett.*, **1969**, 1533.
129. K. Heusler and J. Kalvoda, *Helv. Chim. Acta*, **46**, 2020 (1963).
130. H. Ueberwasser, K. Heusler, J. Kalvoda, Ch. Meystre, P. Wieland, G. Anner, and A. Wettstein, *Helv. Chim. Acta*, **46**, 344 (1963).
131. A. Bowers, R. Villotti, J. A. Edwards, E. Denot, and O. Halpern, *J. Amer. Chem. Soc.*, **84**, 3204 (1962).
132. J. Tadanier, *J. Org. Chem.*, **28**, 1744 (1963).
133. K. Heusler, J. Kalvoda, Ch. Meystre, G. Anner, and A. Wettstein, *Helv. Chim. Acta*, **45**, 2161 (1962).
134. J. Kalvoda, K. Heusler, G. Anner, and A. Wettstein, *Helv. Chim. Acta*, **46**, 618 (1963).
135. L. Velluz, G. Muller, R. Bardoneschi, and A. Poittevin, *C. R. Acad. Sci., Paris*, **250**, 725 (1960).
136. Ch. Meystre, K. Heusler, J. Kalvoda. P. Wieland, G. Anner, and A. Wettstein, *Helv. Chim. Acta*, **45**, 1317 (1962).
137. Baker Castor Oil Co., British Patent 759,416 (1956); cf. *Chem. Zentr.*, **1961**, 1009.
138. T. Koga and M. Tomoeda, *Tetrahedron*, **26**, 1043 (1970).
139. P. Roller and C. Djerassi, *J. Chem. Soc., C*, **1970**, 1089.
140. C. Meystre, J. Kalvoda, G. Anner, and A. Wettstein, *Helv. Chim. Acta*, **46**, 2844 (1963).
141. S. Brois, *J. Amer. Chem. Soc.*, **92**, 1079 (1970).
142. F. Carey and L. Hayes, *J. Amer. Chem. Soc.*, **92**, 7614 (1970).

The Lead Tetraacetate Oxidation of Olefins

ROBERT M. MORIARTY

Department of Chemistry, The University of Illinois at Chicago Circle, Chicago, Illinois

I. INTRODUCTION

The major portion of this review deals with the lead tetraacetate (LTA) oxidation of the carbon–carbon double bond. Acetylenic compounds and allenes are mentioned briefly. The emphasis is on the mechanism and stereochemistry of the reaction. The participation of neighboring groups is discussed and similarities that the LTA oxidation reaction shares with the Hg(II), and Tl(III) oxidations is treated in certain relevant cases. Criegee (1), the single most outstanding contributor to this area, has written a superb summary of research on this versatile reagent.

In fact, this versatility in reaction pathway and products is the focal point of the present review. An attempt is made to show that the manifold reactions of LTA may be incorporated within the framework of unifying mechanistic categories encountered in the electrophilic addition reactions of olefins.

II. CYCLIC OLEFINS

Since some of the most pertinent mechanistic work has been carried out on cyclic olefins, this area of the oxidative reactions will be treated before the acyclic examples.

No systematic study of the behavior of members of a homologous series of cyclic olefins towards oxidation with LTA has been reported. The oxidation of cyclohexene (1) has been studied in detail by several groups (2–5). The products of reaction illustrate the major pathways that occur in the LTA oxidation of olefins; namely, addition, allylic substitution, allylic rearrangement, and, to a minor extent, pinacolic-type ring contraction. Anderson and Winstein (3) report formation of both cis- and trans-1,2-diacetoxycyclohexanes (2) as well as 37% of 3-acetoxycyclohexene (3) and two ring-contracted products 4 and 5 (eq. 1).

A slight change in the relative yield of the various products is observed upon changing the solvent from acetic acid to benzene. Yields of allylic substitution product are higher in benzene than in acetic acid. Kabbe (5) reports a 44% yield of cyclopentenylcarboxaldehyde diacetate (4) from the oxidation of cyclohexene in acetic acid.

Allylic rearrangement in the formation of 3-acetoxycyclohexene (3) has been demonstrated (4). The isotopic distribution observed in the LTA oxidation of cyclohexene-^{13}C in either benzene or acetic acid indicates the intervention of a symmetrical intermediate. This observation, coupled with the fact that 1-methylcyclohexene reacts only slightly faster than cyclohexene, suggests the occurrence of the steps shown in eq. 2.

The initial addition is written as reversible in order to accord with the small rate difference observed between cyclohexene and 1-methylcyclohexene in competition reactions with LTA. The rate-limiting step could be either reaction of the bridged organolead intermediate with acetate or the cleavage of the carbon–lead bond. It may be inferred that the cleavage of the carbon–lead bond is an extremely facile process. Furthermore, in the case of an allylic lead intermediate of Type B, the cleavage reaction leading to an allylic carbonium ion would be expected to occur very rapidly. This point is

$$(2)$$

further discussed later. Intermediate A may be the precursor of the ring-contracted products which can be considered to arise from a pinacolic-type reaction (eq. 3).

$$(3)$$

Cis- and trans-diacetates may derive via several possible modes of decomposition. Assuming a trans stereochemistry for Intermediate A, intervention of an acetoxonium ion 6 could account for the formation of trans-diacetate 7 (eq. 4).

$$(4)$$

Displacement by acetate with inversion of configuration at the carbon–lead bond could account for the formation of cis-diacetate.

The course of thallium acetate oxidation of cyclohexene (1) is very similar to that observed with LTA. Anderson and Winstein (3) found 40 to 50% cis and trans 2 and 40 to 50% 4 as well as 3% of 3.

At this stage some discussion is appropriate regarding the present status of organolead reaction intermediates of Types *A* and *B* mentioned above. Such representations are frequently encountered throughout the literature on the reactions of LTA, and they have been invoked extensively by Criegee (1,6,7) in discussing the LTA oxidation of olefins. Although these intermediates are convenient representations, and have been accepted largely through usage, it should be emphasized that in no acyclic or monocyclic olefinic case have such organolead compounds been isolated or even indicated as intermediates by any physical method. The analogy for such metalloorganic intermediates is based upon the behavior of olefins towards Hg(II) salts where electrophilic attack by Hg(OAc)$_2$ yields isolable organomercurinium compounds (8,8a,8b,9,10). Thus, eq. 5 represents norbornene (**8**) → **9**.

$$\text{8} \qquad + \; Hg(OAc)_2 \; \longrightarrow \qquad \overset{HgOAc}{\underset{OAc}{}} \qquad (5)$$

8   **9**

The oxidation of cyclohexene with mercuric acetate in methanol yields a stable adduct which has been shown by means of X-ray diffraction studies to have *trans* stereochemistry (eq. 6) (8b).

$$\text{1} \quad \xrightarrow[CH_3OH]{Hg(OAc)_2} \quad \overset{OCH_3}{\underset{HgOAc}{}} \qquad (6)$$

1

Stable intermediates have been isolated in the oxythallation reaction of norbornene (**8**) and norbornadiene (eq. 7) (11,12).

$$\text{8} \quad \xrightarrow[CHCl_3]{Tl(OAc)_3} \quad \overset{Tl(OAc)_2}{\underset{OAc}{}} \qquad (7)$$

8   **9**

The fact that Hg(II) and Tl(III) afford isolable organometallic adducts whereas none is obtained with lead may be due in part to the greater electron affinity of Pb(IV). Organolead triacetates have been prepared in the aryl series as in eq. 8 (13,14).

LTA has been shown to react with anisole to give both *p*-methoxyphenyllead triacetate and two (*o* and *p*-) of the three acetoxyanisoles (14a, 14b). Lead tetraacetate is thought to behave as an ambient electrophile

$$(8)$$

reacting with the nucleophilic aromatic ring to yield both acetoxylation and plumbylation. Electrophilic plumbylation is readily reversible and electrophilic substitution occurs *via* a Se 2 mechanism.

Treatment of aliphatic organomercury compounds with LTA yields alkyl acetates (eq. 9) (7).

$$R-Hg-C_6H_5 + Pb(OAc)_4 \longrightarrow R-OAc + Pb(OAc)_2 + C_6H_5-HgOAc \quad (9)$$

An intermediate of the type $R-Pb(OAc)_3$ has been assumed in the reaction of alkyl mercurials with LTA (7). Obviously a valuable contribution to our understanding of the LTA oxidation of olefins would be the synthesis of a compound such as $RCHOAc-CHRPb(OAc)_3$. The behavior of such a compound under the conditions of the oxidation reaction would serve to clarify the postulated role of these species as intermediates. It is interesting to note that treatment of dineopentyl mercury with LTA yields *t*-amylacetate. Assuming intermediacy of $R-Pb(OAc)_3$, it appears that a carbonium ion is involved at some stage in its conversion to ROAc. It should be noted further that carbon radicals may be oxidized by Pb(III) and Pb(IV) (39). This aspect of the oxidative reaction is discussed later.

The oxidation of $(+)$-1-*p*-menthene (10) has been studied by Kergomard (15) and Aratani (16) who found that the allylic acetate formed along with diacetate 12 was racemic (10 → 11) (eq. 10).

$$(10)$$

Wiberg (4) showed that the starting olefin and allylic acetate were both optically stable under the reaction conditions. The symmetrical intermediate (15) is indicated for this oxidation (eq. 11). This is in agreement with the results obtained in the oxidation of ^{13}C-cyclohexene mentioned earlier.

$$\text{(11)}$$

13   **14**   **15**

The oxidation of 2,4,4-trimethylcyclohex-1-ene (**16** → **17**) with LTA in acetic acid may likewise be interpreted in terms of such allylic rearrangement reaction of an organolead intermediate (**17**). None of the alternative, but sterically more crowded, isomer **18** was isolated in this oxidation (eq. 12).

$$\text{(12)}$$

16   **17**   **18**

Furthermore, no methyl group migration was reported although it appears to be a likely possibility at the allylic carbonium ion stage. Mercuric acetate oxidation of **16** yields 24% of **17** (17a).

Several reports of the LTA oxidation of various unsaturated terpenes exist. The oxidation of α-pinene (**19**) is particularly instructive (18–20): treatment with LTA in benzene yielded *cis*-2-acetoxypin-3-ene (**20**), which could be converted under very mild conditions with acetic acid to the product of allylic rearrangement; namely, *trans*-verbenyl acetate (**21**) (eq. 13).

$$\text{(13)}$$

19   **20**   **21**

Earlier, Criegee (20) reported that oxidation of α-pinene (**19**) in benzene yielded *trans*-verbenyl acetate (**21**) whereas oxidation in acetic acid gave, in addition to this product, sobrerol diacetate (**22**) and verbenene. *Trans*-verbenyl acetate (**21**) is a secondary product derived from initially formed *cis*-2-acetoxypin-3-ene (**20**) (18).

These results may be discussed in terms of the mechanistic steps of eq. 14 involving ionic complex *A*. This mechanism is somewhat different

$$(14)$$

from that outlined for cyclohexene in that the initially formed complex does not undergo bimolecular reaction with acetate anion but rather decomposes by intramolecular acetoxyl group transfer. Alternatively, one might propose bimolecular attack but this would occur in a *cis* fashion in order to agree with the observed stereochemical outcome of the reaction. Such *cis* products do in fact occur in the oxymercuration (21) and oxythallation reactions of nor-bornene (11,12). However, *cis* addition appears to be restricted to addition reactions of these reagents to strained bicyclic olefins (10,21,21a).

Formation of sorbrerol diacetate (**22**) may result from the decomposition indicated by eq. 15.

$$(15)$$

The reaction of α-pinene (**19**) with mercuric acetate (4) gives results different from those observed with LTA even though an initial ionic complex analogous to Intermediate *A* may intervene (eq. 16).

Oxidation of apopinene (**23**) follows a similar course (22). The principal

$$(16)$$

product is apoverbenyl acetate (**24**) along with a small amount of apopinane diacetate (**25**) (eq. 17).

$$(17)$$

LTA oxidation of β-pinene (**26**) (23) offers an interesting parallel with both α-pinene and the pair of double bond isomers 2-methylnorbornene (**32**) and 2-methylenenorbornane (**29**) (24). In benzene, LTA oxidation leads to about equal amounts of *trans*-pinocarvyl acetate (**27**) and myrtenyl acetate (**28**) (eq. 18).

$$(18)$$

In acetic acid, however, the yield of *trans*-pinocarvyl acetate (**27**) rose to 46%. Both isomers were stable to acetic acid under conditions even more vigorous than those of the actual LTA oxidation. Again, a bridged organo-lead complex may be involved initially (eq. 19).

$$(19)$$

27   **28**

The allylic oxidation of 5,6-unsaturated steroids with lead tetraacetate occurs to yield a mixture of the 7α- and 7β- acetates (23a). Thus cholesteryl acetate upon treatment with LTA undergoes allylic acetoxylation to the extent of 50% (eq. 19a).

$$(19a)$$

The proposed mechanisms are summarized in eq. 19b. Mechanism A proceeds through a cyclic transition state. Mechanism B involves solvolysis of the allylic organolead intermediate to yield an allylic carbonium ion which coordinates with solvent at C_7 in a nonstereospecific way to yield a mixture of the two epimers.

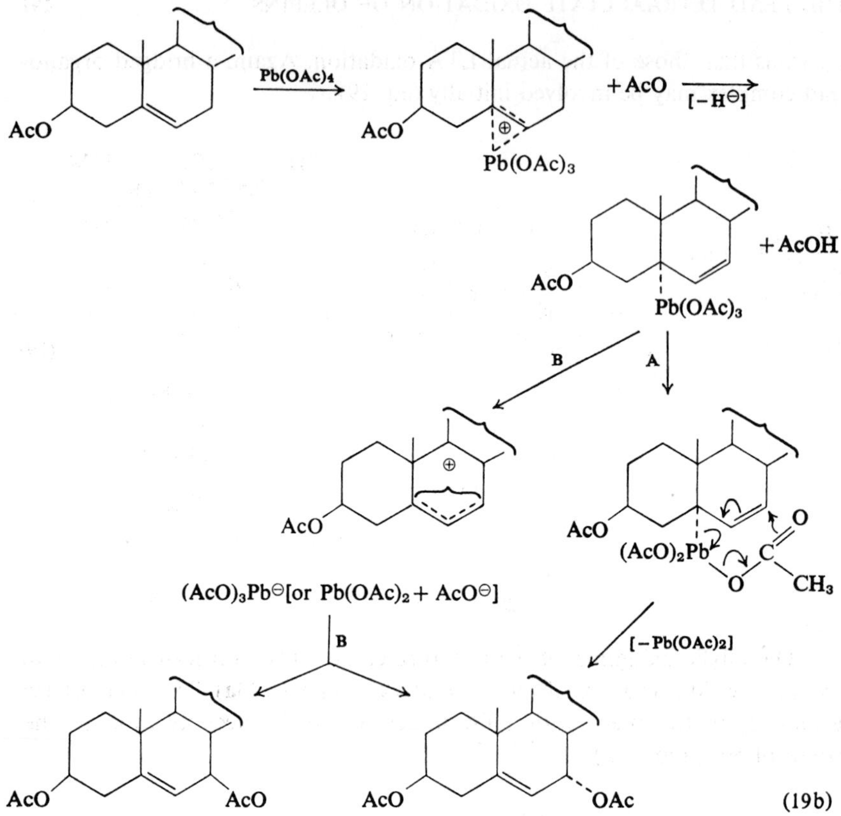

Pb(OAc)4
AcO
+AcO
[−H⊖]
Pb(OAc)3
+AcOH
B
A
(AcO)2Pb
O
C
CH3
(AcO)3Pb⊖[or Pb(OAc)2 + AcO⊖]
[−Pb(OAc)2]
OAc

(19b)

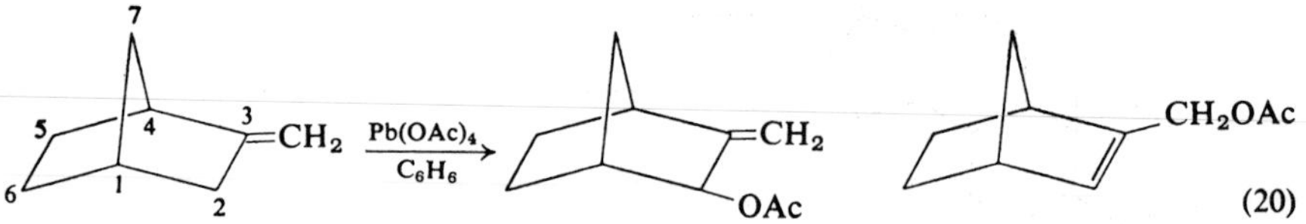

7
5
4
3
6
1
2
CH2
Pb(OAc)4
C6H6
OAc
CH2OAc

(20)

29 **30** (93%) **31** (7%)

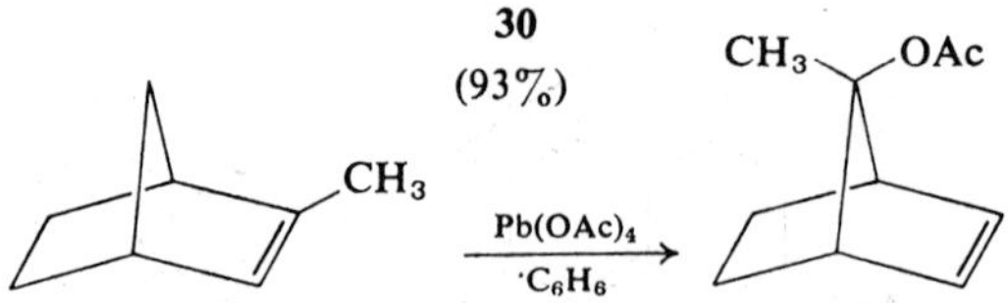

CH3
Pb(OAc)4
C6H6
OAc

32 **33** (35%) (21)

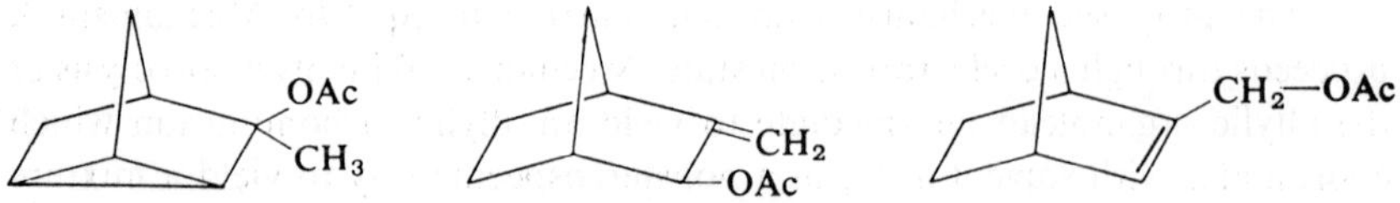

OAc
CH3
CH2
CH2—OAc

34 (56%) **30** (90%) **31** (trace)

Oxidation in benzene solution of the isomeric pair 2-methylenenor-bornane (**29**) (eq. 20) and 2-methylnorbornene (**32**) (eq. 21) gave products in over-all yields of 26 and 35%, respectively (24).

The first intermediate is considered by Erman (24) to arise via electro-philic addition to the double bond to yield the norbornyl cation (eq. 22).

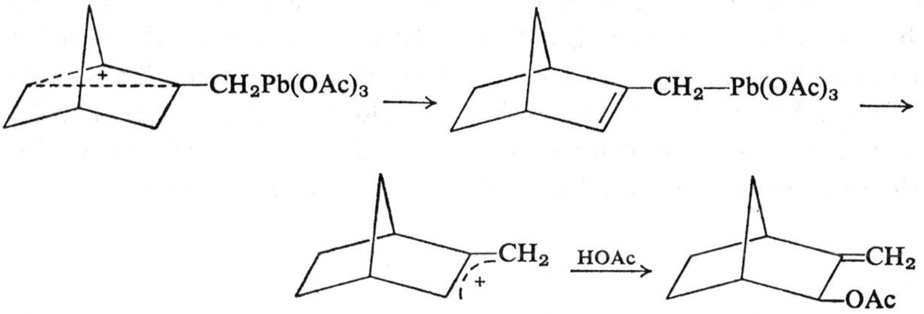

$$(22)$$

30

A concerted rearrangement could be involved as shown in eq. 23.

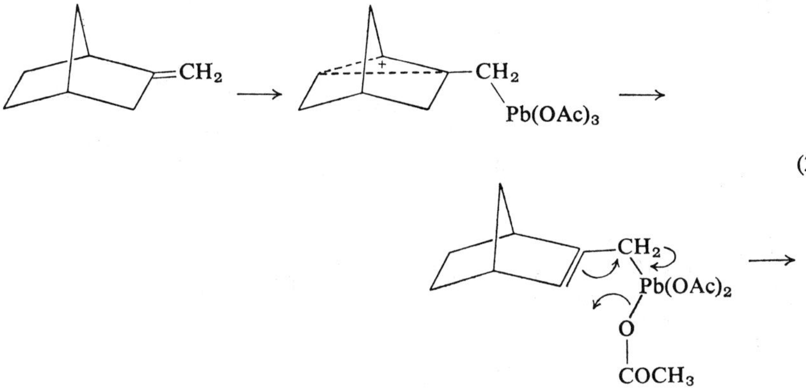

$$(23)$$

$\longrightarrow$ **30**

It is worth noting that no 2,7-diacetoxyl derivative **35** is observed in the LTA oxidation of 2-methylnorbornene (**32**) (eq. 24). This mode of reaction is

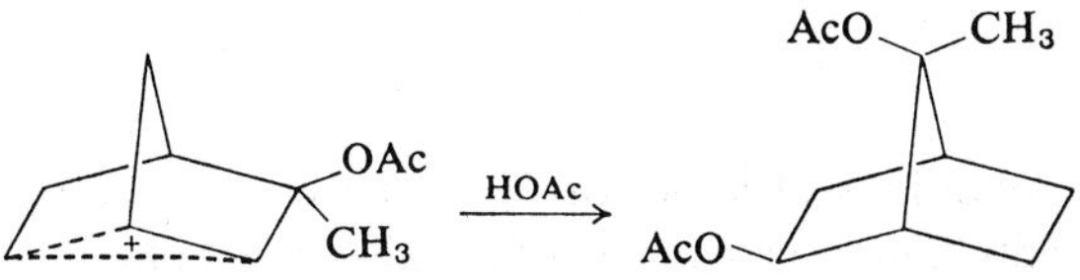

$$(24)$$

35

a major one in the LTA oxidation of norbornene (**8**) under comparable conditions (28,29).

The intermediary allylic carbonium ion is formed by heterolytic cleavage of the C—Pb bond. Predominant attack at C_3 in the *exo* direction is probably due to a combination of torsional effects (25) and the relative stability of the exocyclic double bond compared to the endocyclic arrangement in the norbornyl system.

Similar results as far as the structures of products have been obtained in the bromination of 2-methylnorbornene (**32**) and 2-methylenenorbornane (**29**) with *N*-bromosuccinimide (eq. 25). *Exo*-3-bromo-2-methylenenorbornane (**36**) and 2-methyl-3-bromonortricyclane (**38**) were obtained in a 3:1 ratio but no 2-bromomethylnorbornene-2 (**37**) was detected (26). The allylic cation appears to undergo reaction at both electron deficient positions. Thus treatment of *exo*-2-bromo-2-methylenenorbornane (**36**) with silver acetate in 50% aqueous acetone gave a mixture *exo*-3-hydroxy-2-methylenenorbornane (**39**) and 2-hydroxy-methylnorborn-2-ene (**40**) in a 2:1 ratio (eq. 26) (27).

Since the conditions of ionic solvolysis (silver acetate, aqueous acetone 50%) undoubtedly lead to generation of the allylic carbonium, it is surprising that the same allylic carbonium ion postulated in the LTA reaction of 2-methylenenorbornane (**29**) does not result in a similar *exo-endo*cyclic ratio of acetates. In this sense, the outcome of the lead tetraacetate appears to suggest a mechanism that closely resembles the radical pathway observed in the *N*-bromosuccinimide reaction.

The oxidation of 2-methylnorbornene (**32**) by LTA is thought to involve the stages shown in eq. 27 (24).

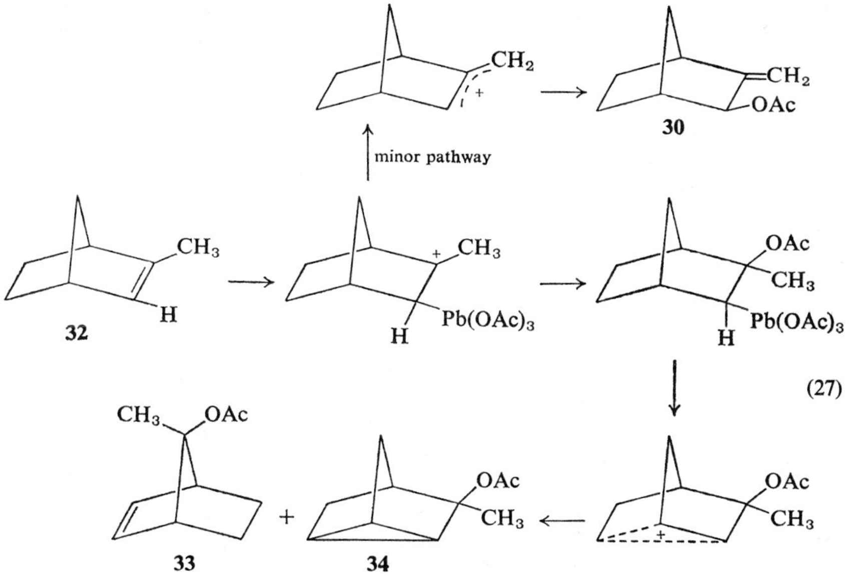

(27)

The oxidation of norbornene by LTA in various solvents has been studied by Alder and co-workers (28) and also by the present author (29) (eq. 28). The principal difference between the results of the two investigations (28,29) is the isolation of substantial amounts of the *anti*-2,7-norbornanediol diacetate (**41**) (28% in acetic acid and 25% in benzene) in the more recent work (29).

Alder suggested initial electrophilic addition followed by rearrangement to the C$_7$-lead derivative (eq. 29).

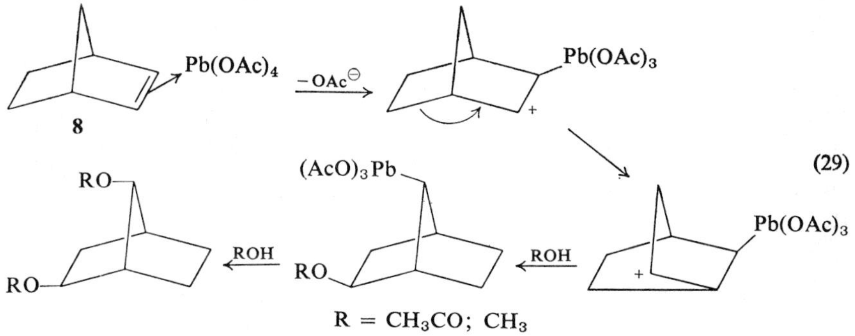

(29)

$$R = CH_3CO; \ CH_3$$

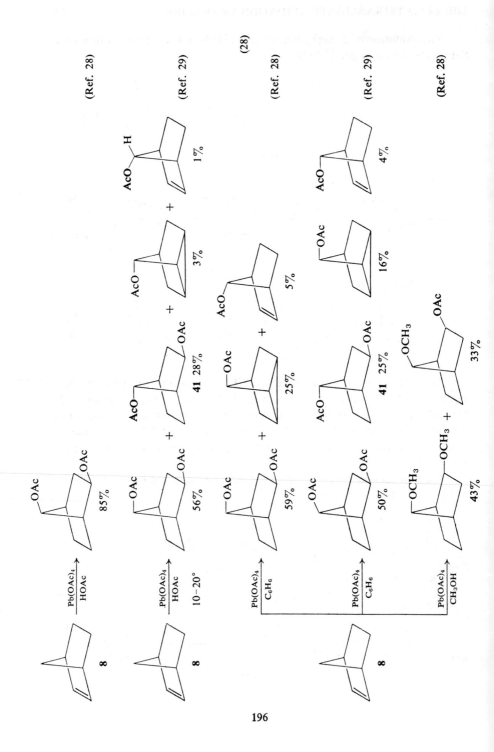

Erman (24) has queried this description on the basis of the well-known course of solvolysis of 7-tosyloxynorbornyl derivatives. The rate of solvolysis would be expected to be very slow and also ring-contraction would be anticipated. This, of course, assumes that the LTA oxidation involves a simple ionization at C_7 to yield a carbonium ion. These objections would be invalidated were a concerted process to occur proceeding with only a small amount of charge separation in the transition state. Also the possibility of a homolytic cleavage of the carbon–lead bond cannot be dismissed (eq. 30).

$$(30)$$

In order to circumvent the C_7 cleavage of the intermediary organolead derivative, Erman (24) suggests the alternative pathway of eq. 31.

$$(31)$$

Comparison of the results of LTA oxidation of camphene (**42**) (30,31) with 2-methylenenorbornane (**29**) is interesting. The major product in this reaction is the enol acetate derivative of homocamphenilone (**43**). The steps may be written as in eq. 33. The first intermediate is analogous to the one proposed in the norcamphene oxidation; however, no β-proton is present. Thus, the formation of an allylic system via deprotonation is precluded.

Naffa and Ourisson (32) have shown that longifolene (**44**) behaves

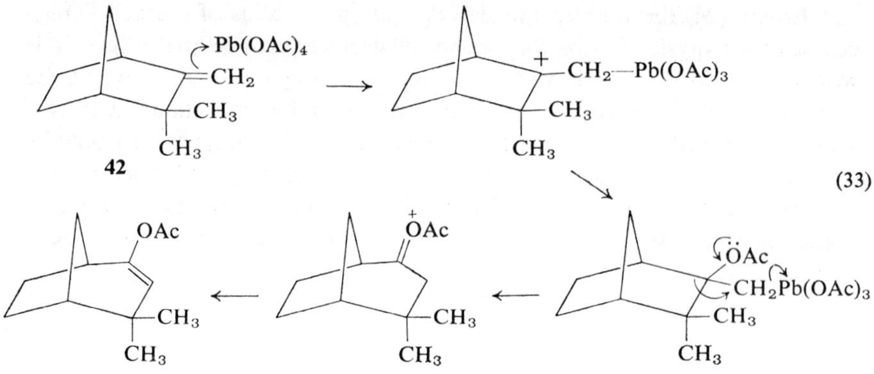

(33)

analogously (**44** → **45**) (eq. 34). Other LTA oxidations of bridged bicyclic

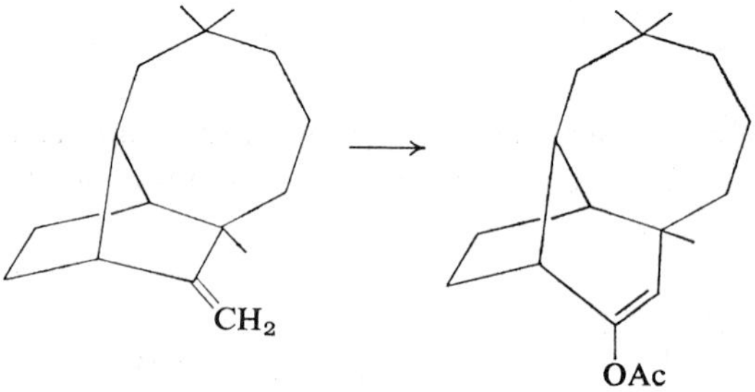

(34)

44   **45**

olefins have been studied by Alder and co-workers (28) (**46** → **47** and **48**) (eq. 35).

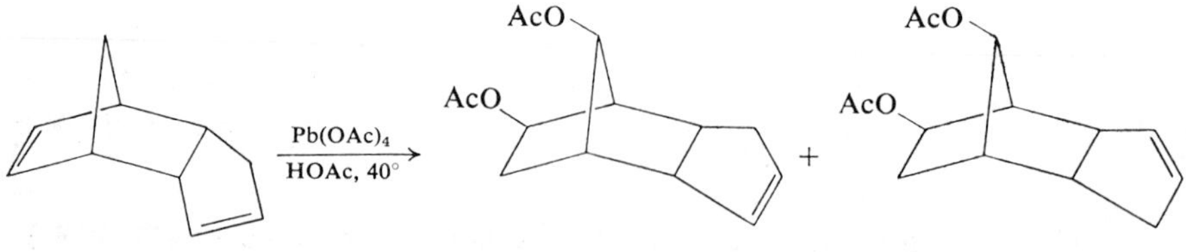

46   **47**   **48**

(35)

Preferential attack at the bicyclo(2.2.1) double bond contrasts with the behavior of dicyclopentadiene (**46**) towards oxidation with perbenzoic acid in which oxidation occurs with equal readiness at both double bonds (33).

Formation of the 2,7-disubstituted product is in agreement with the typical behavior of norbornene towards a large variety of electrophilic reagents.

Norbornadiene upon LTA oxidation was shown to yield both diacetoxy-nortricyclyl and 2,7-diacetoxynorbornenyl derivatives in ratios which were strongly solvent dependent (28) (eq. 36).

$$(36)$$

As mentioned earlier, norbornene as well as norbornadiene undergo cis-exo-oxythallation to yield stable adducts (11,12). Reduction with sodium borohydride proceeds with deoxythallation to yield the parent olefin, and reduction with sodium amalgam in water yields the alcohol (eq. 37).

$$(37)$$

Cocton and DePaulet (34) studied the reaction of 3-methyl-Δ^2-cholestene with Tl(OAc)$_3$ and obtained a 50% yield of a mixture of 4α- and 4β-acetoxy-3-methyl-Δ^2-cholestenes (eq 38).

The authors favor a mechanism for the Tl(OAc)$_3$ oxidation similar to that invoked for the LTA oxidation (eq. 39).

The oxidation of methylene cyclopropane proceeds smoothly *via* ring opening to the unsaturated diacetate (Eq. 39a).

$$(38)$$

50%   4α- and 4β-OAc

$$(39)$$

products

$$(39a)$$

(a) R^1 = R^2 = R^3 = R^4 = H

(b) R^1 = Ph, R^2 = R^3 = R^4 = H

(c) R^1 = R^2 = R^3 = H, R^4 = Ph

(d) R^1 = R^2 = Me, R^3 = R^4 = H

(e) R^1 = R^2 = H, R^3 = R^4 = Me

(a) R^1 = R^2 = R^3 = R^4 = H

(b) R^1 = Ph, R^2 = R^3 = R^4 = H

(c) R^1 = R^2 = Me, R^3 = R^4 = H

A mechanism involving a bridged organolead intermediate is proposed:

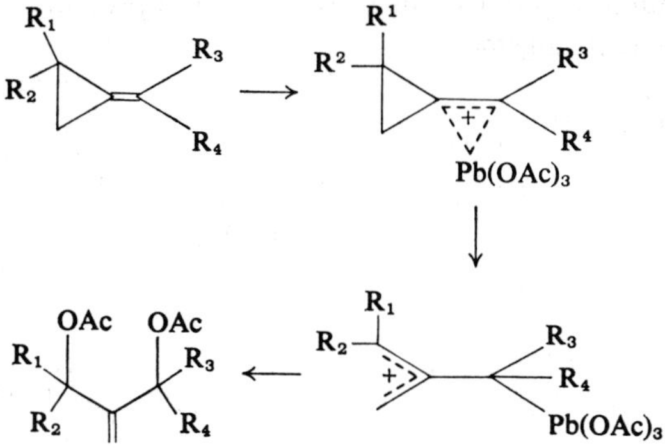

III. ACYCLIC SYSTEMS

Relatively little work has been published on the behavior of simple acyclic olefins towards LTA. Hexene-2,3-methyl-pentene-2 (35) and *cis*-octene-4 (36) undergo allylic substitution in relatively low yield; namely, 7, 6, and 27%, respectively.

p-Methoxystyrene (**49**) reacts with LTA at room temperature to give 90% of the geminal diacetate **50** (6,37) (eq. 40).

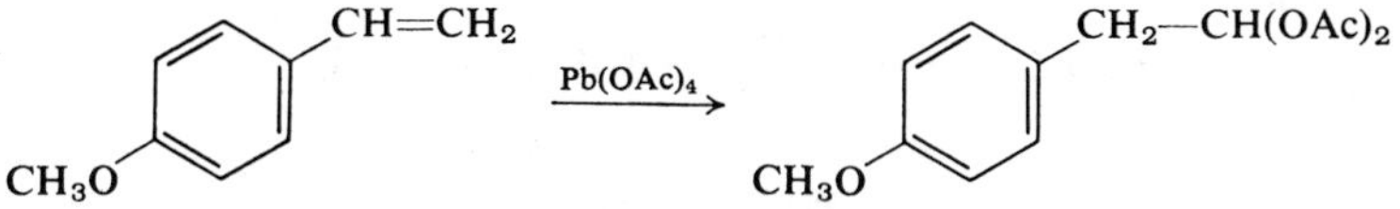

$$(40)$$

Two separate lines of evidence indicate that the *p*-methoxyphenyl group may undergo 1,2-migration in this system. First, *p*-methoxy-α-methylstyrene (**51**) upon oxidation with lead tetraacetate gives a 48% yield of *p*-methoxy-phenyl acetone (**52**) in addition to the vicinal diacetate (**53**) (eq. 41). Second,

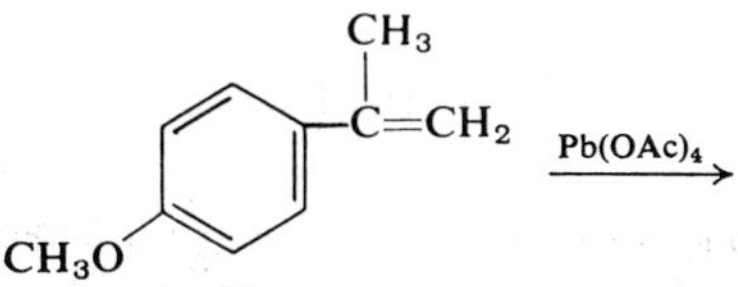

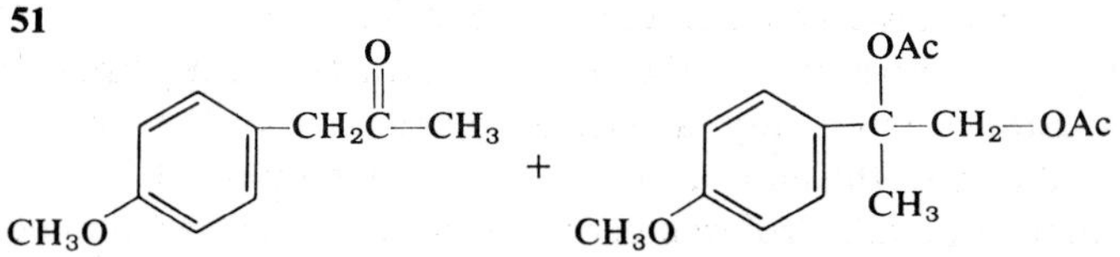

$$(41)$$

oxidation of p-methoxystyrene-α-^{14}C yields terminally labeled products (37) (eq. 42). The authors propose an interesting cyclopropane intermediate to account for the rearrangement.

$$(42)$$

The reaction of styrene and 1,1-diphenylethylene with LTA in benzene at reflux is quite unusual in that methylation of the double bond occurs (eqs. 43 and 44).

$$C_6H_5CH{=}CH_2 \xrightarrow[C_6H_6]{Pb(OAc)_4} C_6H_5\underset{\underset{OAc}{|}}{C}HCH_2CH_3 \qquad (43)$$

$$(C_6H_5)_2C{=}CH_2 \xrightarrow[C_6H_6]{Pb(OAc)_4} (C_6H_5)_2\underset{\underset{OAc}{|}}{C}CH_2CH_3 \qquad (44)$$

A free radical addition of $CH_3\cdot$ to the double bond probably occurs in these reactions. A detailed mechanism for the reaction of $trans$-methylstyrene with LTA has been offered by Heiba et al. (39) (eq. 45) and also by Kropf et al. (39a).

$$C_6H_5CH{=}CH_2 \xrightarrow{CH_3} C_6H_5CH{-}CH_2CH_3 \xrightarrow{Pb(OAc)_4} C_6H_5CHOAcCH_2CH_3$$
$$(45)$$

A further interesting reaction of styrene with LTA has been reported recently by Ichikawa and Uemura (38). Styrene reacts with LTA to produce dihydrofuran derivatives in the presence of a β-diketone, possibly by the route illustrated in eq. 46; namely, oxyplumbation of the olefin and subsequent addition to the enol form of the diketone followed by cyclization to yield ultimately a dihydrofuran derivative **54**. A similar reaction occurs using mercuric acetate.

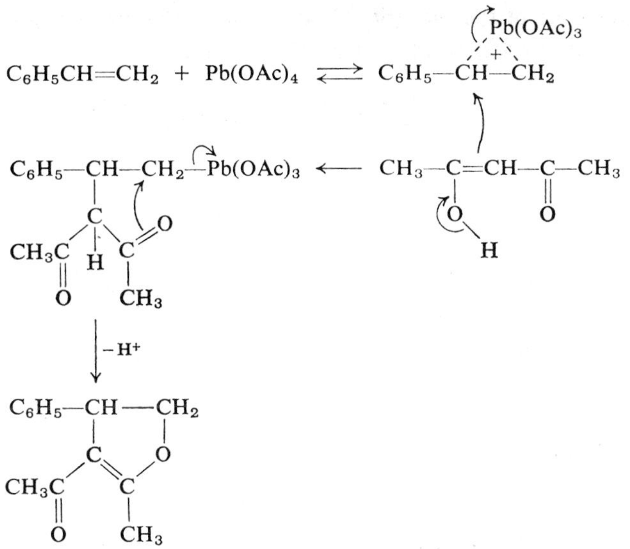

$$C_6H_5CH{=}CH_2 \;+\; Pb(OAc)_4 \;\rightleftharpoons\; C_6H_5{-}\overset{+}{\underset{}{CH}}{-}CH_2\;\Big(Pb(OAc)_3$$

$$C_6H_5{-}CH{-}CH_2{-}Pb(OAc)_3 \;\longleftarrow\; CH_3{-}C{=}CH{-}\underset{O}{\overset{\parallel}{C}}{-}CH_3$$

(46)

$$\downarrow -H^+$$

54

p-Methoxystyrene yields 93% of the dimethyl acetal of p-methoxy-phenylacetaldehyde upon oxidation with mercuric acetate in methanol (5).

Thallium acetate behaves similarly towards styrene (5) (eq. 47).

$$C_6H_5CH{=}CH_2 \xrightarrow[\text{CH}_3\text{OH}]{\text{Tl(OAc)}_3} C_6H_5CH_2CH(OCH_3)_2 \qquad (47)$$

At room temperature the oxythallation adduct may be isolated from the reaction of styrene with thallium acetate in methanol. It decomposes slowly at reflux temperature in methanol but is very quickly converted upon treatment with nitric acid in equimolar quantity to phenylacetaldehyde dimethyacetal.

Isobutylene upon reaction with LTA in acetic acid at 40° yields principally the glyoxylic acid derivative **55** along with a small amount of the 1,2-diacetoxy compound. In the presence of water the principal product obtained

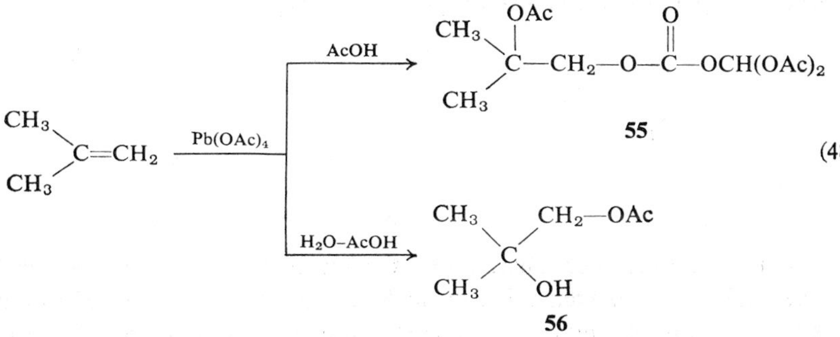

$$\underset{CH_3}{\overset{CH_3}{\diagdown}}C{=}CH_2 \xrightarrow{Pb(OAc)_4}$$

$$\xrightarrow{AcOH} \underset{CH_3}{\overset{CH_3}{\diagdown}}\underset{|}{\overset{OAc}{\underset{|}{C}}}{-}CH_2{-}O{-}\overset{O}{\overset{\parallel}{C}}{-}OCH(OAc)_2$$

55

(48)

$$\xrightarrow{H_2O\text{-}AcOH} \underset{CH_3}{\overset{CH_3}{\diagdown}}\underset{OH}{\overset{CH_2{-}OAc}{\diagup}}C$$

56

is the monoacetate of the diol (56) (36a) (eq. 48). The mechanism of eq. 49 may be invoked in order to rationalize these results.

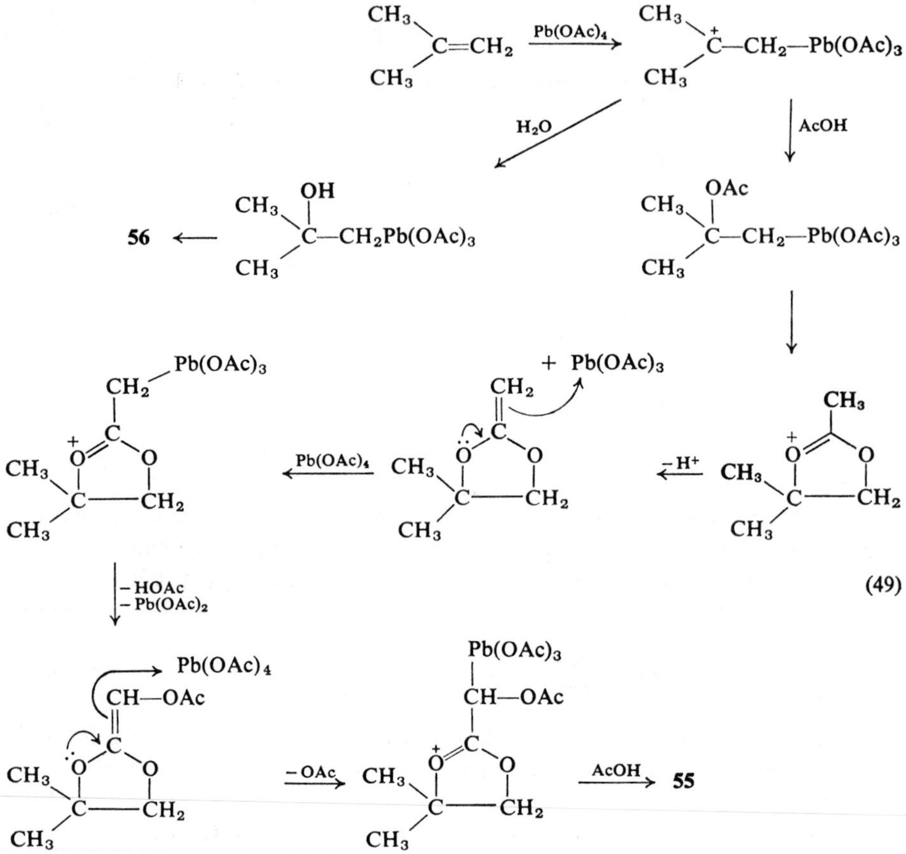

(49)

Thallium acetate oxidation of isobutylene in acetic acid yields as main product (27%) 1,2-diacetoxy-2-methylpropane (5).

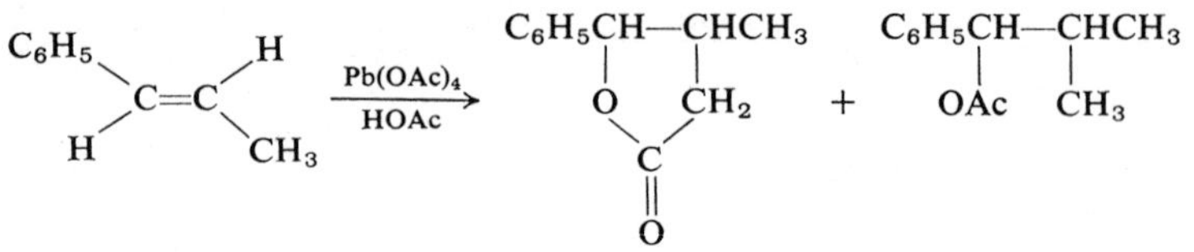

(50)

γ-Lactones have been obtained by the reaction of LTA with octene-1, styrene, and β-methylstyrene. For example, trans-α-methylstyrene (57) yields mainly the lactone 58 and a product (59) that may be considered as formally derived from addition of methyl acetate to the double bond (39) (eq. 50). Minor products were the *erythro* and *threo* 1,2-diacetates **60**, the rearranged 1,1-diacetate **61**, allylic acetate **62** and olefin **63** (eq. 51).

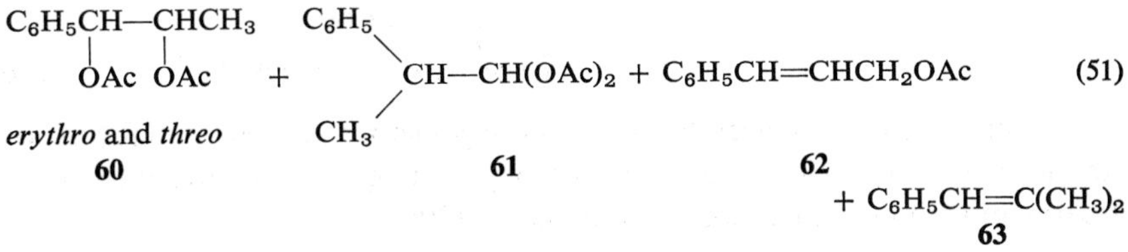

$$
\underset{\substack{erythro \text{ and } threo \\ \mathbf{60}}}{\underset{\underset{\text{OAc } \text{OAc}}{|\quad|}}{C_6H_5CH-CHCH_3}}
+
\underset{\mathbf{61}}{\underset{\substack{CH_3 \\ /}}{\overset{C_6H_5}{\underset{\backslash}{}}}CH-CH(OAc)_2}
+
\underset{\mathbf{62}}{C_6H_5CH=CHCH_2OAc}
\tag{51}
$$

$$+ \; C_6H_5CH=C(CH_3)_2$$
$$\mathbf{63}$$

The authors propose a free radical chain mechanism involving propagation via

$$Pb(III)(OAc)_3 \rightarrow Pb(II)(OAc)_2 + CH_3{}^{\cdot} + CO_2$$

Termination occurs via oxidation of carbon radicals by Pb(III) to carbonium ions. Thus $CH_3{}^{\cdot}$ and ${}^{\cdot}CH_2COOH$ are reasonable radical intermediates because of their relatively high ionization potentials, and eqs. 52–55 may account for the observed products.

$$Pb(OAc)_4 \xrightarrow[\text{or } h\nu]{\Delta} Pb(III)(OAc)_3 + CH_3{}^{\cdot} + CO_2 \tag{52}$$

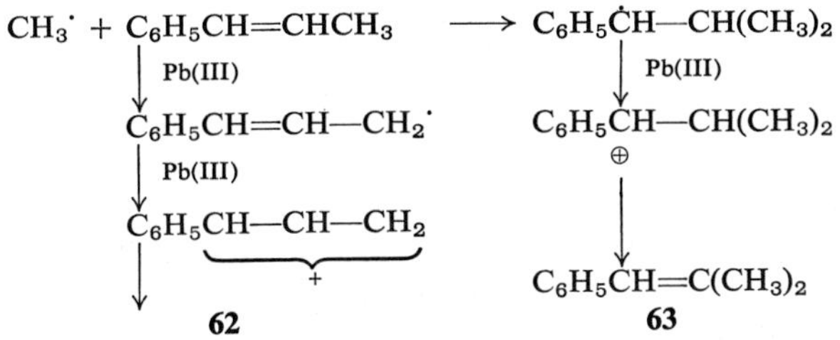

$$CH_3{}^{\cdot} + C_6H_5CH=CHCH_3 \longrightarrow C_6H_5\overset{\cdot}{C}H-CH(CH_3)_2$$

(53)

$$CH_3{}^{\cdot} + CH_3COOH \longrightarrow {}^{\cdot}CH_2COOH + CH_4 \tag{54}$$

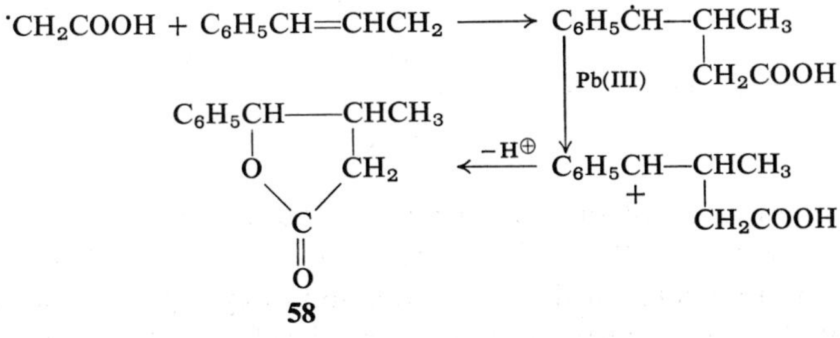

$${}^{\cdot}CH_2COOH + C_6H_5CH=CHCH_2 \longrightarrow C_6H_5\overset{|}{\underset{CH_2COOH}{\overset{\cdot}{C}H-CHCH_3}}$$

(55)

$$\underset{\mathbf{58}}{C_6H_5CH-CHCH_3}$$

Mercuric acetate oxidation of styrene in methanol yields an adduct which is decomposed by the action of boron trifluoride in methanol to yield phenylglycoldimethylether (39a)(eq. 56).

$$C_6H_5CH{=}CH_2 \xrightarrow[\text{CH}_3\text{OH}]{\text{Hg(OAc)}_2} C_6H_5CH(OCH_3)CH_2HgOAc$$

$$\downarrow \text{BF}_3,\text{CH}_3\text{OH}$$

$$C_6H_5CH(OCH_3)CH_2OCH_3 \qquad\qquad (56)$$

Similarly mercuric acetate oxidation of *p*-methoxystyrene initially yields an adduct which forms 40% *p*-methoxyphenylglycoldimethylether upon treatment with boron trifluoride in methanol (39b).

IV. CONJUGATED DIENES AND POLYENES

The LTA oxidation of cyclopentadiene in acetic acid at 20° was reported by Criegee (2) to yield a mixture of the mixed esters of cyclopentenediol and acetylglycolic acid. Cyclohexadiene behaves similarly (40). A mechanism may be written involving steps similar to those invoked (*vide supra*) to explain the formation of glyoxylic acid esters in the oxidation of isobutylene (37) (eq. 57).

An intermediate of the type mentioned above may play a role in the interesting "wet" and "dry" LTA oxidation of cyclopentadiene (41). The stereochemical outcome of the LTA oxidation of cyclopentadiene is solvent dependent. In glacial acetic acid–water a 75–80% yield of the *cis* monoacetate of cyclopentene 1,2-diol of 93% stereochemical purity was obtained.

In dry acetic acid containing potassium acetate a 44% yield of *trans*-3,4-diacetoxycylopentene in 97% stereochemical purity was obtained (eq. 58).

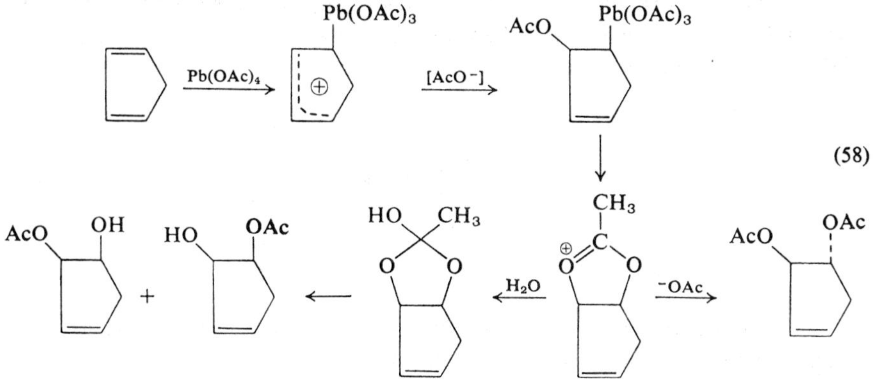

(58)

The stereochemistry of the reaction is completely analogous to that observed in the Prevost reaction or its modification according to Woodward (42). The difference in stereochemistry depends upon the addition of water to the bridged acetoxonium ion to yield the ortho ester, or displacement at the carbon by the acetate anion to yield the *trans*-diacetate.

Recently 1,4 addition of LTA to conjugated dienes has been observed (41a). Typically 1,3-butadiene, isoprene, and 1,3-cyclohexadiene undergo this type addition with LTA and also with Tl(OAc)$_3$. Initial addition of LTA to

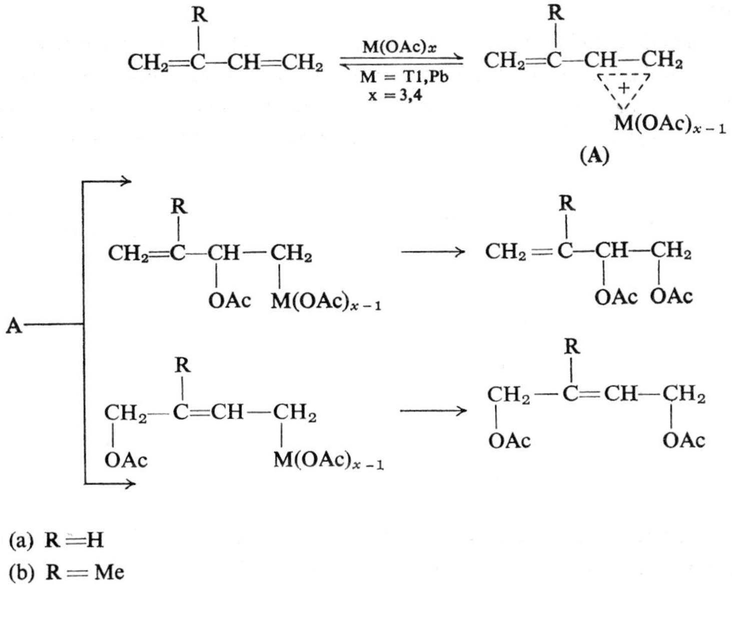

(A)

(58a)

(a) R = H
(b) R = Me

one double bond to form a bridge intermediate (**A**) has been proposed to yield products of 1,2 and 1,4 addition (eq. 58a).

Finkelstein (43) has reported that cyclooctatetraene (**64**) upon oxidation with LTA in acetic acid yields bicyclo(4.2.0)octa-2,4-diene-7,8-diol diacetate (**65**) in 38% yield (eq. 59).

$$(59)$$

The same diacetate was obtained earlier by Cope and co-workers (44) from the reaction of cyclooctatetraene and mercuric acetate in acetic acid. The stereochemistry of the product of lead tetraacetate has not been elucidated although by analogy with recent work on the halogenation of cyclooctatetraene one would assume that it is the *trans*-diacetate. Also, if one assumes that the allowed disrotatory thermal ring closure occurs, then a *cis* ring fusion results (45). The product of oxidation of cyclooctatetraene with mercuric acetate in acetic acid obtained by Cope and co-workers (45) was related by chemical methods to *trans*-bicyclo(4.2.0)octane-7,8-diol.

Furthermore, as in the halogenation studies, the *trans*-diacetate may not be a primary product but perhaps the final product in a multistep process (46–49).

Finkelstein (43) also reported that LTA oxidation of cyclooctatetraene (**64**) in methanol yielded the dimethyl acetal of 2,4,6-cycloheptatriene-1-carboxaldehyde (**66**), while oxidation in the presence of boron trifluoride etherate in acetic acid yielded the corresponding diacetoxy acetal **66a** (eq. 60).

$$(60)$$

Cycloheptatriene carboxylic acid undergoes decarboxylation upon treatment with LTA to yield the tropylium cation (50) (eq. 61).

$$(61)$$

Windaus and Riemann (51) have shown that 3β-acetoxycholestadiene-6,8 (**67**) undergoes dehydrogenation upon treatment with LTA in chloroform acetic acid (**67** $\rightarrow$ **68** and **69**) (eq. 62).

$$(62)$$

Analogous dehydrogenation of p-nitrobenzoyloxycholestadiene-6,8 was reported by Windaus and co-workers (52) when mercuric acetate was used as the oxidizing agent. It is clear that these products could derive from an intermediate formed by initial electrophilic addition to the C_6—C_7 double bond (eq. 63).

$$\rightarrow \mathbf{69} \quad (63)$$

The driving force for the dehydrogenation reactions may be the relief of steric strain in the ring system and also the stability of the resulting dienes and trienes.

In the aromatic series related dehydrogenation reactions occur, the driving force undoubtedly being the aromatic stabilization of the product (eqs. 64 and 65). Again these reactions may be rationalized in terms of the

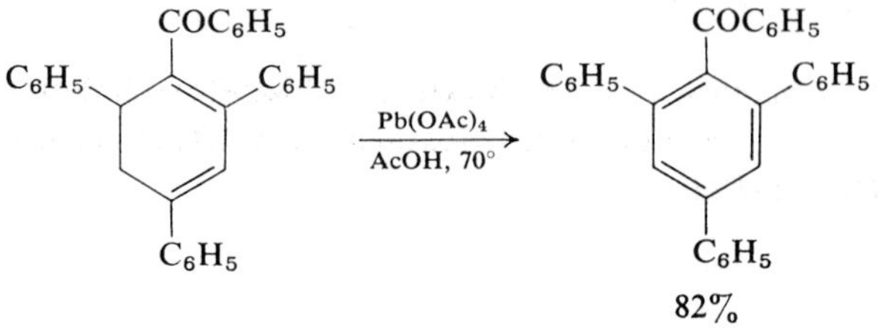

(Ref. 53)   (64)

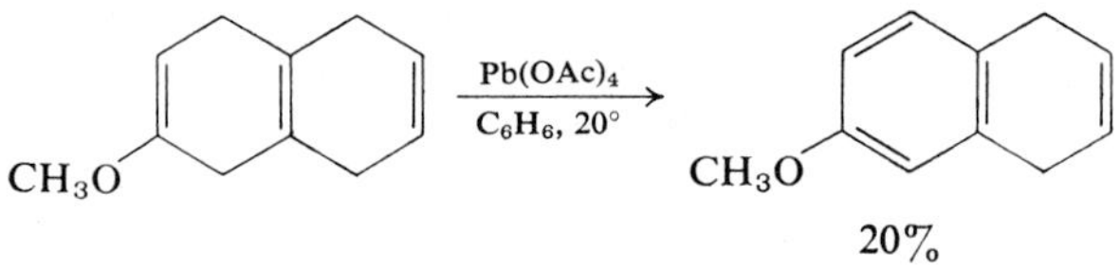

(Ref. 54)   (65)

addition-elimination mechanism mentioned above.

V. NEIGHBORING GROUP PARTICIPATION

Participation of neighboring acetoxyl groups has been postulated earlier in connection with the LTA oxidation of cyclopentadiene and isobutylene. The reactions to be treated now are those in which the neighboring group becomes permanently bonded to the carbon at the site of initial addition of the LTA. In this sense the neighboring group assumes the role of the external nucleophile.

A. Neighboring Hydroxyl

The first example of neighboring group participation by the hydroxyl group in the LTA oxidation of the olefinic group was in the steroidal example (55a) (eq. 66).

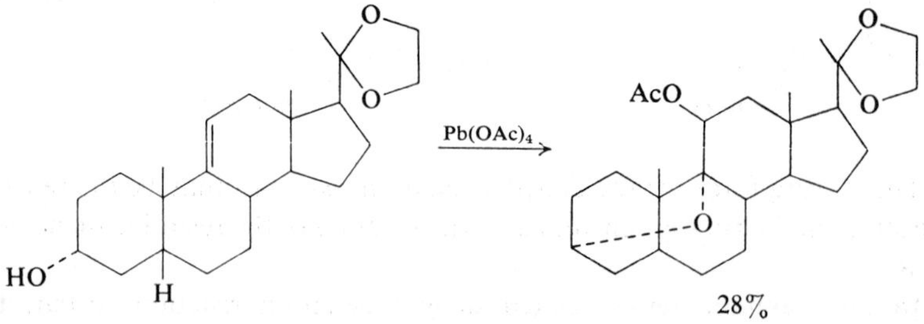

(66)

The reaction was also demonstrated in the bridged bicyclic series in the oxidation of 2-*endo*-hydroxymethylnorbornene (**70**) (55) (eq. 67). The analogous reactions **72** → **73** and **74** → **75** were also observed (eqs. 68 and 69).

$$\text{(67)}$$

$$\text{(68)}$$

$$\text{(69)}$$

An earlier report of this type reaction existed (56) but a structure for the product was not proposed until the above example was published (57). This reaction involved the treatment of coriamyrtin (**76**) with LTA in acetic acid at 20° (eq. 70).

$$\text{(70)}$$

Moon and co-workers (58) have studied several acyclic examples of the cyclization reaction of unsaturated alcohols. No cyclic product was observed

in the case of 3-buten-1-ol (**77**) (eq. 71), but 4-penten-1-ol (**78**) yielded both
five- and six-membered acetoxy oxides (eqs. 72 and 73). Moon and Haynes

30%		10%
+		
13%		60%
+		(71)
20%		13%
+		
32%		0%
Total yield	75%	83%

26%	14%	16%

(Ref. 59)

C_6H_6	37%	0%	12%
C_6H_{12}	21%	0%	21%
C_6H_5N	0%	19%	20%

(59) also studied the alicyclic systems of eq. 74. These workers propose that
the observed products derive from the processes shown in eqs. 75 and 76.

$$ (74) $$

C_6H_{12}	70%	5%	5%
	36%	5%	17%
C_5H_5N	0%	52%	4%

$$ (75) $$

$$ (76) $$

Conceptually the same products may derive from initial addition to the double bond followed by intramolecular interaction of the hydroxyl group.

Intramolecular acetoxy oxide formation was also observed in the case of saturated alcohols (60) (eqs. 77–79). The acetoxy oxide probably derives from

$$ (77) $$

optically active, ca. 4.4%

racemic, ca. 19%   racemic, ca. 15.5 and   racemic, ca. 19.7%
                    17.8% (diastereoisomers)

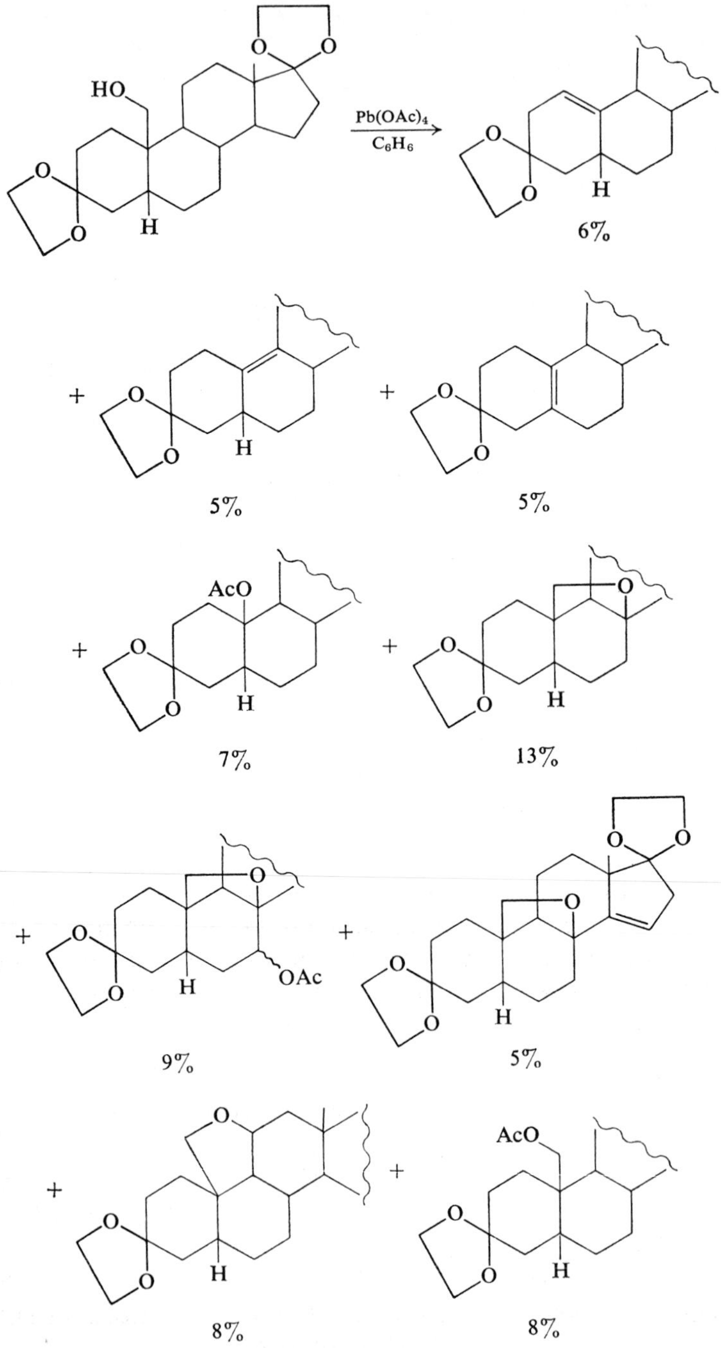

$$\xrightarrow[\text{C}_6\text{H}_6]{\text{Pb(OAc)}_4}$$

6%

+ 5%

+ 5%

+ 7%

+ 13%

(78)

+ 9%

+ 5%

+ 8%

+ 8%

214

a process involving abstraction of the C_8 axial hydrogen by prior cleavage of the lead-oxygen bond (eq. 80). These steps generate the unsaturated alcohol which reacts further with LTA to yield the acetoxy oxide.

A mechanistically similar process occurs in the LTA oxidation of 2-cyclopentylethanol (61) (eq. 81).

Two pathways appear to be possible in the LTA intramolecular oxidative cyclization of unsaturated alcohols. It is known, of course, that both the

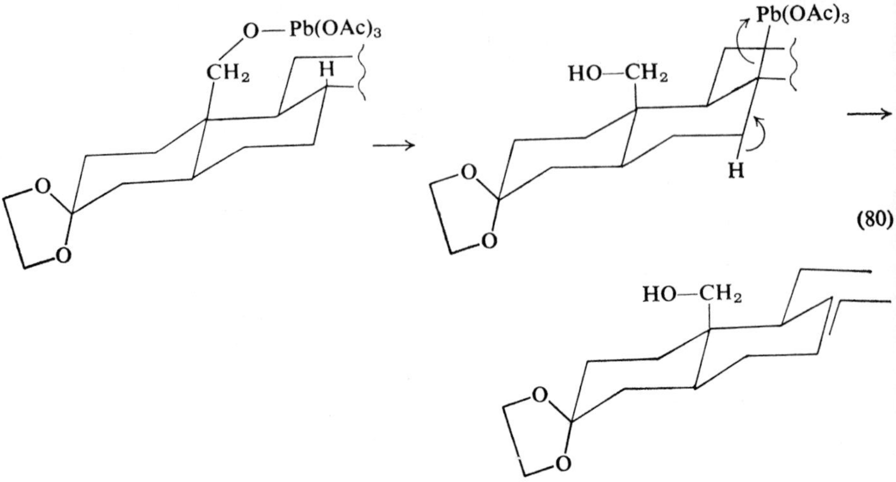

(80)

hydroxyl group and carbon–carbon double bond react with LTA. Consequently, initial reaction of the double bond might lead to a bridged organolead intermediate which could undergo further reaction by intramolecular

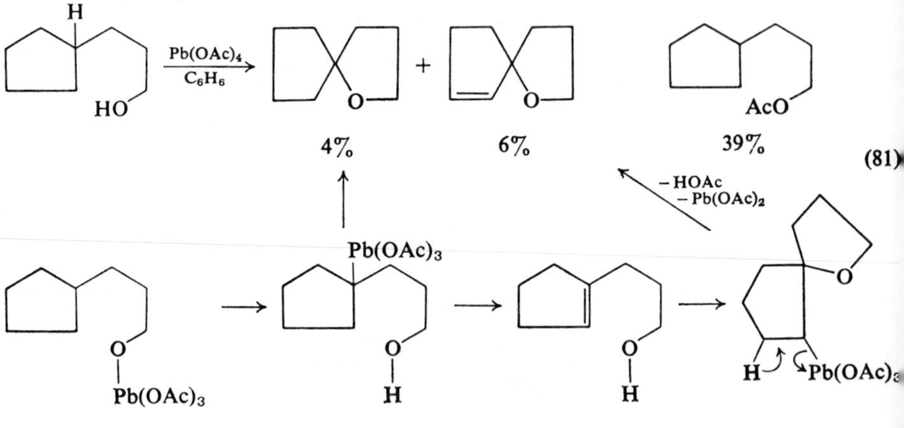

(81)

participation involving the hydroxyl group (Pathway *B*) (eq. 82). Alternatively, the initial reaction might involve the hydroxyl group and subsequent
neighboring group participation by the double bond might lead to acetoxy
oxide (Pathway *A*). An additional uncertainty exists concerning the cleavage
of the carbon–lead or carbon–oxygen bond. This could occur either homolytically or heterolytically. As mentioned earlier, carbon radicals may be
oxidized to the corresponding carbonium ion by lead tetraacetate (39).

Applied to the LTA oxidation of *endo*-2-hydroxymethylnorbornene (**70**), eq. 82 expresses these possibilities. The course of this reaction in terms of a

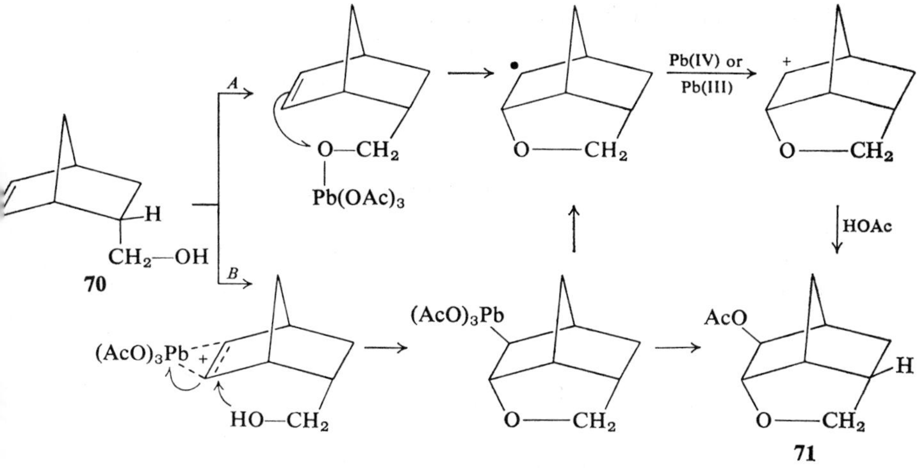

(82)

preference for either of these pathways has not as yet been established firmly. However, with other neighboring groups, a choice has been made in the case of certain bicyclic systems in favor of initial electrophilic addition to the

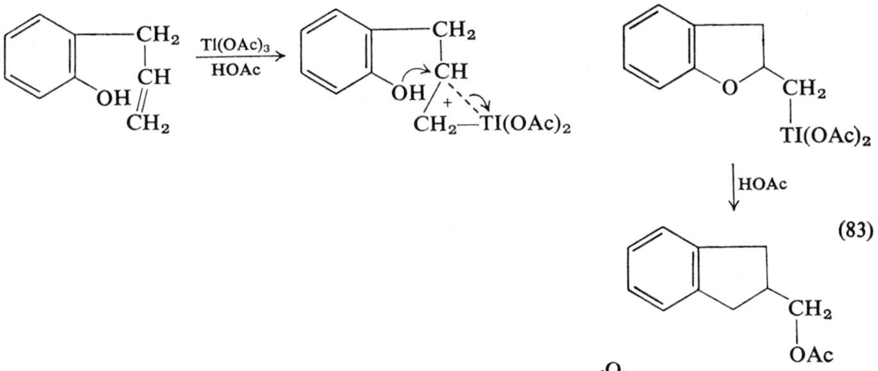

(83)

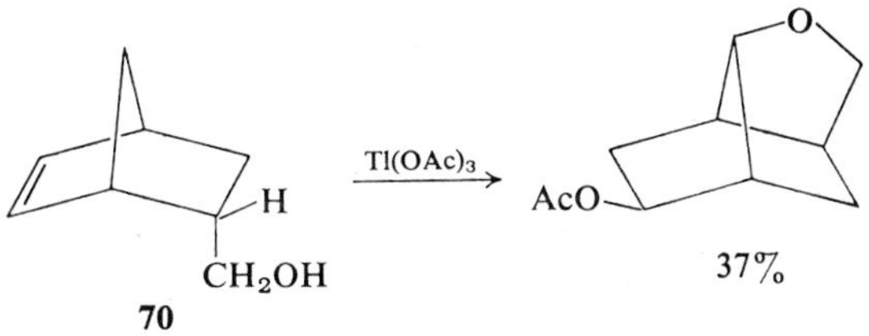

(Ref. 62)   (84)

double bond (*vide infra*) followed by interaction of the neighboring functional group.

Neighboring group participation occurs in the thallium acetate oxidation of phenols (5) and bicyclic olefins (62) (eqs. 83 and 84). For example, thallium acetate oxidation of α-allylphenol initially yields the oxythallation product which upon warming to 80° yields α-acetoxymethylcoumarin in 48% (5).

B. Neighboring Carboxylate

Recently a number of examples have been described of neighboring group participation by the carboxylic acid group in the LTA oxidation of

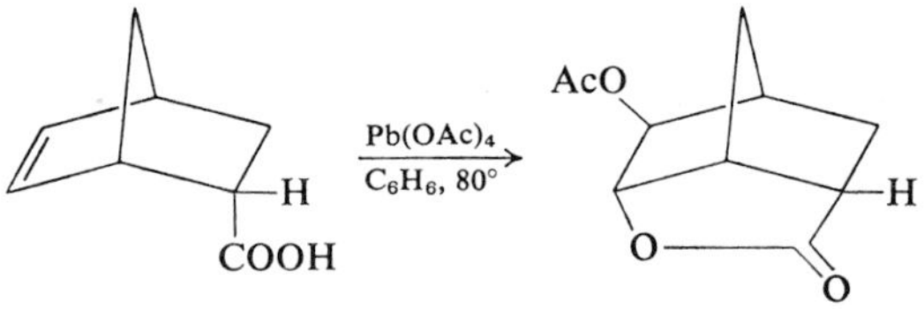

$$\text{(85)}$$

80   **81** (65%)

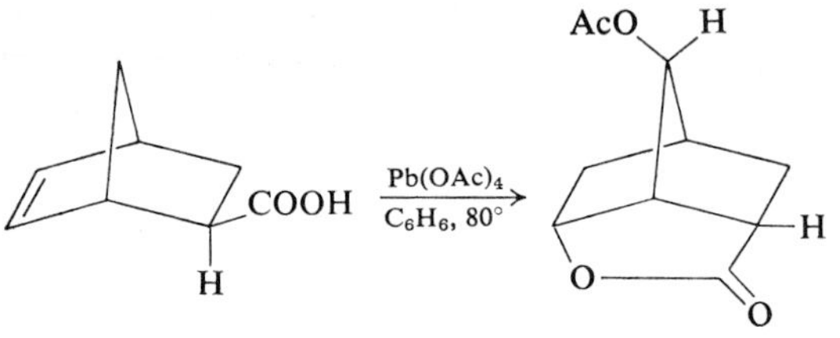

$$\text{(86)}$$

55% (65% in HOAc)

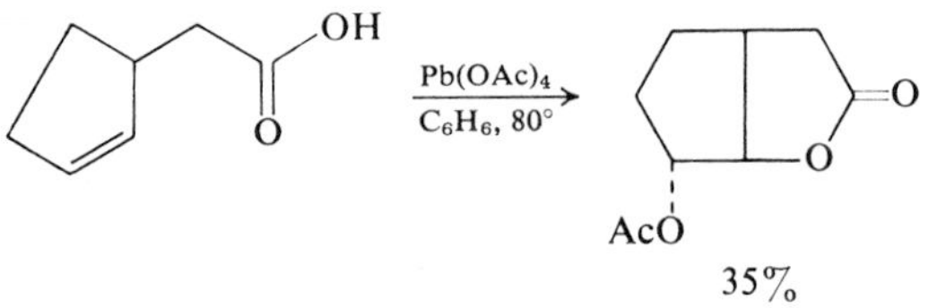

$$\text{(87)}$$

35%

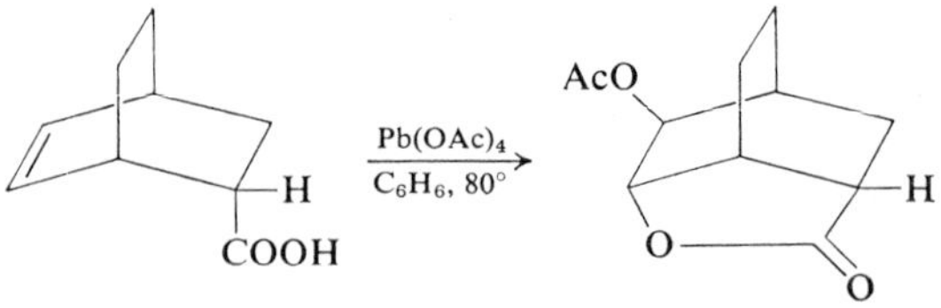

$$\text{(88)}$$

20%

olefins (63,63a). The examples of eqs. 85–89 serve to point up the apparent generality of this reaction.

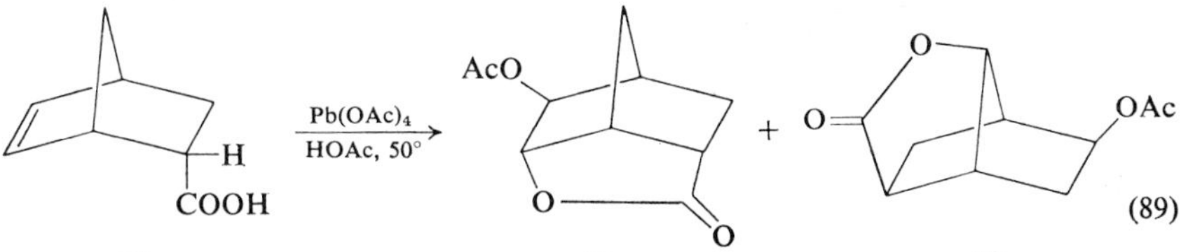

(89)

80   **81**   **82**

In the case of two carboxyl groups within the same molecule, the interesting cagelike bis-lactones were formed [**83** → **84** (64), **85** → **86** (65)] (eqs. 90

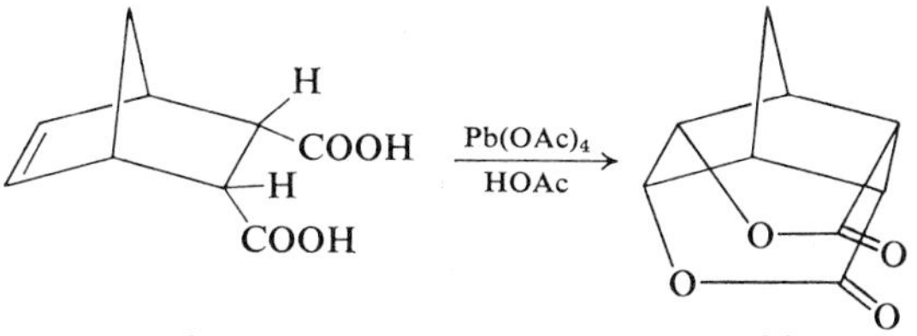

(90)

83   **84**

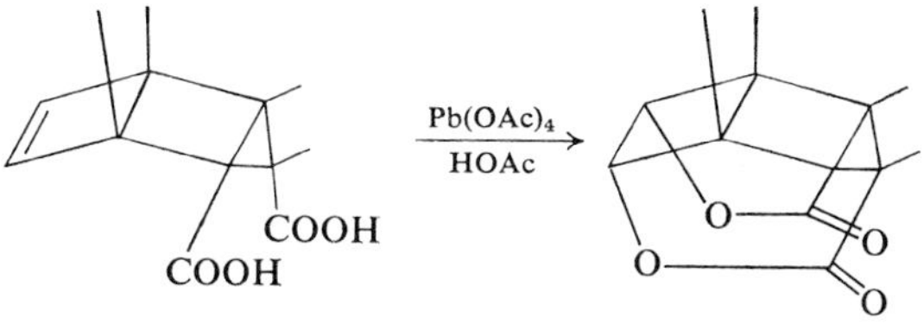

(91)

85   **86**

and 91). A surprising feature of these reactions is that oxidation with LTA does not cause decarboxylation. It is known, for example, that both saturated monocarboxylic and vicinal dicarboxylic acids undergo oxidative decarboxylation upon treatment with LTA (66–72).

As mentioned above for the LTA oxidation of unsaturated alcohols, two pathways may be written for the oxidative acetoxy lactone formation. The first involves initial attack at the double bond (as proposed earlier by Alder et al. for norbornene) (28) followed by intramolecular participation of the carboxylate group in a second step (**80** → **87** → **88** → **81**). The second possible pathway involves participation of the carbon–carbon double bond in the cleavage of the lead–oxygen bond (**89** → **90** → **91** → **81**) (eq. 92).

A priori, the second route, which involves participation of the π-electrons of the double bond in the cleavage of the lead–oxygen bond, appears reasonable since the carboxyl group alone reacts with LTA. Furthermore, a

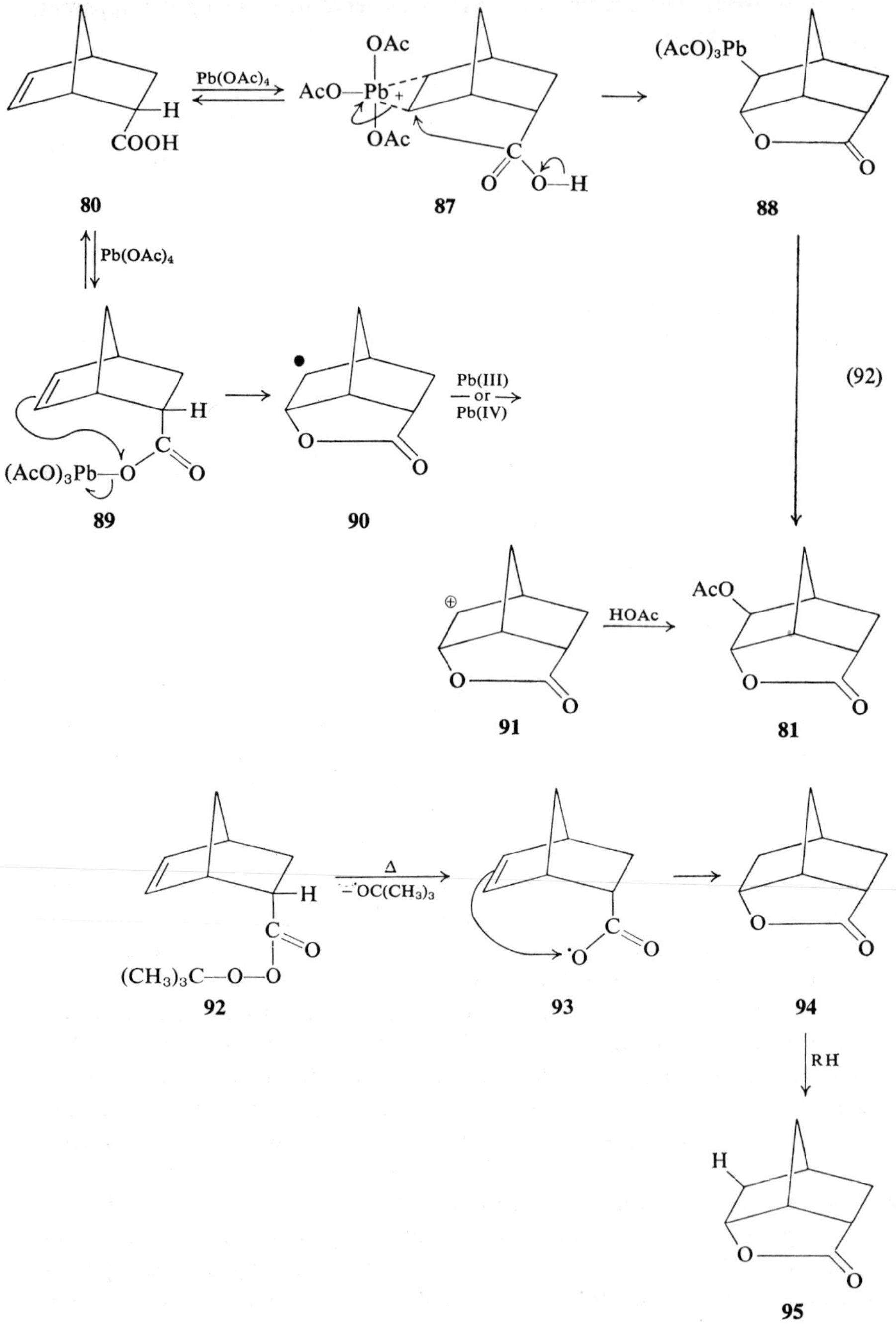

80

87

88

(92)

89

90

91

81

92

93

94

95

similar reaction has been observed in the thermal decomposition of nor-bornene *t*-butyl peroxycarboxylate (92). This result was explained in terms of attack of the double bond on the carboxylate radical (73) (92 → 93 → 94 → 95) (eq. 93).

That the LTA oxidation of 2-*endo*-norbornene carboxylic acid (80) does, in fact, proceed via initial attack at the double bond is indicated by the following observations. First, 2-*exo*-norbornene carboxylic acid (96) is readily oxidized to an acetoxy lactone 99 by LTA in benzene, possible via the steps 96 → 97 → 98 → 99 (eq. 94). Thus the double bond and carboxylate function need not bear a close spatial relationship. Furthermore, the methyl

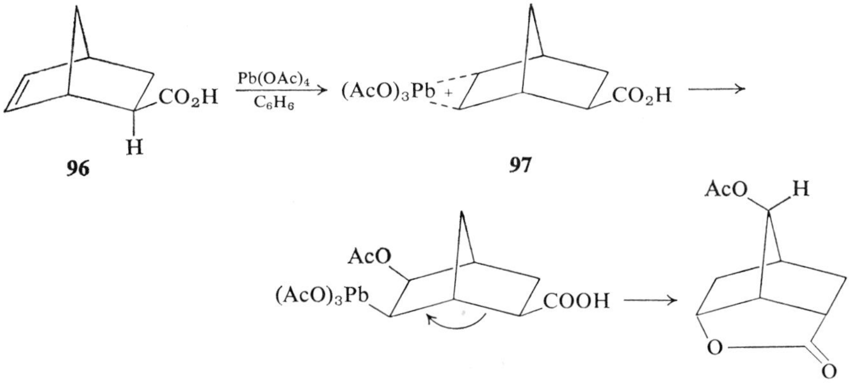

ester of 2-*endo*-norbornene carboxylic acid (100) upon reaction with LTA in benzene yields the same product as that obtained from the free acid. More-over, the saturated ester (101) is completely impassive to LTA (eq. 96) under the conditions of the reaction 100 → 81 (eq. 95). Finally, the *N*-methyl-

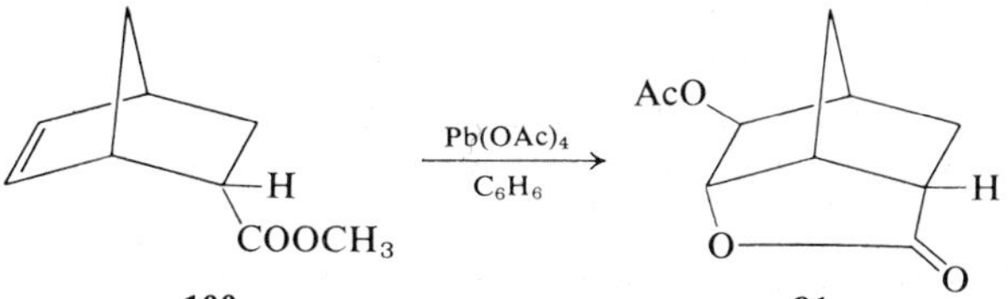

(95)

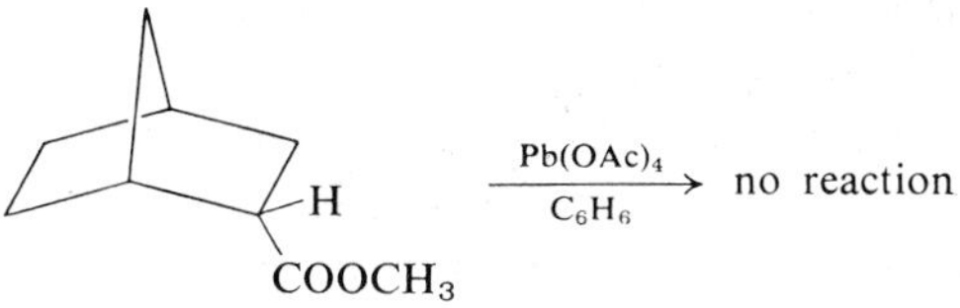

(96)

carboxamido derivatives of the 2-*endo*- and 2-*exo*-norbornene carboxylate, **102** and **103**, respectively, also yield acetoxy lactone upon treatment with LTA (eqs. 97 and 98).

81 (45%)   (97)

81 (25%)   **82** 25%

103

99 (45% in C_6H_6)
(58% in HOAc)   (98)

The question arises as to whether, in general, the initial attack of lead tetraacetate upon an olefinic acid, amide, ester, or alcohol invariably takes place at the double bond and the functional groups intervene at some secondary stage of the reaction. This question cannot be answered with certainty on the basis of the data available. Clearly, in the case of norbornene carboxylic acids, attack at the double bond is the primary step, but norbornene is an atypical olefin in the sense that it possesses a highly strained double bond. With less strained olefins, the rate of electrophilic addition of lead tetraacetate may be sufficiently slow that reaction of the carboxylate function may take precedence.

In the case of lead tetraacetate oxidation of unsaturated alcohols, Moon postulates that the initial reaction is at the alcohol function (58). Thus he writes the eq. (99) mechanism as typical for the LTA oxidation of unsaturated alcohols in which acetoxy oxide is formed.

(99)

No really compelling evidence has been offered on which to base a firm choice between this mechanistic sequence and the one involving initial reaction at the double bond. Thus the reason offered by Moon for favoring the above mechanistic pathway is that no products were obtained that resulted from the addition of lead tetraacetate to the double bond alone.

Some other more complex examples of participation are encountered in the LTA oxidation of unsaturated dicarboxylic acids. These reactions were found in the attempted oxidative decarboxylation. Two groups of workers have investigated the LTA oxidation of methylfumaropimarate (104) (74,75). Ayer and McDonald (74) found the products (105–108) in addition to the expected diene, which was formed in 10% yield (eq. 100).

It was observed that the cyclopropyl lactone reacted with acetic acid to yield the addition product (105 → 107). When the LTA oxidation of methyl fumaropimarate (104) is carried out in acetic acid, the principal product is the same acetoxy lactone. Zalkow and Brannon (75) reported that the cyclopropyl lactone is the major product in this reaction (105).

Ayer and McDonald (74) call attention to the hindered nature of the double bond in methyl fumaropimarate (104) and they therefore propose a mechanism of oxidation involving initial attack of lead tetraacetate at the carboxylate function.

Another interesting example of such lactone formation in the LTA oxidation of a carboxylic acid was encountered by Kitahonoki and Takano (76) in the oxidation of a benzobicyclo(2.2.2) derivative 109 → 110 (eq. 101). This reaction, as well as the previous one (104 → 105), may be viewed as being initiated by generation of a carbonium ion formed by decarboxylation (eq. 102).

The driving force for the (2.2.2) → (3.2.1) change is the formation of a benzylic carbonium ion although the migration of the C_4—C_8 bond would also yield a benzylic carbonium ion.

A superior method for the transformation of vincinal dicarboxylic acids into the corresponding olefin is via electrolytic bisdecarboxylation (77,78). This process avoids the undesired lactone formation so frequently observed in the case of unsaturated systems.

An interesting and synthetically useful variation of the LTA oxidative decarboxylation employs LTA in conjunction with a lithium halide (79,80). The product of this reaction, when applied to an alkyl carboxylic acid, is the

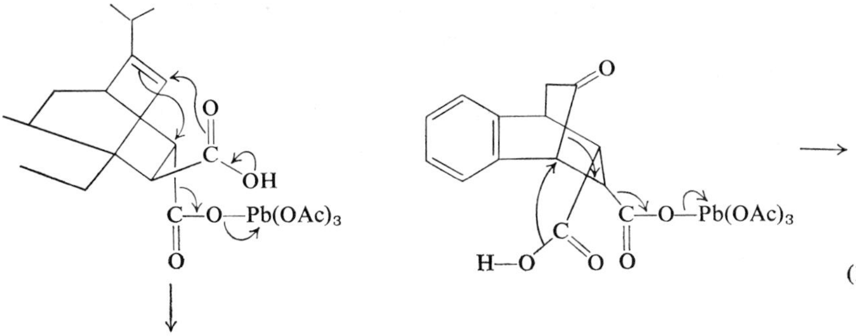

(102)

105 or 106

lower alkyl halide. The process has been termed "halodecarboxylation" and it has been studied in detail by Kochi.

Attempted halodecarboxylation of 2-*endo*-norbornene carboxylic acid (80) using LTA–lithium chloride in 1.5:10 molar ratio in benzene caused no decarboxylation but a high yield of an acetoxy lactone isomeric with that formed in the oxidation using LTA alone was obtained (eq. 103).

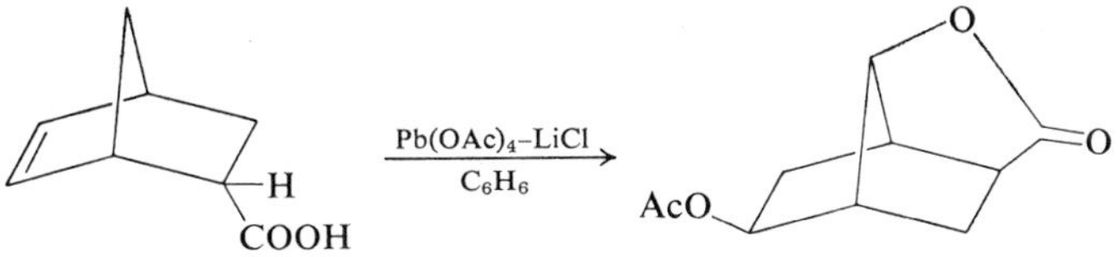

(103)

80   82 (70%)

This lactone may be regarded formally as a product of Wagner-Meerwein rearrangement of the carbonium ion 111a and this was validated by showing that acetolysis of the related *p*-toluene-sulfonate ester 111 did in fact yield the rearranged acetoxy lactone (81) (eq. 104).

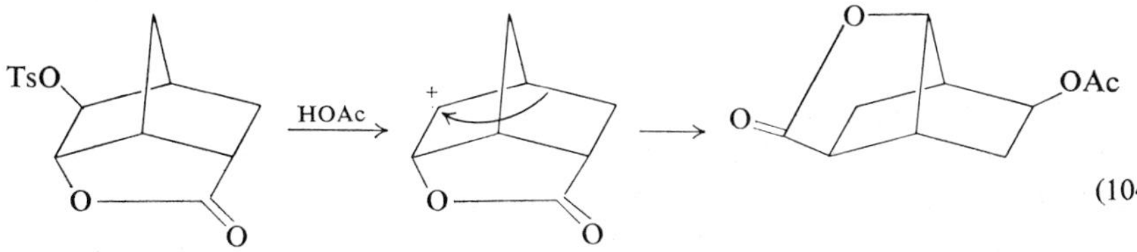

(104)

111   111a   82

The reasonable assumption may be made that lithium chloride somehow induces the formation of the carbonium ion in the LTA oxidation of 80. It is possible that Cl coordinated to the lead atom in the organolead intermediate could stabilize the transition state for carbonium ion formation via a direct heterolytic cleavage of the carbon–lead bond.

This mechanism is supported by the observation that LTA oxidation of *endo*-norbornene-2-carboxylic acid (**80**) in acetic acid (relative to benzene) yields **82** in addition to **81** in about a 1:1 ratio.

Analogous intramolecular participation occurs in the thallium acetate oxidation of norbornene derivatives (eqs. 105–107).

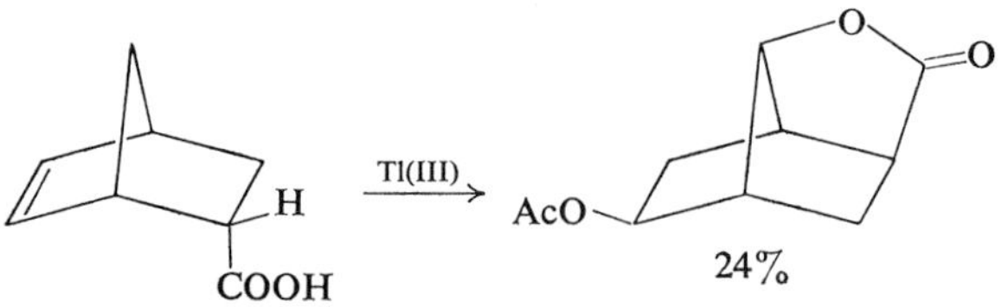

(105)

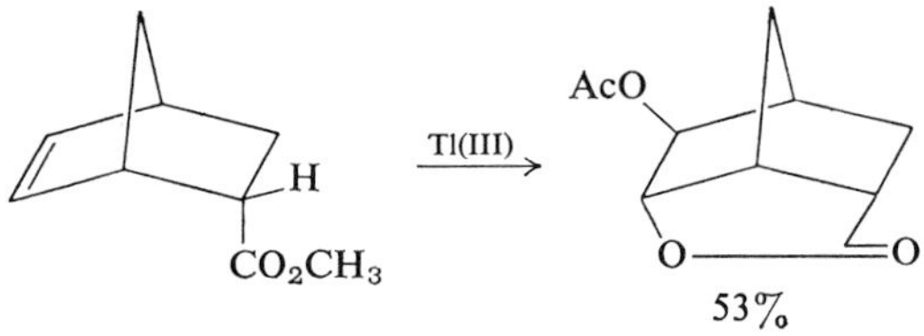

24%

(106)

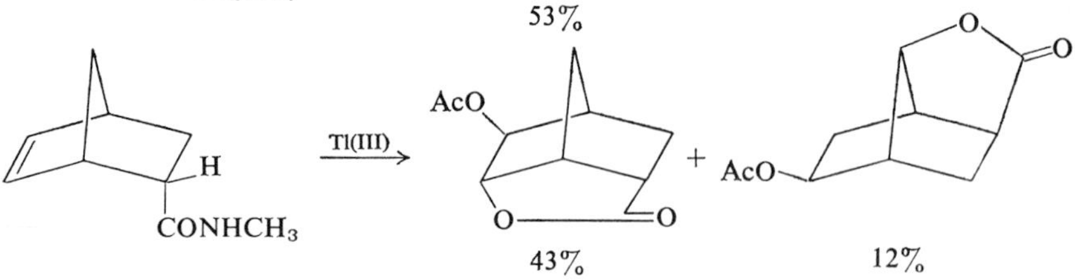

53%

(107)

43%   12%

VI. OXIDATION OF ALLENIC AND ACETYLENIC COMPOUNDS

La Forge and Acree (82) reported on the reaction of 1-phenyl-1,2-butadiene, 1-cyclohexyl-2,3-pentadiene, and 2,3-pentadiene with LTA. These workers reported that 1-phenyl-1,2-butadiene yielded as principal product a crystalline compound whose formula corresponded to the addition of two acetoxyl groups to one of the double bonds of the allenic system. The two possible isomers were designated by the authors as either $C_6H_5CH{=}COAc{-}CHOAcCH_3$ or $C_6H_5CHOAcCOAc{=}CHCH_3$ with no choice expressed with regard to stereochemistry about the double bond. Catalytic hydrogenation yielded a dihydro derivative. Subsequent chemical transformations reported by these workers do not enable one to decide upon the structure of the initial product, and this problem merits investigation. Products from the LTA oxidation of the other two allenes mentioned above were not identified in this study.

Moon and Campbell (83) recently have reported on the LTA oxidation of phenylacetylene (**112**). They find methylphenylacetylene (**113**) as the major product (24%) along with minor amounts of products derived formally from addition of methyl acetate to the double bond as well as ketone formation, **114**, **115**, and **116**, respectively (eq. 108). Because this reaction proceeds in

$$\tag{108}$$

pyridine these authors tend to reject a free-radical mechanism although an intermediate such as $C_6H_5\overset{\cdot}{C}HCH_3$ certainly would appear to be an attractive one for rationalization of the products obtained.

Oxidation of phenylbenzylacetylene leads to benzylic acetoxylation as well as formation of phenylbenzoylacetylene (84).

VII. OXIDATION OF CARBONYL COMPOUNDS

The product of LTA oxidation of a compound containing the CH_2CO group results from replacement of an α-hydrogen by an acetoxyl group. Catalysis by Lewis acids and the course of reaction of cholestane-3-one (**117**) and cholestane-2-one (**119**) as discussed below indicate that prior enolization

$$\tag{109}$$

of the carbonyl compound may be involved. In this sense the reaction may be considered to involve addition of LTA to the double bond. Also, reaction between LTA and enolic hydroxyl group must be considered (eq. 109).

The reaction has been applied to numerous ketonic systems including cyclic and steroidal examples as well as β-diketonic ones (85–89). Henbest and co-workers (89) have shown that boron trifluoride exerts a strong catalytic effect upon the α-acetoxylation reaction. Cholestan-2- and 3-one give the 3α-acetoxy and 2α-acetoxy ketones at room temperature upon oxidation with LTA–BF₃ (**119** → **120**; **117** → **118**), respectively (eqs. 110 and 111). By comparison, temperatures of 70–80° are needed for α-acetoxylation of cyclohexanone. The more easily enolized β,γ-unsaturated ketones such as cholest-5-ene-3-one react at 15–20° to give 4α-acetoxy compounds.

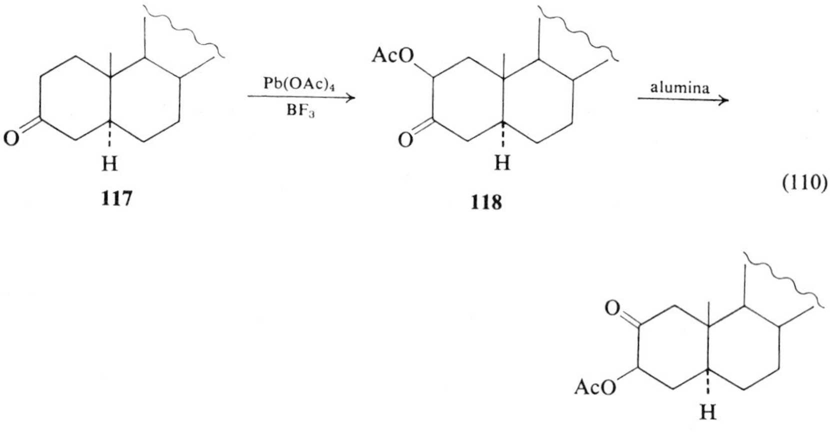

(110)

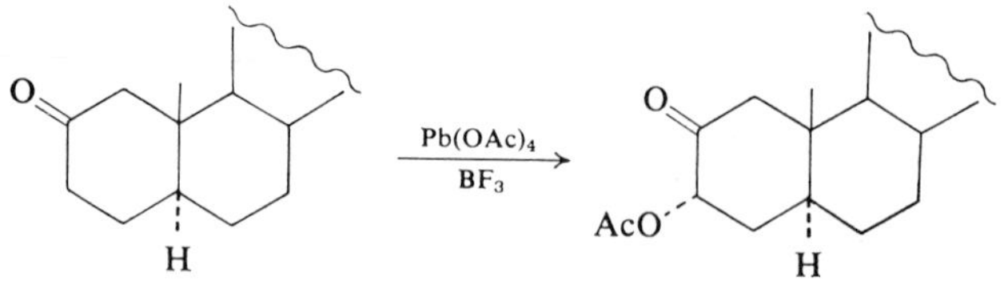

(111)

The stereochemical outcome of these reactions is completely analogous to results observed in the bromination of the corresponding ketones, i.e., attack from the less hindered α-side (90). The isomerization of the 2α-acetoxycholestan-3-one to 3β-acetoxycholestan-2-one, **118** → **118a**, probably proceeds via enolization and α-protonation.

LTA oxidation of 3-oxo-4,5-oxido steroids causes acetoxylation in the

2α-position. These products rearrange under very mild conditions to the 2,3-dioxo-Δ^4 steroids (eq. 111a) (95a).

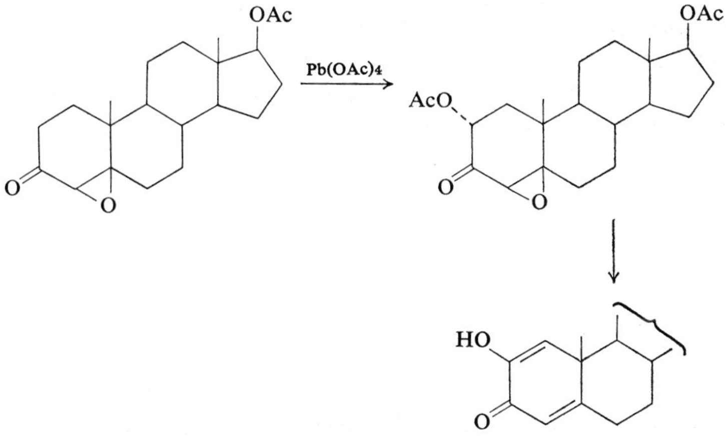

(111a)

An interesting example of neighboring group participation is available in the behavior of norquassic acid (121) toward oxidation with LTA (91) (eq. 112).

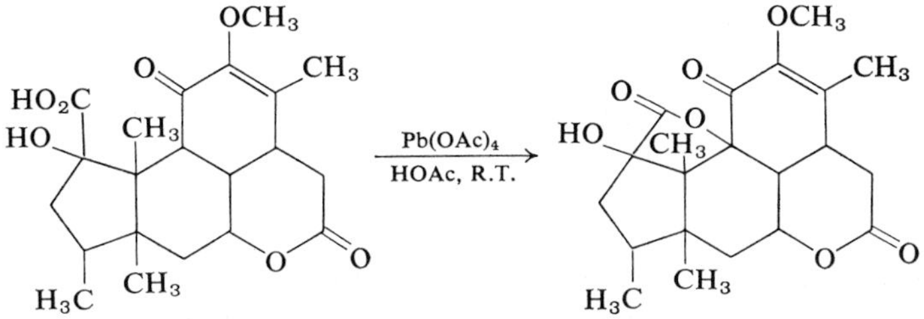

(112)

No less than four different functional groups which may react with lead tetraacetate are present in 121; namely, hydroxyl, carboxyl, enol ether, and ketonic carbonyl. Indeed, product 122 corresponds to perhaps the least

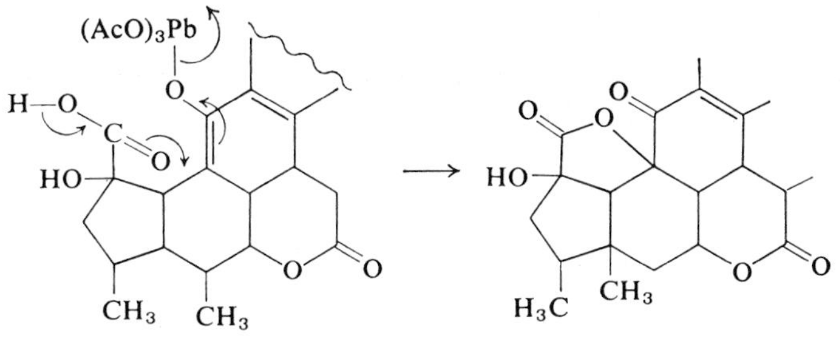

(113)

expected reaction. The carboxyl group may intercept an intermediate in the oxidation of the ketonic carbonyl group (eq. 113).

Such an enol lead ester may intervene in the α-acetoxylation reaction of ketones with LTA as mentioned above.

In the case of the α,β-unsaturated ketone pulegone **123**, the double bond is the site of addition yielding *cis* and *trans*-4-acetoxyisopulegone (**124**) (92) (eq. 114). Only a minor amount of 2-acetoxypulegone (**125**) was formed.

$$(114)$$

123	**124**	**125**
	Major	Minor

This contrasts with the behavior of pulegone upon oxidation with mercuric acetate in which case **123** yields **125**. The product may arise by the addition-elimination scheme shown in eq. 115.

$$(115)$$

| **123** | | **124** |

LTA oxidation of isophorone probably also involves reaction of the enol form although better yields were obtained using benzene alone in the absence of Lewis acid catalysts such as BF_3 (eq. 115a) (92a).

$$(115a)$$

It has been shown that LTA acetoxylates the enolate anion rapidly and in high yield (eq. 115b) (92b).

$$(115b)$$

Enol ethers undergo rapid reaction with lead tetraacetate (6). A reasonable mechanism for this reaction involves initial electrophilic attack upon the double bond followed by addition of HOAc to the oxonium ion and solvolysis of the lead ester (eq. 116). It was found by Johnson et al. that with enol

$$RO-CH=CH_2 \longrightarrow R\overset{+}{O}=CH-CH_2-Pb(OAc)_3$$

$$Pb-(OAc)_3 \quad \Big\downarrow_{-OAc}$$

$$\overset{OAc}{\underset{|}{C}} \qquad\qquad (116)$$

$$\underset{OAc}{\overset{OAc}{\underset{|}{RO-CH-CH_2-Pb(OAc)_3}}} \longrightarrow \underset{}{\overset{OAc}{\underset{|}{RO-CH-CH_2OAc}}}$$

esters the product of oxidation with LTA results from cleavage of the ester group (93). Acyl transfer to acetate ion may yield the anhydride and the organolead intermediate which may ultimately yield the acetoxy aldehyde (eq. 117).

$$\overset{O}{\overset{||}{RC}}-O-CH=CHR \xrightarrow{Pb(OAc)_4} \overset{O}{\overset{||}{R-C}}-\overset{+}{O}=CR-\overset{Pb(OAc)_3}{\underset{|}{CHR}}$$

$$AcO^-$$

$$(117)$$

$$O=CR-\underset{Pb(OAc)_3}{\overset{\cdot}{\underset{|}{CHR}}} \longrightarrow O=CRCHR \underset{OAc}{\overset{}{\underset{|}{}}}$$

Reaction of enol acetates with LTA also leads to α-acetoxylation of ketones. Nambara and Fishman (94) synthesized 5α-androstane-3β,16α-diol-17-one diacetate by this method.

(118)

Enamines are oxidized by LTA by apparent electrophilic addition to the —N—C=C— system. The transformation of **126** → **127** and **128** among yohimbinoid alkaloids represents an example of this reaction (95). Product **127** losses HOAc and the resulting pyroline is further oxidized by LTA to the pyridinium structure (eq. 118).

Recently the α-acetoxylation of lactones has been demonstrated (95a) (eq. 118a). This reaction occurs under rather forcing conditions.

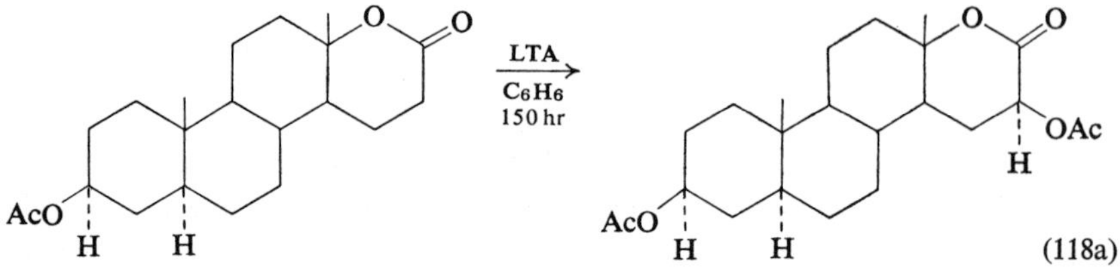

(118a)

VIII. FRAGMENTATION REACTION

No acetoxy lactone formation was observed in the LTA oxidation of a β,γ-unsaturated acid **129** (96) (eq. 119). The possibility of formation of an allylic carbonium ion is undoubtedly an important factor in these reactions.

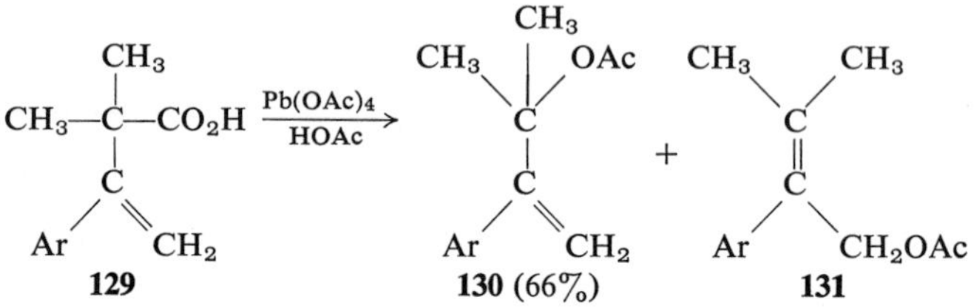

(119)

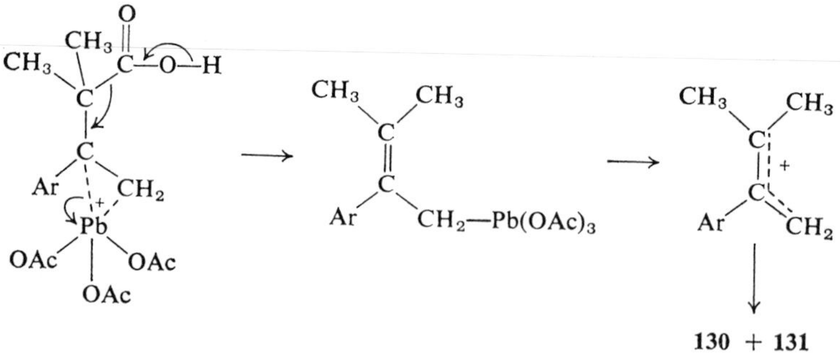

(120)

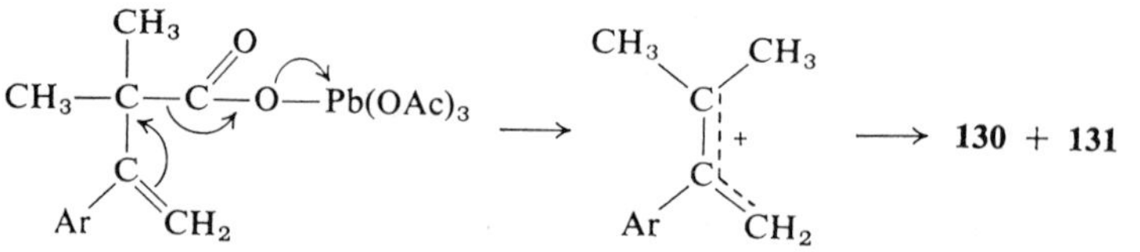

130 + 131

(121)

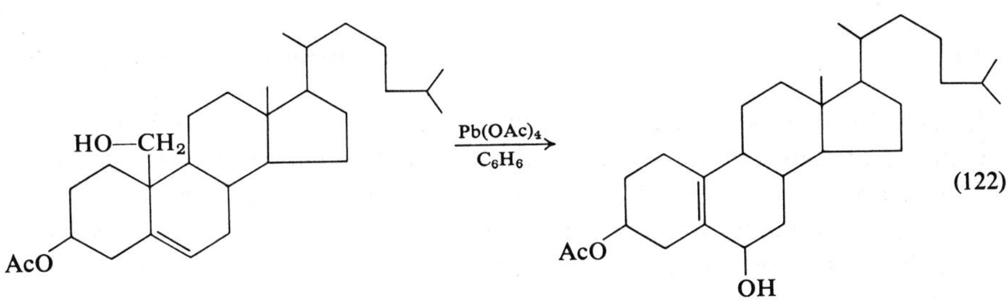

(122)

132   **133**

(123)

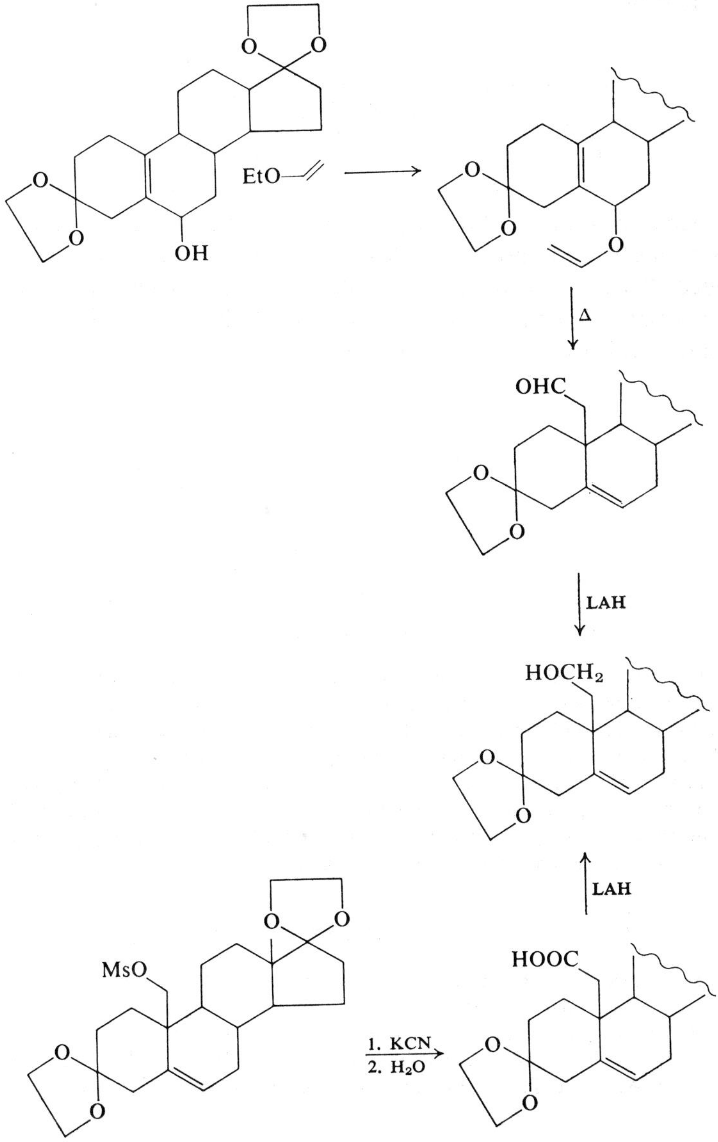

233

In fact, this may be considered as a fragmentation initiated by electrophilic addition of lead tetraacetate to either the double bond or carboxyl group followed by loss of carbon dioxide to yield the allylic ion (eqs. 120 and 121).

Among steroidal unsaturated alcohols, analogous fragmentation reactions have also been observed (55) (eq. 122). The 6β-stereochemistry of the hydroxyl group for a related example was demonstrated by the elegant route of eq. 123 (100). A related example of the fragmentation reaction is $134 \rightarrow 135$ (eq. 124) (97).

(124)

134   **135**

Simpler examples of this type of fragmentation have also been observed. It was mentioned earlier that lead tetraacetate oxidation of 3-buten-1-ol yielded 30% allyl acetate as well as other products which may result from an intermediary allyl carbonium ion (58). Similarly, oxidation of 1-allylcyclohexanol yields cyclohexanone (99) (eq. 125).

(125)

80%

Acknowledgment

The author wishes to thank Professor Guy Ourisson, Institute of Chemistry, University of Strasbourg, for his hospitality during the writing of this work. Thanks also go to the Petroleum Research Fund for support of my sabbatical year at Strasbourg and to the National Institutes of Health for support of the portion of the author's research presented in this review. Special gratitude is owed to Dr. D. A. V. Awang, Chicago Circle, for a critical reading of the entire manuscript, and Mrs. J. Johnson for typing and technical drawing. Thanks are also due to Professor Kurt Schaffner, University of Geneva, for making unpublished results available to me.

References

1. R. Criegee, in K. B. Wiberg, Ed., *Oxidation In Organic Chemistry*, Academic Press, New York, 1965, pp. 335–351.
2. R. Criegee, *Justus Liebigs Ann. Chem.*, **481**, 263 (1930).
3. C. B. Anderson and S. Winstein, *J. Org. Chem.*, **28**, 605 (1963).
4. K. B. Wiberg and S. D. Nielsen, *J. Org. Chem.*, **29**, 3353 (1964).

5. H. J. Kabbe, *Justus Liebigs Ann. Chem.*, **656**, 204 (1962).
6. R. Criegee, P. Dimroth, K. Noll, R. Simon, and C. Weis, *Chem. Ber.*, **90**, 1070 (1957).
7. R. Criegee, P. Dimroth, and R. Schempf, *Chem. Ber.*, **90**, 1337 (1957).
8. A. M. Birks and G. F. Wright, *J. Amer. Chem. Soc.*, **62**, 2412 (1940). (a) J. Romeyn and G. F. Wright, *J. Amer. Chem. Soc.*, **69**, 697 (1947). (b) A. G. Brook and G. F. Wright, *Acta Crystallogr.*, **4**, 50 (1951); *Chem. Abstr.*, **45**, 4321 (1951).
9. M. Malaiyandi and G. F. Wright, *Can. J. Chem.*, **41**, 1493 (1963).
10. For recent work on the stereochemistry of such adducts, see A. Factor and T. G. Traylor, *J. Org. Chem.*, **33**, 2607 (1968).
11. S. Winstein and K. C. Pande, *Tetrahedron Lett.*, **46**, 3393 (1964).
12. F. A. L. Anet, *Tetrahedron Lett.*, **46**, 3399 (1964).
13. V. I. Lodochnikova, E. M. Panov, and K. A. Kocheshkov, *Zh. Obshch. Khim.*, **34**, 4022 (1964).
14. E. M. Panov, V. J. Lodochnikova, and K. A. Kotscheschkow, *Dokl. Akad. Nauk SSSR*, **111**, 1042 (1956).
14a. D. R. Harvey and R. O. C. Norman, *J. Chem. Soc.*, **1964**, 4860.
14b. R. O. C. Norman and C. B. Thomas, *J. Chem. Soc.* (*B*), **1970**, 421.
15. A. Kergomard, *Ann. Chim.* (Paris), **8**, 153 (1953).
16. T. Aratani, *Nippon Kagaku Zasshi*, **78**, 1534 (1957); *Chem. Abstr.*, **54**, 1587 (1960).
17. I. Alkonyi, *Chem. Ber.*, **96**, 1873 (1963). (a) I. Alkonyi, *Chem. Ber.*, **95**, 279 (1962).
18. G. H. Whitham, *J. Chem. Soc.*, **1961**, 2232.
19. Y. Matsubara, *J. Chem. Soc. Jap.*, **78**, 907, 909 (1957); *Chem. Abstr.*, **506**, 22056 (1956).
20. R. Criegee, *Angew. Chem.*, **70**, 173 (1958).
21. T. G. Traylor and A. W. Baker, *Tetrahedron Lett.*, **19**, 15 (1959). (a) For a good discussion of such *cis* addition, see T. G. Traylor, *Accounts Chem. Res.*, **2**, 152 (1969).
22. J. A. Retamar and C. Fernandez, *Rev. Fac. Ing. Quim Univ. Nac. Litoral, Santa Fe Arg.*, **33–34**, 25–31 (1964–65); *Chem. Abstr.*, **66**, 95222 (1957).
23. L. E. Grunewald, *J. Org. Chem.*, **30**, 1673 (1965).
23a. M. Stefanović, A. Jokić, Z. Maksimović, Lj Lorenc, and M. Lj Mioailović, *Helv. Chim. Acta*, **53**, 1895 1970
24. W. F. Erman, *J. Org. Chem.*, **32**, 765 (1967).
25. P. v. R. Schleyer, *J. Amer. Chem. Soc.*, **89**, 701 (1967).
26. C. W. Jefford and W. Wojnarowski, *Tetrahedron Lett.*, **34**, 3763 (1968).
27. C. W. Jefford and W. Wojnarowski, *Chem. Commun.*, **1968**, 129.
28. K. Alder, F. H. Flock, and H. Wirtz, *Chem. Ber.*, **91**, 609 (1958).
29. Unpublished results of R. M. Moriarty and H. Walsh, 1966.
30. S. Wakabayashi, *Kogyo Kagaku Zasshi*, **63**, 627 (1960); *Chem. Abstr.*, **56**, 7165 (1962).
31. W. Hückel and H. G. Kirschner, *Chem. Ber.*, **80**, 41 (1947).
32. P. Naffa and G. Ourisson, *Bull. Soc. Chim. Fr.*, **1954**, 1115.
33. K. Alder and G. Stein, *Justus Leibigs Ann. Chem.*, **485**, 234 (1931).
34. B. Cocton and A. C. DePaulet, *Bull. Soc. Chim.*, fascicule 9, **1966**, 2947.
34a. R. Noyori, Y. Tsuda, and H. Takaya, *Chem. Comm.*, **1970**, 1181.
35. E. Detilleux and J. Jadot, *Bull. Soc. Roy. Sci. Liege*, **24**, 366 (1955).
36. E. Hahl, Diplomarbeit, Technical University of Karlsruhe, Germany, 1956, quoted by R. Criegee on p. 339 of Ref. 1. (a) E. Hahl, Dissertation, Technical University of Karlsruhe, Germany, 1958.

37. Y. Yukawa and N. Hayashi, *Bull. Chem. Soc. Jap.*, **39**, 2255 (1966).
38. K. Ichikawa and S. Uemura, *J. Org. Chem.*, **32**, 493 (1967).
39. E. I. Heiba, R. M. Dessau, and W. J. Koehl, Jr., *J. Amer. Chem. Soc.*, **90**, 2706 (1968). (a) G. F. Wright, *J. Amer. Chem. Soc.*, **57**, 1993 (1935). (b) R. Schempf, Dissertation, Technische Hochschule, Karlsruhe, 1957.
39a. H. Kropf, J. Gelbrich, and M. Ball, *Tetrahedron Lett.*, **39**, 3427 (1969).
40. W. Hückel and K. Kümmerle, *J. Prakt. Chem.*, [2] **160**, 74 (1942).
41. F. V. Brutcher, Jr., and F. J. Vara, *J. Amer. Chem. Soc.*, **78**, 5695 (1956).
41a. S. Uemura A. Tabata, and M. Okano, *Chem. Comm.*, **1970**, 1630.
42. For a review of the Prevost and Woodward methods, see C. V. Wilson, *Org. Reactions*, **9**, 332–387 (1957), pp. 350–352.
43. M. Finkelstein, *Chem. Ber.*, **90**, 2097 (1957).
44. A. C. Cope, N. A. Nelson, and D. S. Smith, *J. Amer. Chem. Soc.*, **76**, 1100 (1954).
45. R. Hoffmann and R. B. Woodward, *Accounts Chem. Res.*, **1**, 17 (1968).
46. R. Huisgen and G. Boche, *Tetrahedron Lett.*, **1965**, 1769.
47. R. Huisgen, G. Boche, W. Hechtl, and H. Huber, *Angew. Chem.*, **78**, 595 (1966); *Angew. Chem. Int. Ed. Engl.*, **5**, 585 (1966).
48. G. Boche, W. Hechtl, H. Huber, and R. Huisgen, *J. Amer. Chem. Soc.*, **89**, 3344 (1967).
49. R. Huisgen, G. Boche, and H. Huber, *J. Amer. Chem. Soc.*, **89**, 3345 (1967).
50. M. J. S. Dewar, C. R. Gannellin, and R. Pettit, *J. Amer. Chem. Soc.*, **79**, 1767 (1957); C. R. Gannellin and R. Pettit, *J. Chem. Soc.*, **1958**, 55.
51. A. Windaus and U. Riemann, Hoppe-Seyler's *Z. Physiol. Chem.*, **274**, 206 (1942).
52. A. Windaus, U. Riemann, and G. Zühlsdorff, *Justus Liebigs Ann. Chem.*, **552**, 135, 142 (1942).
53. H. Meerwein, *Chem. Ber.*, **77**, 227 (1944).
54. A. J. Birch, A. R. Murray, and H. Smith, *J. Chem. Soc.*, **1951**, 1945.
55. R. M. Moriarty and K. Kapadia, *Tetrahedron Lett.*, **1964**, 1165. (a) H. Immer, M. Lj. Mihailovic, K. Schaffner, D. Arigoni, and O. Jeger, *Experientia*, **16**, 530 (1960). *Helv. Chim. Acta*, **45**, 753 (1962).
56. T. Koriyone and N. Kawano, *J. Pharm. Soc. Jap.*, **71**, 924 (1951).
57. T. Okuda and T. Yoshida, *Chem. Ind.* (London), **1965**, 37.
58. S. Moon and J. M. Lodge, *J. Org. Chem.*, **29**, 3453 (1964).
59. S. Moon and L. Haynes, *J. Org. Chem.*, **31**, 3067 (1966).
60. D. Hauser, K. Heusler, J. Kalvoda, K. Schaffner, and O. Jeger, *Helv. Chim. Acta*, **47**, 1961 (1964).
61. S. Moon and B. H. Waxman, *J. Org. Chem.*, **34**, 288 (1969).
62. Unpublished result of R. M. Moriarty and H. Gopal, 1969.
63. R. M. Moriarty, H. G. Walsh, and H. Gopal, *Tetrahedron Lett.*, **36**, 4363 (1966). (a) R. M. Moriarty, H. Gopal, and H. G. Walsh, *Tetrahedron Lett.*, **36**, 4369 (1966).
64. K. Alder and S. Schneider, *Justus Liebigs Ann. Chem.*, **524**, 189 (1936).
65. R. Criegee, H. Kristinsson, D. Seebach, and F. Zanker, *Chem. Ber.*, **98**, 2331 (1965).
66. G. Büchi, R. E. Erickson, and N. Wakabayashi, *J. Amer. Chem. Soc.*, **83**, 927 (1961).
67. E. J. Corey and J. Casanova, Jr., *J. Amer. Chem. Soc.*, **85**, 165 (1963).
68. C. A. Grob, M. Ohta, and A. Weiss, *Angew. Chem.*, **70**, 343 (1958).
69. C. A. Grob, M. Ohta, E. Renk, and A. Weiss, *Helv. Chim. Acta*, **41**, 1191 (1958).
70. W. A. Mosher and C. L. Kehr, *J. Amer. Chem. Soc.*, **75**, 3172 (1953).
71. J. K. Kochi, *J. Amer. Chem. Soc.*, **87**, 1811 (1965).

72. N. A. LeBel and J. E. Huber, *J. Amer. Chem. Soc.*, **85**, 3193 (1963).
73. M. Martin and D. C. DeJongh, *J. Amer. Chem. Soc.*, **84**, 3526 (1962).
74. W. A. Ayer and C. E. McDonald, *Can. J. Chem.*, **43**, 1429 (1965).
75. L. H. Zalkow and D. R. Brannon, *J. Chem. Soc., Suppl. I,* **1964**, 5497.
76. K. Kitahonoki and Y. Takano, *Tetrahedron Lett.*, **1963**, 1597.
77. P. Radlick, R. Klem, S. Spurlock, J. J. Sims, E. E. van Tamelen, and T. Whitesides, *Tetrahedron Lett.*, **49**, 5117 (1968).
78. H. H. Westberg and H. J. Dauben, Jr., *Tetrahedron Lett.*, **49**, 5123 (1968).
79. J. K. Kochi, *J. Amer. Chem. Soc.*, **87**, 1811 (1965).
80. J. K. Kochi, *J. Org. Chem.*, **30**, 3265 (1965).
81. R. M. Moriarty, C. R. Romain, and T. O. Lovett, *J. Amer. Chem. Soc.*, **89**, 3927 (1967). (a) Unpublished results of H. Gopal and R. M. Moriarty, 1968.
82. F. B. La Forge and F. Acree, Jr., *J. Org. Chem.*, **6**, 208 (1941).
83. S. Moon and W. J. Campbell, *Chem. Commun.*, **1966**, 470.
84. J. Jadot and M. Neuray, *Bull. Soc. Roy. Sci. Liege*, **30**, 247 (1962).
85. G. W. K. Cavill and D. H. Solomon, *J. Chem. Soc.*, **1955**, 4426.
86. E. Detilleux and J. Jadot, *Bull. Soc. Roy. Sci. Liege*, **29**, 208 (1960); *Chem. Abstr.*, **65**, 7275 (1961).
87. R. Criegee and K. Klonk, *Justus Liebigs Ann. Chem.*, **564**, 1 (1949).
88. L. Fieser and R. Stevenson, *J. Amer. Chem. Soc.*, **76**, 1728 (1954).
89. H. B. Henbest, D. N. Jones, and G. P. Slater, *J. Chem. Soc.*, **1961**, 4473.
90. C. Djerassi and T. Nakano, *Chem. Ind.* (London), **1960**, 1385.
91. Z. Valenta, A. H. Gray, S. Papadopoulos, and C. Podesva, *Tetrahedron Lett.*, **20**, 25 (1960).
92. L. H. Zalkow and J. W. Ellis, *J. Org. Chem.*, **29**, 2626 (1964).
92a. J. W. Ellis, *J. Org. Chem.*, **34**, 1154 (1969).
92b. J. W. Ellis, *Chem. Comm.*, **1970**, 406.
93. W. S. Johnson, B. Gastambide, and R. Pappo, *J. Amer. Chem. Soc.*, **79**, 1991 (1957).
94. T. Nambara and J. Fishman, *J. Org. Chem.*, **27**, 2131 (1962).
95. N. Finch, C. W. Gemenden, I. H-C Hsu, and W. I. Taylor, *J. Amer. Chem. Soc.*, **85**, 1520 (1963).
95a. M. Stefanović, A Djarmati, and M. Gašić, *Tetrahedron Lett.*, **32**, 2769 (1970).
96. J. Jacques, C. Weidmann, and A. Horeau, *Bull. Soc. Chim. Fr.*, **1959**, 424.
97. H. Immer, M. Lj. Mihailovic, K. Schaffner, D. Arigoni, and O. Jeger, *Experientia*, **16**, 530 (1960); *Helv. Chim. Acta*, **45**, 753 (1962).
98. M. Amorosa, L. Caglioti, G. Cainelli, H. Immer, J. Keller, H. Wehrli, M. Lj. Mihailovic, K. Schaffner, D. Arigoni, and O. Jeger, *Helv. Chim. Acta*, **45**, 2674 (1962).
99. E. A. Braude and O. H. Wheeler, *J. Chem. Soc.*, **1955**, 320.
100. J. Hill, J. Iriarte, K. Schaffner, and O. Jeger, *Helv. Chim. Acta*, **49**, 292 (1966).

Stereochemical Features of Vinylic Radicals

LAWRENCE A. SINGER

*Department of Chemistry, University of Southern California,
Los Angeles, California*

Vinyl radicals have been investigated by a number of workers during the last few years. The approaches to the subject are often different, but when taken together they provide an emerging picture of the stereochemical capabilities of these transient species. Other recent summaries are by Bentrude (1) and Simamura (2).

I. GENERAL CONSIDERATIONS

Vinyl radicals can assume either linear (sp-hydribized) (**1**) or bent (sp^2-hybridized) (**2**) configurations. Basic questions about any particular vinyl radical are (*1*) what is the stable configuration, **1** or **2**?; and if it is **2**, (*2*) what is the rate of the inversion process **2** $\rightleftharpoons$ **3**?

Unfortunately, direct structural information is difficult to obtain on these highly reactive transient species. Some available epr data are presented

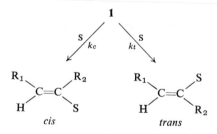

later in this chapter. In large part, our present picture of the stereochemical capabilities of vinyl radicals is based on chemical and kinetic data.

II. CHEMICAL APPROACH

The olefin products derived from the vinyl radical intermediates provide useful information for answering questions (*1*) and (*2*), at least in part. If a linear intermediate is formed, both *cis*- and *trans*-olefins can be formed with the olefin ratio being determined by *stereoselectivity* features in the scavenging steps (k_c, k_t) (Case 1, below). If bent intermediates are formed and inversion is *fast* relative to scavenging, the product ratio is determined by the relative populations of the two bent forms and again stereoselectivity features in the scavenging steps (Case 2, below). If bent intermediates are involved but inversion is *slow* relative to scavenging, the olefin product ratio is *stereospecifically* determined—the degree of specificity depending on the ratio of k_{scav}/k_{inver} (Case 3, below) (*3*).

Case 1. Linear vinyl radical, where S is some scavenger.

Case 2. Bent vinyl radicals, fast inversion.

Case 3. Bent vinyl radicals, slow inversion. As in Case 2, but k_i, k_{-i} are comparable to or smaller in magnitude than k_c, k_t.

A. Vinyl Radicals as Intermediates in Radical Additions to Acetylenes

Vinyl radicals are formed as intermediates in free radical additions to acetylenes as shown in eq. 1. This process is depicted as a free radical chain where the propagating steps are (*1*) X· adding to the acetylene to give a transient vinyl radical **4**, and (*2*) **4** abstracting Y· from covalent XY.

$$X\cdot + R_1C{\equiv}CR_2 \longrightarrow \underset{\textbf{4}}{\overset{R_1}{\underset{X}{\diagup}}C{=}\overset{\cdot}{C}{\sim}R_2} \overset{X-Y}{\longrightarrow} \overset{R_1}{\underset{X}{\diagup}}C{=}C\overset{R_2}{\underset{Y}{\diagdown}} + X\cdot$$

and/or   (1)

$$\overset{R_1}{\underset{X}{\diagup}}C{=}C\overset{Y}{\underset{R_2}{\diagdown}} + X\cdot$$

In this approach, the stereochemistry of the initial addition Step (*1*) must be considered if **4** is a bent radical and if inversion is slower than scavenging (Case 3, above). For example, if Step (*1*) is a *trans* addition, and if Case 3 is involved, an over-all *trans* addition product will be observed. However, if Case 1 or Case 2 is involved, the stereochemistry of the products is independent of the initial addition step.

B. Addition of Hydrogen Bromide to Acetylenes

Addition of two equivalents of hydrogen bromide to 2-butyne under free radical conditions gives exclusively *d,l*-2,3-dibromobutane (4). Since it has been shown that *cis*- and *trans*-2-bromo-2-butene add hydrogen bromide homolytically in an *anti*-manner (5), it can be concluded that the initial addition to 2-butyne must be *anti* (eq. 2).

$$CH_3C{\equiv}CCH_3 + 2HBr \longrightarrow$$

(2)

Abell and Piette found that photolysis of a mixture of 2-butyne and hydrogen bromide at 77°K rapidly produced an epr signal of seven equally spaced lines with a hyperfine constant of 8.3 G (6). This pattern indicates equal electron density at the two methyl groups and is consistent with a bridged structure for the intermediate radical (5).

5

The epr spectra similarly obtained with *cis*- and *trans*-2-butene revealed seven line patterns with splittings of 10.8 and 12.6 G, respectively (6). Again, bridged structures are consistent with these results. Such bridged structures conveniently explain the observed *anti*-additions of hydrogen bromide to 2-butyne and the 2-butenes.

Skell and Allen (7) found that hydrogen bromide added to propyne in the liquid phase at −78° to give *cis*-1-bromo-1-propene. They suggest that bromine atom adds stereospecifically *trans* yielding a 1-methyl-2-bromovinyl radical (6) which is scavenged in a hydrogen atom transfer step prior to inversion (Case 3) (eq. 3). However, they point out that another explanation is a bridged radical species that necessarily reacts with hydrogen bromide in a stereospecific manner.

(3)

C. Addition of Thiyl Radicals to Acetylenes

The addition of thiyl radicals to acetylenes has been extensively studied and much of this work is reviewed by Oswald and Griesbaum (8a) and more recently by Kellog (8b). Although this approach to the study of vinyl radicals is complicated by (*1*) the potential isomerization of the initial adduct products by thiyl radicals, (*2*) the potential over-all addition of two molecules of thiol to the acetylene bond, and (*3*) the potential *cis-trans* isomerization of the thioethenes under often used photolysis conditions (9), careful work by several groups has provided useful stereochemical information.

Oswald and coworkers (10) studied the addition of thiol compounds to phenylacetylene and observed mainly *cis*-olefins in the product mixtures. Rapid *cis-trans* adduct isomerization does occur in the presence of catalytic amounts of aromatic thiol compound so the observed product ratio depends on the reaction conditions. The subsequent product isomerization can be minimized, however, by running the addition in the presence of a large excess of phenylacetylene over thiol compound (22-fold excess). In this way it was found that benzenethiol adds to phenylacetylene at 0° to give $>95\%$ *cis*-adduct (eq. 4).

$$C_6H_5SH + HC{\equiv}CC_6H_5 \xrightarrow[0°]{} \underset{>95\% \ cis}{C_6H_5SCH{=}CHC_6H_5} \qquad (4)$$

Kampmeier and Chen (11) observed that the addition of thiolacetic acid to 1-hexyne (hexyne/thiolacetic acid $\gtrsim 6$) leads to an adduct ratio of *cis-trans* = 82/18. They determined the equilibrium distribution of be *cis/trans* $\approx$ 50/50 so that their kinetic ratio is enriched in the thermodynamically less stable isomer.

These stereochemical results (10,11) are explained by the scheme shown in eq. 5. The predominance of *cis*-adduct in the kinetic product mixture results from stereoselective capture of the transient vinyl radicals ($k_c > k_t$). These studies could not distinguish between Cases 1 and 2. Kampmeier and Chen exclude Case 3 on the basis that for that scheme to be operative in their system, the initial addition of thiyl radical to 1-hexyne would be required to yield an 82/18 mixture of the radical precursors of the *cis*- and *trans*-olefins (11a). Although definitive information on the stereochemistry of such radical additions is scarce, a nonstereospecific addition seems unlikely.

$$R_1S{\cdot} + HC{\equiv}CR_2$$

(5)

cis-Adduct   trans-Adduct

Very recently, Wedegaertner and Kampmeier (11b) have examined the free radical addition of ethyl mercaptan to ethoxyacetylene in pentane. An excess of acetylene was maintained in order to suppress product isomerization. The *cis/trans* distribution of the 1-ethoxy-2-(ethylthio)ethane products was determined as a function of per cent reaction. Samples at the lowest per cent conversions show the kinetically controlled product ratio to be *cis/trans* > 100 over a mercaptan concentration range of 1.01 to 8.1 × 10^{-4} M. The thermodynamic mixture of products is *cis/trans* = 0.37.

The very high *cis/trans* kinetic ratio can be accounted for by a *trans* addition of a thiyl radical to give a *cis*-vinyl radical which chain transfers with mercaptan more rapidly than it isomerizes (Case 3). Were Case 2 to apply, the data would require an unreasonably large difference in the free energies of the transition states for the transfer steps (13). A linear radical intermediate (Case 1), however, might provide sufficient stereoselectivity to account for the data. Theoretical calculations indicate a bent structure for α-oxyvinyl radicals (see Section V); thus, Case 3 is the most attractive interpretation

Thus equation 5 cannot be generalized to all thiyl radical–acetylene additions. In any particular system, if bent vinyl radicals are intermediates, a spectrum of stereochemical results are possible ranging from 100% stereoselectively to 100% stereospecifically determined product ratios depending on the relative magnitude of the rate processes involving inversion and scavenging.

D. Other Additions to Acetylenes

The benzoyl peroxide catalyzed additions of carbon tetrachloride (12) and chloroform (12,13) to 1-hexyne and 1-octyne yield predominantly *cis*-olefins by a stereoselective reaction. The proposed scheme (12,13) is similar to that described above for the additions of thiyl radicals to terminal acetylenes.

$$Cl_3C\cdot \ + \ HC{\equiv}C(CH_2)_3CH_2R \ \longrightarrow \ Cl_3CCH{=}\overset{\cdot}{C}(CH_2)_3CH_2R$$

cis- and *trans*-Olefins

(6)

An intramolecular δ-hydrogen atom abstraction competes with the bimolecular scavenging process in these reactions and leads preferentially to cyclopentane products (eq. 6). The relative reactivity of primary to secondary hydrogens is 1:22 so that while this competing process is important in the 1-octyne system, it is relatively unimportant in the 1-hexyne system (13). Heiba and Dessau observe a similar cyclization in the free radical addition of carbon tetrachloride to propargyl esters (14).

Benkeser and co-workers (15) report that the benzoyl peroxide catalyzed additions of silicochloroform to terminal acetylenes lead to *cis/trans* olefin ratios of $\sim 3/1$ (eq. 7). Their ratios probably are representative of the kinetic product distributions since they report only a 10% change in the olefin ratios in samples removed periodically during a run.

$$\text{RC}\equiv\text{CH} + \text{SiHCl}_3 \xrightarrow{\text{Benzoyl Peroxide}} \text{RCH}=\text{CHSiCl}_3 \qquad (7)$$

$$
\begin{array}{cc}
 & cis/trans \\
\text{R} = \text{C}_3\text{H}_7 & 79/21 \\
\text{C}_4\text{H}_9 & 77/23 \\
\text{C}_5\text{H}_{11} & 75/25
\end{array}
$$

Free radical additions of tin hydrides to terminal acetylenes are reported to yield *trans*-olefins (16) but facile isomerization of the olefin products may be occurring via an addition-cleavage mechanism with tin radicals (17).

III. STEREOSPECIFIC GENERATION OF VINYL RADICALS FROM ISOMERIC ACYLPEROXY SOURCES

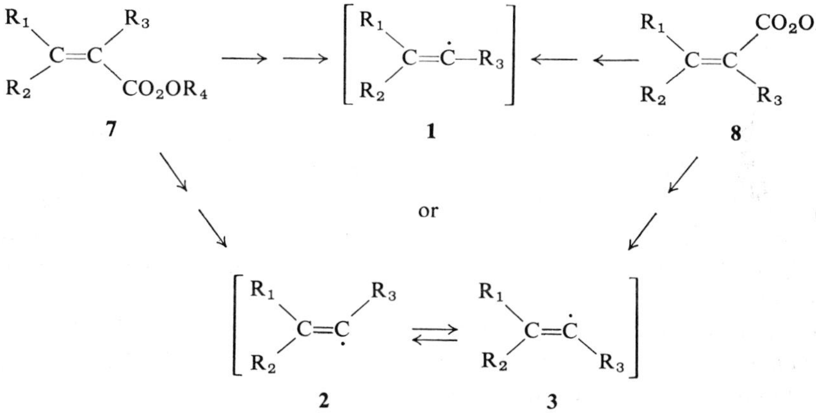

Vinyl radicals can be generated in a stereospecific manner from the homolytic decomposition of isomeric acylperoxy sources (7,8). Both *t*-butyl peroxyesters and diacylperoxides have been utilized. This approach has the potential advantage of a clear stereochemical entry point into the vinyl radical system ($7 \rightarrow 2$, $8 \rightarrow 3$) which is not true with the previously discussed acetylene systems. Of course, if the intermediate vinyl radical has a linear configuration (1), both 7 and 8 yield 1 (eq. 8).

Although these peroxyesters (9) are one-bond cleavage systems, meaning that acyloxy radicals (10) are initially formed upon oxygen–oxygen bond homolysis (18,19), significant decarboxylation to the vinyl radicals occurs (eq. 9).

$$
\begin{array}{c}
\overset{\displaystyle O}{\overset{\|}{RC}}\text{O—OBu-}t \longrightarrow RCO_2{}^{\boldsymbol{\cdot}} + {}^{\boldsymbol{\cdot}}\text{OBu-}t \\
\textbf{9 (where R is} \qquad \textbf{10} \\
\text{a vinyl group)}
\end{array}
$$

$$\longrightarrow R{}^{\boldsymbol{\cdot}} + CO_2$$

(9)

A. α-Alkyl and α-Arylvinyl Radicals

Thermal decomposition of either the *cis*- or *trans*-isomer of the *t*-butyl peroxyesters of α-methylcinnamic acid (11) (20), α,β-dimethylcinnamic acid (12) (21), α-phenylcinnamic acid (13) (20), and α-phenylcrotonic acid (14) (21) in hydrocarbon solvents yields the same olefin product mixture. Under these decomposition conditions the solvent serves as the scavenger of the transient vinyl radicals in a hydrogen atom transfer reaction. The olefin product ratios are shown in Table 1.

$$
\begin{array}{cc}
\begin{array}{c}
C_6H_5 \\
\diagdown \\
C{=}C(CH_3)CO_3Bu\text{-}t \\
\diagup \\
H \\
\textbf{11}
\end{array}
&
\begin{array}{c}
C_6H_5 \\
\diagdown \\
C{=}C(CH_3)CO_3Bu\text{-}t \\
\diagup \\
CH_3 \\
\textbf{12}
\end{array}
\end{array}
$$

$$
\begin{array}{cc}
\begin{array}{c}
C_6H_5 \\
\diagdown \\
C{=}C(C_6H_5)CO_3Bu\text{-}t \\
\diagup \\
H \\
\textbf{13}
\end{array}
&
\begin{array}{c}
CH_3 \\
\diagdown \\
C{=}C(C_6H_5)CO_3Bu\text{-}t \\
\diagup \\
H \\
\textbf{14}
\end{array}
\end{array}
$$

Since the same olefin ratio is observed from either isomeric precursor, the olefin products are stereoselectively determined, meaning either Case 1 or Case 2 is operative in each system. Further, the data reveal olefin ratios distinctly different from the expected thermodynamic values. For example, at equilibrium at 110°, the *cis/trans* propenylbenzene ratio is ~0.33 (20) while the kinetic product ratios range from 0.76 with toluene as scavenger to 1.55 with cumene. In all cases the olefin product ratio is richer in the thermodynamically *less stable* isomer than is found in the equilibrium mixture. Also, the bulkier the scavenger, the stronger this trend.

TABLE 1

Olefin Products from the Thermal Decompositions of *t*-Butyl Peroxycinnamates in Hydrocarbon Solvents

System	Scavenger (temp.)	Olefin products	Expected equilibrium olefin ratio (*cis/trans*)	Observed olefin ratio (*cis/rans*)	Ref.
cis- or *trans*-**11**	Toluene (110°)	Propenylbenzenes	~0.33	0.76	
	Cyclohexene (110°)			0.84	20
	Cumene (110°)			1.55	
cis- or *trans*-**12**	Cumene (110°)	2-Phenyl-2-butenes	~4	1.1–1.2	21
cis- or *trans*-**13**	Cyclohexene (110°)	Stilbenes	~0.1	~4	20
	Cumene (110°)			~6	
cis- or *trans*-**14**	Toluene (105°)	Propenylbenzenes	~0.33	1.39	
	Cyclohexene (105°)			1.58	22
	Cumene (105°)			2.45	

These results reflect the steric differences between the transition states leading to the isomeric products. In the case of the α-methyl-β-phenylvinyl radical system (11), the transition state leading to the thermodynamically more stable *trans*-propenylbenzene requires the scavenger to approach the radical site on the side bearing the bulkier group (phenyl vs. hydrogen). The resulting greater steric interaction between the incoming scavenger and phenyl, as compared to phenyl and hydrogen, is an important factor in determining the relative energies of the two isomeric transition states. Consequently, transfer of hydrogen atom so that it ends up *trans* to the bulkier group on the β-carbon is enhanced, which leads to an enrichment of the product mixture in the thermodynamically less stable isomer. A more quantitative analysis of this effect is possible using olefin product ratios obtained at different temperatures.

The temperature dependencies of the olefin product ratios in the α-methyl-β-phenylvinyl (15) and α-phenyl-β-methylvinyl (16) radical systems in toluene, cyclohexene, and cumene have been determined (20,22) and are shown in Table 2.

$$C_6H_5\underset{H}{\overset{}{\diagdown}}C=\dot{C}{\sim}CH_3 \qquad\qquad CH_3\underset{H}{\overset{}{\diagdown}}C=\dot{C}{\sim}C_6H_5$$

15   16

In system 15 the diacyl peroxide was used as a precursor in addition to the peroxyester. Both 15 and 16 were generated by photolysis as well as thermolysis of the precursors. Since the initial bond breaking process in these systems involves only oxygen–oxygen bond fission, acyloxy radicals are intermediates and are the *immediate* precursors of the vinyl radicals. The acyloxy radicals are long-lived enough for interception by hydrogen atom transfer from solvent molecules. Accordingly, they must have reached thermal equilibrium before decarboxylation. Therefore, the vinyl radicals eventually formed by the photolysis route are energetically equivalent to those resulting from the thermolysis route. The internal consistency of the data in Table 2 further support this position.

The data in Table 2 are displayed in Figures 1 and 2 as plots of Log (*cis/trans*) vs. $1/T$. If Case 1 prevails, a plot of log [(*cis*-olefin)/(*trans*-olefin)] vs. $1/T$ gives a slope defined by $\Delta\Delta H^{\ddagger}_{t-c}/(2.3R)$ where $\Delta\Delta H^{\ddagger}_{t-c}$ is the difference between the enthalpy of activation terms for the transition states leading to the *trans*- and *cis*-olefins. This term shows the extent of stereoselectivity in the system and is expected to be positive, since approach by the donor in the transition state leading to *trans*-olefin involves greater nonbonding inter-

TABLE 2

Olefin Product Ratios at Various Temperatures in the α-Methyl-β-phenylvinyl
and α-Phenyl-β-methylvinyl Radical Systems (20,22)

Radical system	Scavenger	Precursor[a]	Mode of decomposition	Temp. (°C)	Olefin[b] ratio (cis/trans)
15	Toluene	P	Thermal	110	0.76
		DAP	Thermal	90	0.68
		DAP	Thermal	79	0.65
		DAP	Photo	1	0.50
	Cyclohexene	P	Thermal	110	0.84
		DAP	Thermal	80	0.77
		DAP	Photo	1	0.59
		DAP	Photo	−75	0.38
	Cumene	P	Thermal	110	1.55
		DAP	Thermal	79	1.44
		DAP	Photo	1	1.34
		DAP	Photo	−75	1.22
16	Toluene	P	Thermal	105	1.39
		P	Photo	50	1.42
		P	Photo	22	1.43
		P	Photo	0	1.45
	Cyclohexene	P	Thermal	105	1.58
		P	Photo	48	1.77
		P	Photo	22	1.87
		P	Photo	0	2.10
	Cumene	P	Thermal	105	2.45
		P	Photo	50	3.15
		P	Photo	25	3.70
		P	Photo	0	4.55

[a] P = peroxyester, DAP = diacyl peroxide.
[b] Propenylbenzenes.

actions with the groups on the β-carbon compared to the other transition
state.

If Case 2 prevails, the slope in the plot mentioned above is defined by
$(\Delta\Delta H^{\ddagger}_{t-c} + \Delta H^{\circ}_{t-c})/(2.3R)$, where the second enthalpy term is the difference
between the enthalpies of formation of the two isomeric radicals.

Negative slopes are observed in system 15 with toluene, cyclohexene,
and cumene as scavengers with the calculated enthalpy differences being
−0.74, −0.64, and −0.21 kcal/mole, respectively (20). It was suggested that

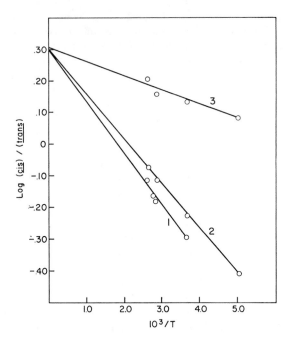

Figure 1. Log (*cis/trans*) for propenylbenzenes derived from α-methyl-β-phenylvinyl radical (15) vs. 1/T. Curve 1, toluene; Curve 2, cyclohexene; Curve 3, cumene.

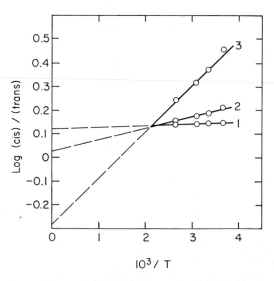

Figure 2. Log (*cis/trans*) for propenylbenzenes derived from α-phenyl-β-methylvinyl radical vs. 1/T. Curve 1, toluene; Curve 2, cyclohexene; Curve 3, cumene.

TABLE 3
Stereoselectivity Parameters for Hydrogen
Atom Transfer to Vinyl Radicals

H-Atom donor	System	
	15	**16**
	$\Delta\Delta\Delta H^{\ddagger}_{t-c}$ (kcal/mole)	
Cumene	0.53	0.83
Cyclohexene	0.10	0.18
Toluene	(0.00)	(0.00)

Case 2 operates in this system and that the negative slopes reflect the domination of the slope term by ΔH°_{t-c} which is negative if the *trans*-vinyl radical is
more stable than the *cis*. As the bulk of the donor increases, the $\Delta\Delta H^{\ddagger}_{t-c}$
term increases, resulting in an over-all less negative enthalpy term.

In system **16**, positive slopes are observed with toluene, cyclohexene,
and cumene as donors (22). It is suggested that Case 1 operates here so that
the slope values reflect only the one enthalpy term $\Delta\Delta H^{\ddagger}_{t-c}$.

Table 3 shows the stereoselectivity parameters for systems **15** and **16**
standardized to toluene as donor ($\Delta\Delta\Delta H^{\ddagger}_{t-c}$). Implicit in this treatment is the
assumption that the equilibrium distribution between *cis*- and *trans*-vinyl
radicals is unchanged through the solvent series toluene, cyclohexene, and
cumene.

The results reveal greater stereoselectivity parameters in vinyl radical
system **16**. Kopchik and Kampmeier (13) point out that greater steric interaction might be expected between an incoming hydrogen atom donor and the
groups on the β-carbon in a linear vinyl radical than in a bent vinyl radical
due to the different lines of approach that the donors take (**17, 18**).

These stereoselectivity results suggest that the α-methyl-β-phenylvinyl
radical is sp^2-hybridized while the α-phenyl-β-methylvinyl is sp-hybridized.

An sp-hybridized 1-phenylvinyl radical has been suggested in the past (23–25) to explain the enhanced rates of free radical additions to phenylacetylene. Phenylacetylene captures methyl radicals 16 times faster than methylacetylene (24). The very large stereoselectivity features of the α-phenyl-β-phenylvinyl radical (see Table 1) can be understood within the context of a linear radical (13). It is tempting to extrapolate this result to α-arylvinyl radicals in general. Results of calculations on the α-phenylvinyl radical, presented later in this chapter, suggest appreciable stabilization of the linear over the bent form.

A different opinion on the structure of the α-phenyl-β-phenylvinyl radical has been put forth by Simamura (2,26) and Tokumaru (27). Simamura reports that decomposition of t-butyl cis- and $trans$-α-phenylperoxycinnamates in carbon tetrachloride at 110° leads to the same cis- and $trans$-chlorostilbene mixture ($cis/trans = 76/24$). However, at 80°, the $trans$-peroxyester yields $cis/trans = 84/16$ while the cis-peroxyester yields a 77/23 ratio. Simamura concludes that the α-phenyl-β-phenylvinyl radical must be bent to account for the observed partial specificity.

Tokumaru has reported that the decomposition of the diacyl peroxides (**19** and **20**) at 60° gives stilbene products as follows: in chloroform, **19** yields $cis/trans = 89/11$; **20** yields $cis/trans = 20/80$; in cumene, **19** yields $cis/trans = 88/12$; **20** yields $cis/trans = 6/94$.

Tokumaru's results in cumene appear to be incompatible with results from our laboratory at 110° where stereoselective capture of the intermediate radicals occurs (20). We estimate as a lower limit $k_{scav}/k_{inver} \lesssim 10^{-3}$ in order for equilibration to precede hydrogen atom transfer (at 110°) and $k_s/k_i \approx 1$ where partial stereospecificity is observed (at 60°). These values lead to a *minimum* difference in the enthalpies of activation for inversion and scavenging ($\Delta H_i^{\ddagger} - \Delta H_s^{\ddagger}$) of about 15 kcal/mole. A reasonable lower limit estimate for the enthalpy of activation for hydrogen atom transfer from cumene to the transient vinyl radical is about 5 kcal/mole which requires $\Delta H_i^{\ddagger} \gtrsim 20$ kcal/mole. Unless the inversion process has an entropy of activation that is positive and *very large* (ca. $+25$ e.u.), the absolute rate constant for the inversion process is very *small* relative to the scavenging process. Since such a value for $\Delta S_i^{\ddagger}$ appears unreasonable, we conclude that the data at 110 and 60° are inconsistent.

The results of **19** and **20** in chloroform are similarly inconsistent with the results in cumene at 110° to the extent that chloroform and cumene might be expected to be equally effective as scavengers. This comparison is based on the similarity of their chain transfer constants for styrene polymerization (at 60°, cumene, $C = 0.82$; chloroform, $C = 0.5$) (28a) and their similar reactivities in hydrogen atom transfer to phenyl radical (chloroform/cumene ~3.5) (28b).

The results of Simamura (2,26) are not as obviously incompatible with the cumene results at 110° since carbon tetrachloride might be expected to be a better scavenger than cumene (based on the relative chain transfer constants for styrene polymerization) (28). However, the experimental details of these investigations (26,27) apparently have not been published. Since important control experiments are *necessary* in these systems, in this writer's opinion, these results must be taken with some reservations until the details are published.

B. α-Halovinyl Radicals

Thermal decomposition of the *cis*- and *trans*-isomers of *t*-butyl α-chloro- and α-bromoperoxycinnamates (**21, 22**) in the presence of cumene and dihydroanthracene (DHA) as scavengers leads to different olefin product ratios (29,30). These results can be explained by Case 3 where bent vinyl radicals are formed with comparable inversion and hydrogen atom scavenging rates.

$$C_6H_5 \diagdown \atop H \diagup C{=}C(Cl)CO_3Bu\text{-}t \qquad\qquad C_6H_5 \diagdown \atop H \diagup C{=}C(Br)CO_3Bu\text{-}t$$

$$\textbf{21} \qquad\qquad\qquad\qquad \textbf{22}$$

Analysis of the kinetic scheme depicted in Case 3 reveals that starting from *cis*-peroxyester (**23**), the olefin ratio is given by (*trans*-olefin)/(*cis*-olefin)

(10)

TABLE 4

Stereospecificity Parameters in Hydrogen Atom Transfer to the α-Chloro-β-phenyl-vinyl Radical (29)

H-Atom donor	k_c/k_{-i} (M^{-1})	k_t/k_i (M^{-1})
Cumene	0.1	0.2
DHA	4.3	12.1

$= (k_t k_{-i})/(k_c k_i) + (k_t[SH])/k_i$ while from the *trans*-peroxyester (**24**), the olefin ratio is given by $(cis\text{-olefin})/(trans\text{-olefin}) = (k_c k_i)/(k_t k_{-i}) + (k_c[SH])/k_{-i}$ (eq. 10).

It is predicted and observed for system **21** that the extent of specificity in the olefin product mixture is directly proportional to the scavenger concentration. The stereospecificity parameters are given in Table 4.

In addition to showing that α-chloro- and α-bromovinyl radicals are bent species, these results reveal that DHA is 40–60 times more efficient as a scavenger than cumene towards these radicals. Unfortunately, the absolute rate constants for neither the scavenging nor inversion steps are known at this time.

C. Other Stereospecific Reactions with Vinyl Radicals

It has been reported that the *cis*- and *trans*-β-phenylvinyl radicals generated by thermal decomposition of diacylperoxides **25** and **26** are trapped with bromotrichloromethane before complete equilibration (31) (eq. 11).

The products derived from the reduction of *cis*- and *trans*-3-chloro-3-hexene with sodium naphthalenide show stereospecificity (32) (eq. 12). Sargent and Brown suggest that this reaction initially yields vinyl radicals and chloride ions upon electron transfer from the naphthalene radical anions. The vinyl radicals are scavenged by a second electron transfer from the radical anions to give carbanions before complete equilibration at the vinyl radical stage. The carbanions are believed to be configurationally stable under the reactions conditions.

$$
\underset{H}{\overset{C_2H_5}{>}}C=C\underset{Cl}{\overset{C_2H_5}{<}} \quad \xrightarrow[DME]{[\text{naphthalene}]^{\cdot-} Na^+} \quad \underset{H}{\overset{C_2H_5}{>}}C=C\underset{H}{\overset{C_2H_5}{<}} \quad + \quad \underset{H}{\overset{C_2H_5}{>}}C=C\underset{C_2H_5}{\overset{H}{<}}
$$

44%   56%

(12)

$$
\underset{H}{\overset{C_2H_5}{>}}C=C\underset{C_2H_5}{\overset{Cl}{<}} \quad \xrightarrow[DME]{[\text{naphthalene}]^{\cdot-} Na^+} \quad 18\% \qquad 82\%
$$

Fry and Mitnick (33) in a similar study on the 3-iodo-3-hexenes, using electrochemical reduction, report partial stereospecificity in the 3-hexene product mixture. The scheme proposed again involves electron transfer to give a transient vinyl radical which is rapidly scavenged by a second electron transfer.

Neuman and Holmes (34) observed different olefin product ratios from the photodecomposition of the *cis*- and *trans*-4-iodo-3-heptenes. However, secondary reactions under the experimental conditions made a definitive analysis of the system difficult.

Kuivila (35) reports partial stereospecific reduction of *cis*- and *trans*-2-bromo-2-butene with tin hydride at $-78°$ (eq. 13). At room temperature, either *cis*- or *trans*-2-bromo-2-butene yields 65% *trans*- and 35% *cis*-2-butene (36). However, equilibration of the olefin products may be occurring via an addition-cleavage mechanism with tin radicals (17).

$$
\underset{H}{\overset{CH_3}{>}}C=C\underset{Br}{\overset{CH_3}{<}} \quad \xrightarrow[-78°]{Bu_3SnH} \quad \underset{H}{\overset{CH_3}{>}}C=C\underset{H}{\overset{CH_3}{<}} \quad + \quad \underset{H}{\overset{CH_3}{>}}C=C\underset{CH_3}{\overset{H}{<}}
$$

57%   43%   (13)

$$
\underset{H}{\overset{CH_3}{>}}C=C\underset{CH_3}{\overset{Br}{<}} \quad \xrightarrow[-78°]{Bu_3SnH} \quad 15\% \qquad 85\%
$$

IV. EPR STUDIES

Epr spectroscopy is potentially the most direct way to obtain information on the structures of vinyl radicals. The magnetic equivalence or nonequivalence of the β-hydrogen coupling constants together with the magnitude of the

α-H, β-H, and α-^{13}C hyperfine splittings can provide detailed structural information. Studies have been carried out on several systems.

Cochran, Adrian, and Bowers (37) studied the epr spectrum of the vinyl radical generated by the photolysis of a matrix composed of 1% hydrogen iodide, 9% acetylene, and 90% argon at 4.2°K. Under these conditions, hydrogen atoms formed from the photolysis of hydrogen iodide added to acetylene (eq. 14).

$$\text{HI} \xrightarrow[\text{2537 Å}]{h\nu} \text{H}^{\cdot} + \text{I}^{\cdot}$$

$$\xrightarrow{\text{C}_2\text{H}_2} \begin{array}{c} \text{H} \\ \diagdown \\ \text{C}=\dot{\text{C}}\text{H} \\ \diagup \\ \text{H} \end{array} \tag{14}$$

The observed epr spectrum under these conditions is complex. The lines are asymmetric and the number of distinguishable peaks is greater than the eight lines expected for the hyperfine interaction of three nonequivalent protons, or the six lines expected for the interaction of one set of two equivalent protons and a single proton. The additional splittings are ascribed to possible anisotropic hyperfine interactions with the three protons and/or an anisotropy in the electronic g factor which can occur in polycrystalline samples.

The gross features of the spectrum suggest an eight line pattern with hyperfine splittings for three nonequivalent protons of 68, 34, and 16 G. Assignments of these splittings to specific protons was possible using additional data obtained when the experiment was repeated using deuterioacetylene (eq. 15).

$$\text{DC}\equiv\text{CD} + \text{H}^{\cdot} \xrightarrow{4.2°\text{K}} \begin{array}{c} \text{D} \\ \diagdown \\ \text{C}=\dot{\text{C}} \\ \diagup \qquad \diagdown \\ \text{H} \qquad\qquad \text{D} \end{array} \tag{15}$$
$$\textbf{27}$$

The spectrum observed in this experiment showed two groups of lines separated by 64 G. There were no lines corresponding to splittings of 34 or 16 G. The authors conclude that one of the β-hydrogen coupling constants must be 68 G and that only *one* of the two possible isomers was present. As the temperature was raised to 32°K, there was no sign of a second isomer appearing. Since theory predicts that the *trans*-isomer (**27**) will have the larger coupling constant, it was concluded only the *trans*-isomer is present at 4.2°K, possibly because of a lower zero point energy for this isomer. A second explanation based on a stereospecific *trans*-addition to yield **27**, and a high barrier for isomer interconversion, was ruled out on the basis of the apparent facile isomer inversion observed at 77°K (38).

Fessenden and Schuler (38) observed an epr signal assigned to the vinyl radical by radiolysis (2.8 MeV electrons) of ethylene at $-180°C$. They found a four line pattern with apparent coupling constants of 102.44 and 13.39 G. They concluded that the four line pattern represents the outer lines of a more complete pattern. The unobserved central lines would be a four line pattern if the β-hydrogens are nonequivalent, and a doublet if they are equivalent. The outer lines are unchanged. The lack of detection of these lines was related to an intermediate rate for interconversion (between 3.3×10^7 and 3.3×10^9 sec^{-1}) which results in line broadening. The inversion barrier was estimated to be ~ 2 kcal/mole which is a minimum value since the lifetimes of the individual forms probably are limited by tunneling through the true barrier.

By assuming that the magnitudes of the two β-hydrogen coupling constants in the vinyl radical are in the same ratio as in the α-methylvinyl radical, they obtain coupling constants of 65, 37, and 13.39 G for the trans-β-H, cis-β-H, and α-H, respectively, which compare with values of 68, 34, and 16 G reported by Cochran, Adrian, and Bowers (37).

Fessenden (39) observed the α-^{13}C splitting in the vinyl radical to be 107.57 G by radiolysis of an ethylene-1-^{13}C-ethane mixture at 178°C. The magnitude of the α-^{13}C splitting was related to the amount of s-character in the orbital containing the unpaired electron by the theory of Karplus and Fraenkel (40). The amount of s-character (6.5%) in turn was related to the angle of bend which was estimated to be 151°.

Kasai and Whipple (41) studied the epr spectrum of the vinyl radical in a neon matrix by the vapor phase photolysis of the iodide and rapid trapping near 4°K. Highly oriented substitution of the vinyl radical into the neon matrix occurs. The β-hydrogens are magnetically equivalent on the time scale of the measurement (10^{-8} sec).

A study on the α-methylvinyl radical was carried out by Fessenden and Schuler (38) by radiolysis of a 2.5 mole % solution of allene in ethane at $-172°C$. A sixteen line pattern made up of four groups of four lines each is observed which is the expected result for a five proton system containing three equivalent protons. The calculated coupling constants are $a^{CH_3} = 19.48$ G, $a_\beta^H = 57.89$ G, and $a_\beta^{H'} = 32.92$ G. A bent structure for the radical is concluded based on the nonequivalence of the β-H coupling constants. The barrier to inversion in this radical is somewhat greater than the 2 kcal/mole estimate for the simple vinyl radical. A larger barrier for the α-methyl system is to be expected if tunneling of the α-H is important in determining the apparent barrier in the vinyl radical.

Fenistein, Marx, Moreau, and Serre (42) examined the epr spectrum of the α-cyanovinyl radical formed by ^{60}Co radiolysis of α-chloracrylonitrile at several temperatures. At 4.2°K the spectrum is composed of four lines which corresponds to two nonequivalent protons with coupling constants of $a_\beta{}^H = 60$ G and $a_\beta{}^{H'} = 42$ G. At 77°K the spectrum appears as a dissymmetric triplet, indicating equivalent protons having a coupling constant of $a_\beta{}^H = 48.2 \pm 0.1$ G. In addition, each line is further split into a triplet by interaction with the nuclear spin of nitrogen with a coupling constant of $a^N = 3.5$ G. The authors conclude that the radical is formed and trapped in a bent configuration at 4.2°K because of a barrier of ca. 2.5 kcal/mole to

TABLE 5
EPR Data on Vinyl Radicals

Vinyl radical	Conditions	Coupling constants (Gauss)	Ref.
H′ ＼ C=Ċ ／ ＼ H H	4.2°K, argon matrix	$a_\alpha{}^H = 16$, $a_\beta{}^H = 68$, $a_\beta{}^{H'} = 34$	37
	93°K, liquid ethylene	$a_\alpha{}^H = 13.39$, $a_\beta{}^H = 65$, $a_\beta{}^{H'} = 37$	38
	4°K, neon matrix	$a_\alpha{}^H = 13.3$, $a_\beta{}^H = a_\beta{}^{H'} = 51.5$	41
H ＼ C=^{13}Ċ ／ ＼ H H	95°K, liquid ethylene–ethane	$a^{13}C = 107.57$	39
H′ ＼ C=Ċ ／ ＼ H CH₃	101°K, liquid ethane	$a^{CH_3} = 19.48$, $a_\beta{}^H = 57.89$, $a_\beta{}^{H'} = 32.92$	38
H′ ＼ C=Ċ⟶CN ／ H	4.2°K, zeolite	$a_\beta{}^H = 60$, $a_\beta{}^{H'} = 42$	42
	77°K, zeolite	$a_\beta{}^H = a_\beta{}^{H'} = 48.2$	
H′ ＼ C=Ċ—C≡CH ／ H	4°K, argon matrix	$a_\beta{}^H = a_\beta{}^{H'} = 43.4$, $a^H = 13.7$	43

reorganization into the linear form. At 77°K the radical rapidly converts into the more stable linear form (eq. 16).

$$CH_2=C{\overset{Cl}{\underset{CN}{<}}} + e^- \longrightarrow CH_2=\dot{C}\!\sim\!CN + Cl^- \tag{16}$$

The structurally similar butatrienyl radical was studied by Kasai, Skattebøl, and Whipple (43) by photolysis of a mixture of 1% hydrogen iodide, diacetylene, and argon during deposition near 4°K (eq. 17). The spectrum was analyzed in terms of a three-proton system containing two equivalent protons in the linear structure. The isotropic coupling constants are $a_\beta^H = a_\beta^{H'} = 43.4$ G and $a_\gamma^H = 13.7$ G.

$$HC\equiv C\!-\!C\equiv CH + H^\cdot \longrightarrow \overset{H'}{\underset{H}{>}}C=\dot{C}\!-\!C\equiv CH \tag{17}$$

The epr data on vinyl radicals discussed above is summarized in Table 5.

V. CALCULATIONS

While simple π-electron Hückel theory is successful in predicting un-paired electron distributions in π-systems (44), the method does not give satisfactory results with σ-systems (those where the hybrid orbital containing the odd electron has some s-character—such as the vinyl radical in the bent form). In general, for the latter, a full valence electron (inclusion of the s-electrons in the calculation) treatment is necessary. In order to make such calculations manageable, approximations are used. These semiempirical calculations usually are based on the Hartree-Fock-Roothaan equations for molecular orbitals taken as a linear combination of atomic orbitals (LCAO) (45) and use the iterative self-consistent field method (SCF) to minimize the total electronic energy and so obtain good approximations of the wave functions. The several methods differ from one another in the variety and value of the interaction terms present as parameters.

Several calculations have been carried out on the vinyl radical (28). Petersson and McLachlan (46) and Drago and Peterson (47) independently carried out extended Hückel-type calculations (48). The latter find an energy minimum at an angle of bend of $\theta = 155°$ using the bond length parameters of ethylene. The barrier to inversion by this calculation is predicted to be 1.4 kcal/mole.

$$
\begin{array}{c}
\text{H}' \\
\diagdown \\
\text{C}=\text{C} \\
\diagup \qquad \diagdown \\
\text{H} \qquad \qquad \text{H}
\end{array}
$$

θ

28

A similar type calculation by Yonezawa and co-workers (49) led to coupling constants of $a_\beta{}^{H'} = 20.98$, $a_\beta{}^H = 40.74$, and $a_\alpha{}^H = 11.18$ G as compared with values of McLachlan (calculated at $\theta = 146°$) fo 32.9, 66.8, and 13.6 G, respectively. A separate LCAO MO approach by Dixon predicts coupling constants of 52.3, 82.0, and 14.5, respectively, at $\theta = 150°$ (50). A similar value for θ is derived from a calculation based on a method developed by Berthier (51).

Pople and co-workers (52) recently reported an SCF-LCAO MO calculation based on an extension of the CNDO (complete neglect of differential overlap) approximation method (53). This variation, called INDO (intermediate neglect of differential overlap), leads to proton coupling constants of 21.2., 55.1 and 17.1 G for β-H', β-H, and α-H, respectively, and ^{13}C coupling constants of $+178$ and -14.5 G for the α and β carbons. The experimental values for the latter are $+107.57$ and -8.55 G (39). This calculation probably is not the best one possible with this method since a standard C—C bond length of 1.40 Å and an angle of 120° were used. Further optimization of these parameters might lead to better agreement with experiment.

Other SCF-LCAO MO calculations on the α-cyanovinyl and butatrienyl radicals (42) give coupling constants in good agreement with the experimental values (42,43) when using linear structures for these systems.

We (54) and Kopchik and Kampmeier (13) have carried out semiempirical calculations in an attempt to predict the effect of various substituents on the structures of vinyl radicals. While we utilized the CNDO method (53), Kopchik and Kampmeier used the extended Hückel approach (48). Computer programs for both methods are available through the Quantum Chemistry Program Exchange, Indiana University.

Our results appear in Table 6 and are graphically summarized in Figure 3. For the simple vinyl radical the CNDO method predicts an energy minimum near 160° and a low barrier to inversion (< 1 kcal/mole). The α-fluorovinyl radical has an energy minimum near 140° with a barrier to inversion of close to 6 kcal/mole. The introduction of a phenyl group in the β-position raises the barrier by a few kcal/mole. The structure with the fluorine and phenyl *trans* to one another is predicted to be more stable than the *cis* by about 1 kcal/mole.

The energy profile of the α-phenylvinyl radical resembles a fairly steep well centered around a linear structure ($\theta = 180°$) having the aromatic

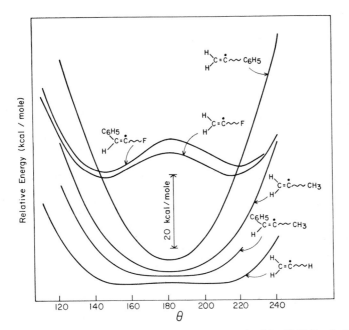

Figure 3. Total energy profiles of vinyl radicals as determined by CNDO calculations.

π-system in good position to overlap with the p-orbital containing the unpaired electron ($\alpha = 90°$). This interaction seems to be significant since rotation around the $C-C_6H_5$ bond shows an energy maximum of 17 kcal/mole when the aromatic π-system is orthogonal to the p-orbital. Even in a bent configuration ($\theta = 120°$), a lower energy is found for $\alpha = 90°$ relative to $\alpha = 0°$ although the difference is less (~ 8 kcal/mole).

These results can be compared to those cited by Simamura (2). He reports that simple Hückel theory indicates that the stabilization gained by delocalization in the structure where $\theta = 180°$ and $\alpha = 90°$ only slightly exceeds the stabilization lost by severance of the conjugation of the phenyl with the double bond. Our results suggest that the delocalization effect is significant.

The above results on the vinyl, α-fluorovinyl, and α-phenylvinyl radicals are in general agreement with those obtained using other methods. Some earlier results on the vinyl radical already have been mentioned (46–51). In addition, Kopchik and Kampmeier (13), using extended Hückel theory (Table 7), find an energy minimum for the α-vinylvinyl radical (a model for the α-phenylvinyl) in a linear structure having good interaction between the p-orbital containing the unpaired electron and the π-system of the α-vinyl

TABLE 6
CNDO Calculations on Vinyl Radicals

System	Bond lengths (Å)		θ (degrees)	Total energy in atomic units (1 a.u. = 27.21 eV)	Relative energy (kcal/mole)
H₂C=Ċ—H (θ)	C=C	1.353	110	−16.03967	220
	C—H	1.071	120	−16.05488	12.4
			130	−16.06518	5.93
			140	−16.07128	1.53
			150	−16.07408	0.40
			160	−16.07472	(0.00)
			170	−16.07441	0.19
			180	−16.07417	0.34
H₂C=Ċ—F (θ)	C=C	1.333	110	−43.04573	22.1
	C—H	1.080	120	−43.06505	10.0
	C—F	1.348	130	−43.07635	2.94
			140	−43.08106	(0.00)
			150	−43.08080	0.16
			160	−43.07752	2.84
			170	−43.07371	4.60
			180	−43.07193	5.70
C_6H_5(H)C=Ċ—F (θ)	C=C	1.333	110	−88.74683	22.60
	C=C	1.393	120	−88.76686	10.10
	C—H	1.080	130	−88.77848	2.85
	C—C_6H_5	1.450	140	−88.78304	(0.00)
	C—F	1.348	150	−88.78222	0.51
			160	−88.77791	3.20
			170	−88.77267	6.45
			180	−88.76957	8.40
			200	−88.77495	5.10
			210	−88.77956	2.18
			220	−88.78127	1.10
			230	−88.77703	3.74

(styrene vinyl radical structure with θ and α)

($\alpha = 0$ when ring is in molecular plane)
C=C   1.340
C=C   1.430 (benzene)
C—H   1.100
C—C_6H_5 1.355

			θ	Total energy	Relative energy
at $\alpha = 90°$			180	−61.825775	(0.00)
			160	−61.815765	6.25
			140	−61.782168	27.4
			120	−61.728791	60.5

Effect of rotation around C—C_6H_5

			θ	Total energy	Relative energy
at $\alpha = 90°$			180	−61.825775	(0.00)
$\alpha = 45°$			180	−61.812675	8.20
$\alpha = 0°$			180	−61.797913	17.40

$\alpha = 90°$	120	−61.768782	35.60
$\alpha = 45°$	120	−61.763201	39.20
$\alpha = 0°$	120	−61.755431	44.00

C—H (CH$_3$) 1.070
C—H (vinyl) 1.100
C—CH$_3$ variable as indicated
C=C variable as indicated

Optimization of C—CH$_3$ using C=C as 1.353 Å,[a] $\theta = 180°$

1.375	−24.83165
1.425	−24.83255
1.458	−24.827950
1.488	−24.821092
1.540	−24.803131

Optimization of C=C using C—CH$_3$ as 1.459 Å,[b] $\theta = 180°$

1.250	−24.835507
1.300	−24.839036
1.353	−24.827950

Effect of bending using C—CH$_3$ as 1.400 Å and C=C as 1.275 Å

180	−24.843009	(0.00)
170	−24.842043	0.61
160	−24.838898	2.56
150	−24.832894	6.35
140	−24.823105	12.35
130	−24.808533	21.6
120	−24.788167	34.4

C—CH$_3$	1.488	120	−70.48426	24.00
C=C	1.353	140	−70.51209	6.25
C—H (vinyl)	1.080	150	−70.51860	2.50
C—H (methyl)	1.101	160	−70.52171	0.60
C—C (benzene)	1.393	170	−70.52271	(0.00)
		180	−70.52271	(0.00)
		190	−70.52248	0.14
		200	−70.52132	0.87
		210	−70.51831	2.76
		220	−70.51064	7.50

[a] A graphical analysis of the data indicates an energy minimum near 1.400 Å.
[b] A graphical analysis of the data indicates an energy minimum near 1.275 Å.

group. Their calculations on the α-fluorovinyl radical indicate the bent structure ($\theta = 120°$) is more stable than the linear by 27 kcal/mole (55). An α-hydroxy substituent gives an energy preference for bent ($\theta = 120°$) over linear geometry of 23 kcal/mole in a similar calculation (55). CNDO calculations of an energy/geometry profile for the α-methoxyvinyl radical reveal an energy minimum at 148° with a difference of 2.4 kcal/mole between the minimum and a linear structure (56).

The CNDO results on the α-methylvinyl system (Table 6) surprised us. This method predicts an energy minimum at 180°. The energy profile appears as a more flat-bottomed well as opposed to the steep well of the α-phenyl system. Introduction of a phenyl group in the β-position flattens the bottom more but the minimum clearly is still at 180°. In this system we went through a bond length optimization procedure for the C=C and C—CH₃ bonds. In contrast, the calculations of Kopchik and Kampmeier (13) reveal an energy minimum near 140° with a predicted barrier to inversion of less than 3 kcal/mole.

The available epr data on the α-methylvinyl radical (38) indicate a bent structure for this radical and our kinetic results suggest a similar structure for the α-methyl-β-phenylvinyl radical (20).

TABLE 7
Extended Hückel Calculations on Vinyl Radicals (13)

System	Bond lengths (Å)		θ (degrees)	Total energy in atomic units (1 a.u. = 27.21 eV)	Relative energy (kcal/mole)
H \\ C=Ċ~H / H	C=C C—H	1.34 1.10	120 135 150 165 180	−6.61965 −6.62567 −6.62732 −6.62714 −6.62688	4.82 1.04 (0.00) 1.13 2.77
H \\ C=Ċ~CH₃ / H	C=C C—H C—CH₃	1.34 1.10 1.54	90 100 110 120 130 140 150 160 170 180	−10.13424 −10.17778 −10.20149 −10.21328 −10.21822 −10.21926 −10.21829 −10.21663 −10.21523 −10.21476	53.44 26.08 11.17 3.76 1.71 (0.00) 0.61 1.66 2.54 2.83
H \\ C=Ċ~CH=CH₂ / H	C=C C—H C—C₂H₃ α 0° 0° 90°	1.34 1.10 1.54	When $\alpha = 0°$, two π bonds are in same plane 120 180 180	 −12.67478 −12.67410 −12.68671	 7.49 7.92 (0.00)

TABLE 8
Experimental and Theoretical Approaches to the Structures of
Selected Vinyl Radicals

System	Experimental		Theoretical	
	EPR	Kinetic	EHM	CNDO
$H_2C{=}\dot{C}{\sim}H$	Bent	—	Bent	Bent
$H_2C{=}\dot{C}{\sim}CH_3$	Bent	Bent[a]	Bent	Linear
$(C_6H_5)(H)C{=}\dot{C}{\sim}CH_3$	—	Bent	—	Linear
$H_2C{=}\dot{C}{\sim}F$	—	—	Bent	Bent
$(C_6H_5)(H)C{=}\dot{C}{\sim}F$	—	Bent[b]	—	Bent
$H_2C{=}\dot{C}{\sim}OR$	—	Bent[c]	Bent[d]	Bent[e]
$H_2C{=}\dot{C}{\sim}C_6H_5$	—	Linear[f]	Linear[g]	Linear

[a] Based on work on the α-ethyl-β-ethylvinyl radical, Refs. 32 and 33.
[b] Based on results of the α-chloro and α-bromo systems, Refs. 29 and 30.
[c] Based on work in Ref. 11b.
[d] R = H.
[e] R = CH_3.
[f] Based on results of the α-phenyl-β-phenyl- and α-phenyl-β-methylvinyl radical systems, Refs. 20 and 22.
[g] Using the α-vinylvinyl radical as a model, Ref. 13.

A comparison of experimental and theoretical approaches to the stable structures of some selected vinyl radicals appears in Table 8. Except for the α-methyl systems, the agreement is good. What is lacking are epr data on vinyl radicals with phenyl and halogen as α-substituents. The theoretical predictions appear to be straightforward in these two cases.

It is not within the cognizance of this writer to comment on why the extended Hückel and CNDO methods give opposite predictions of structure for the α-methylvinyl radical. The latter has a good history for predicting structures and physical parameters for small, closed-shell molecules (53). Until the limitations inherent in calculations on open-shell molecules by semiempirical methods are better defined, we caution users of these available programs to be conservative in their conclusions.

Acknowledgments

I wish to thank my co-workers in this area, Dr. N. P. Kong and Mrs. J. Chen, for their efforts and the National Science Foundation for financial assistance in support of our studies.

References

1. W. G. Bentrude, *Ann. Rev. Phys. Chem.*, **18**, 300 (1967).
2. O. Simamura, in N. L. Allinger and E. L. Eliel, Eds., *Topics in Stereochemistry*, Vol. 4, Wiley-Interscience, New York, 1970, chap. 1.
3. The terms stereoselectivity and stereospecificity are used as defined by H. E. Zimmerman, L. Singer, and B. S. Thyagarajan, *J. Amer. Chem. Soc.*, **81**, 108 (1959), footnote 16.
4. C. Walling, M. S. Kharasch, and F. R. Mayo, *J. Amer. Chem. Soc.*, **61**, 1711 (1939).
5. H. L. Goering and D. W. Larson, *J. Amer. Chem. Soc.*, **81**, 5937 (1959).
6. P. I. Abell and L. H. Piette, *J. Amer. Chem. Soc.*, **84**, 916 (1962).
7. P. S. Skell and R. G. Allen, *J. Amer. Chem. Soc.*, **80**, 5997 (1959); *ibid.*, **86**, 1559 (1964).
8. (a) A. A. Oswald and K. Griesbaum, in N. Kharasch and C. Meyers, Eds., *The Chemistry of Sulfur Compounds*, Vol. 2, Pergamon, New York, 1966, chap. 9.
   (b) R. M. Kellog, in E. Huyser, Ed., *Methods in Free Radical Chemistry*, Dekker, New York, 1969.
9. S. Groen, R. M. Kellog, J. Buter, and H. Wynberg, *J. Org. Chem.*, **33**, 2218 (1968).
10. A. A. Oswald, K. Griesbaum, B. E. Hudson, and J. M. Bregman, *J. Amer. Chem. Soc.*, **86**, 2877 (1964).
11. (a) J. A. Kampmeier and G. Chen, *J. Amer. Chem. Soc.*, **87**, 2608 (1965).
    (b) D. K. Wedegaertner and J. A. Kampmeier, Unpublished results.
12. E. I. Heiba and R. M. Dessau, *J. Amer. Chem. Soc.*, **88**, 1589 (1966); *ibid.*, **89**, 3772 (1967).
13. R. M. Kopchik and J. A. Kampmeier, *J. Amer. Chem. Soc.*, **90**, 6733 (1968).
14. E. I. Heiba and R. M. Dessau, *J. Amer. Chem. Soc.*, **89**, 2238 (1967).

15. (a) R. A. Benkeser and R. A. Hickner, *J. Amer. Chem. Soc.*, **80**, 5298 (1958). (b) R. M. Benkeser, M. L. Burrows, L. E. Nelson, and J. V. Swisher, *J. Amer. Chem. Soc.*, **83**, 4385 (1961).

16. (a) A. J. Leusink and H. A. Budding, *J. Organometal. Chem.*, **11**, 533 (1968). (b) A. A. Leusink, H. A. Budding, and W. Drenth, *J. Organometal. Chem.*, **11**, 541 (1968).

17. R. Sommer and H. G. Kuivila, *J. Org. Chem.*, **33**, 802 (1968).

18. P. D. Bartlett and R. R. Hiatt, *J. Amer. Chem. Soc.*, **80**, 1398 (1958).

19. L. A. Singer, in D. Swern, Ed., *Organic Peroxides*, Wiley-Interscience, New York, 1970, chap. 5.

20. L. A. Singer and N. P. Kong, *Tetrahedron Lett.*, **1966**, 2089; *J. Amer. Chem. Soc.*, **88**, 5213 (1966).

21. J. A. Kampmeier and R. M. Fantazier, *J. Amer. Chem. Soc.*, **88**, 1959 (1966).

22. L. A. Singer and J. Chen, *Tetrahedron Lett.*, **1969**, 4849.

23. K. W. Doak, *J. Amer. Chem. Soc.*, **72**, 4681 (1950).

24. M. Gazith and M. Szwarc, *J. Amer. Chem. Soc.*, **79**, 3339 (1957).

25. G. E. Owen, Jr., J. M. Pearson, and M. Szwarc, *Trans. Faraday Soc.*, **61**, 1722 (1965).

26. N. Wada, K. Tokumaru, and O. Simamura, *Abstracts, 21st Annual Meeting of the Chemical Society of Japan*, Vol. 3, Osaka, Japan, April 1968, p. 2069.

27. As reported by M. J. Perkins, in B. Capon et al., Eds., *Organic Reaction Mechanisms 1968*, Wiley, London, 1969, chap. 9.

28. (a) C. Walling, *Free Radicals in Solution*, Wiley, New York, 1957, p. 152. (b) R. F. Bridger and G. A. Russell, *J. Amer. Chem. Soc.*, **85**, 3754 (1963).

29. L. A. Singer and N. P. Kong, *J. Amer. Chem. Soc.*, **89**, 5251 (1967).

30. L. A. Singer and N. P. Kong, *Tetrahedron Lett.*, **1966**, 5141.

31. O. Simamura, K. Tokumaru, and H. Yui, *Tetrahedron Lett.*, **1966**, 5141.

32. G. D. Sargent and M. W. Brown, *J. Amer. Chem. Soc.*, **89**, 2789 (1967).

33. A. J. Fry and M. A. Mitnick, *J. Amer. Chem. Soc.*, **91**, 6207 (1969).

34. R. C. Neuman, Jr., and G. D. Holmes, *J. Org. Chem.*, **33**, 4317 (1968).

35. H. G. Kuivila, *Accounts Chem. Res.*, **1**, 299 (1968).

36. G. M. Whitesides and C. P. Casey, *J. Amer. Chem. Soc.*, **88**, 4541 (1966).

37. E. I. Cochran, R. J. Adrian, and V. A. Bowers, *J. Chem. Phys.*, **40**, 213 (1964).

38. R. W. Fessenden and R. H. Schuler, *J. Chem. Phys.*, **39**, 2147 (1963).

39. R. W. Fessenden, *J. Phys. Chem.*, **71**, 74 (1967).

40. M. Karplus and G. K. Fraenkel, *J. Chem. Phys.*, **35**, 1312 (1961).

41. P. H. Kasai and E. H. Whipple, *J. Amer. Chem. Soc.*, **89**, 1033 (1967).

42. S. Fenistein, R. Marx, C. Moreau, and J. Serre, *Theor. Chim. Acta*, **14**, 339 (1969).

43. P. H. Kasai, L. Skattebøl, and E. B. Whipple, *J. Amer. Chem. Soc.*, **90**, 4509 (1968).

44. For example, see A. Carrington, *Quart. Rev.*, **17**, 67 (1963).

45. C. C. J. Roothaan, *Rev. Mod. Phys.*, **23**, 69 (1951).

46. G. A. Petersson and A. D. McLachlan, *J. Chem. Phys.*, **45**, 628 (1966).

47. R. S. Drago and H. Peterson, Jr., *J. Amer. Chem. Soc.*, **89**, 5774 (1967).

48. R. Hoffman and W. N. Lipscomb, *J. Chem. Phys.*, **36**, 2179, 3489 (1962).

49. T. Yonezawa, H. Nakatsumi, T. Kawamura, and H. Kato, *Bull. Chem. Soc., Jap.*, **40**, 2211 (1967).

50. W. T. Dixon, *Mol. Phys.*, **9**, 201 (1965).

51. Y. Ellinger, A. Rassat, R. Subra, and G. Berthier, *Theor. Chim. Acta*, **10**, 289 (1968).

52. J. A. Pople, D. L. Beveridge, and P. A. Dobosh, *J. Amer. Chem. Soc.*, **90**, 4201 (1968).

53. J. A. Pople and G. A. Segal, *J. Chem. Phys.*, **43**, S136 (1965); *ibid.*, **44**, 3289 (1966).
54. L. A. Singer and J. Chen, Unpublished results.
55. R. M. Kopchik and J. A. Kampmeier, Unpublished results.
56. D. K. Wedegaertner, Unpublished results.

Stereochemistry of Hofmann Eliminations

JAMES L. COKE

Department of Chemistry, University of North Carolina,
Chapel Hill, North Carolina

I. INTRODUCTION

Few areas of mechanistic chemistry have received as much attention over the years as 1,2-eliminations. Much of the early work was concerned with direction of elimination (i.e., Hofmann vs. Saytzeff orientation) and whether a given reaction followed a unimolecular or bimolecular mechanism. The question of stereochemistry of eliminations was somewhat neglected until conformational concepts had been well developed. This is unfortunate since it has frequently resulted in mechanistic proposals without regard to stereochemistry.

It is the purpose of this chapter to examine some recent evidence on the stereochemistry of leaving groups in Hofmann eliminations of trimethyl-ammonium hydroxides or alkoxides. Because elimination reactions have been reviewed in several places (1–8), the early literature will not be covered. However, it will be necessary to review a few cases that led to the general thinking that prevailed until the last few years. It is hoped that by reviewing the methods and compounds used in recent work and the results obtained, future mistakes may be prevented and some misconceptions may be cleared up. This should allow development of a theoretical concept of the Hofmann elimination mechanism consistent with what is now known about the stereochemistry of reaction.

II. GENERAL CONSIDERATIONS

A. Terminology

Because the geometry of departing groups in a 1,2-elimination is a conformational concept rather than a configurational one, it is more appropriate to use the conformational terms *syn* and *anti* (9) rather than the old terminology *cis* and *trans*. The terms *syn* (**2**) and *anti* (**1**) will be used in this chapter to describe the departing groups and thus the stereochemistry of the

mechanism, while the configurational terms *cis* and *trans* or *erythro* and *threo* will be used to describe the stereochemistry of the starting molecules and the olefin products. This chapter will deal primarily with the stereochemistry of the elimination mechanism, but many recent stereochemical investigations have also shed light on other aspects of the problem.

The possible mechanisms for Hofmann elimination, with regard to molecularity, have been discussed in great detail (1–8). It seems reasonably safe to say that E1 (carbonium ion) mechanisms do not operate in the systems which will be discussed in this chapter. It will be assumed that as long as a kinetic isotope effect for loss of hydrogen is observed and no exchange of deuterium or hydrogen in the starting material is found, that an E1cb (carbanion) mechanism is not operating. These assumptions should not be taken as rigorously proven in all cases and any new system should be examined with caution.

This leaves two fundamental mechanisms to be considered for Hofmann eliminations. The first is the E2 mechanism which is a concerted but not necessarily synchronous breakage of the carbon–hydrogen and carbon–nitrogen bonds. The E2 mechanism can follow *syn* stereochemistry to give *cis* olefin (eq. 1) or *trans* olefin (eq. 2) or *anti* stereochemistry to give *cis* olefin (eq. 3) or *trans* olefin (eq. 4).

The second possibility is the ylid mechanism that involves the reversible removal of an α-proton followed by unimolecular decomposition of the α-carbanion (eq. 5). If the ylid mechanism is intramolecular in the elimination step and if breakage of the carbon–hydrogen and carbon–nitrogen bonds

are concerted, then the stereochemistry of the elimination step must be *syn*. This kind of elimination could still produce *cis* or *trans* olefin products. One major difference between a *syn* E2 and *syn* ylid mechanism is that in the former the β-proton lost in the elimination step is found in the solvent or protonated base after reaction whereas in the latter mechanism the β-proton is found in the trimethylamine. This allows deuterium labeling to distinguish easily between these two mechanisms.

In most elimination reactions it is generally thought that a coplanar conformation of the leaving groups (dihedral angle of 0° or 180°) is more favorable than a noncoplanar arrangement regardless of whether the stereochemistry is *syn* or *anti* (10,11). This is of considerable significance in Hofmann elimination of trimethylammonium salts because there is always a side reaction competitive with olefin formation. This is direct nucleophilic attack by base on the carbon of a methyl group to generate N,N-dimethylalkylamine (eq. 6). Depending on the conditions used for elimination, this

$$
\begin{array}{c}
\text{Me} \\
\mid \\
\text{R---N}^+\text{---Me} \longleftarrow :\text{B} \longrightarrow \text{R---NMe}_2 + \text{Me---B} \\
\mid \\
\text{Me}
\end{array}
\qquad (6)
$$

side reaction may be a minor or major reaction path, but if coplanarity cannot be achieved it may be the only reaction observed.

B. Early Examples

In many molecules the stereochemistry of Hofmann elimination can be determined from the geometry of the olefin product. This is possible when there is one hydrogen and two other different types of substituents on the β-carbon. This particular type of model was the one used in most early work on Hofmann eliminations. Unfortunately, the substituents needed to construct such models introduce severe steric or electronic factors that influence the stereochemistry of elimination in ways difficult to predict. It is becoming increasingly apparent that extreme caution should be used in arriving at general conclusions from a given model compound.

One of the earliest and most widely quoted examples is that of the *erythro* and *threo* isomers of N,N,N-trimethyl-1,2-diphenylpropylammonium ethoxide (12). The *threo* isomer **3** gives only *trans*-α-methylstilbene on elimination (eq. 7) while the erythro isomer **4** gives only *cis*-α-methylstilbene (eq. 8). These results are only consistent with an *anti* mechanism operating

$$ (7) $$

$$ (8) $$

on compounds **3** and **4**. It should be noted that both **3** and **4** would have very severe eclipsing strain in the conformation needed for *syn* elimination. In addition, the phenyl groups introduce electronic effects that make it difficult to extrapolate the above results to simple aliphatic systems.

Two other examples often cited are the *cis* and *trans* isomers of *N,N,N*-trimethyl-2-phenylcyclohexylammonium hydroxide (**5** and **6**, respectively) (13). The *cis* isomer **5** undergoes almost exclusive *anti* elimination to give 1-phenylcyclohexene (eq. 9). The *trans* isomer **6** also gives 1-phenylcyclo-hexene as a product (eq. 10), and this was proposed to be the result of re-

$$\text{(9)}$$

5

$$\text{(10)}$$

6

arrangement of initially formed 3-phenylcyclohexene (13). However, it was shown that 1-phenylcyclohexene is the initial product in eq. 10 and that it must arise by a *syn* mechanism (14,15). The possibility of an ylid mechanism or an E1cb mechanism with reversion of an intermediate carbanion back to starting material in eq. 10 were both excluded using deuterium labeling (16) and it was concluded that compound **6** most likely undergoes a *syn* E2 elimination although with probably a high degree of carbanion character. The observation (17) that compound **5** reacts 133 times as fast as compound **6** tended to support the generally held view that *anti* elimination was the preferred stereochemistry for Hofmann elimination.

Another set of examples in which the electronic effect of the phenyl group is absent is the case of the *cis* and *trans* isomers of *N,N,N*-trimethyl-2-methylcyclohexylammonium hydroxide (**7** and **8**, respectively) (18). The *cis* isomer **7** on heating gives 95% 1-methylcyclohexene (eq. 11) and this can

$$\text{(11)}$$

7

only occur by *anti* elimination. The *trans* isomer **8** gives pure 3-methylcyclo-hexene (eq. 12). The stereochemistry of the mechanism in eq. 12 is unknown but it is quite unlikely that compound **8** undergoes a *syn* or *anti* elimination from a normal chair conformation. The *syn* mechanism would have to be a noncoplanar arrangement and the *anti* mechanism would involve placement of both the methyl group and the trimethylammonium group in axial conformations, to the extent that the transition state resembles the ground state.

$$\text{(12)}$$

It is much more likely that compound **8** reacts in a twist conformation (19) and the mechanism is most likely *anti* although this stereochemistry has not been determined.

The results in eq. 11 are unusual in that the most highly substituted olefin is the product. This result has been rationalized in terms of steric crowding of the trimethylammonium group in an axial conformation next to an equatorial methyl, causing the carbon–nitrogen bond to be much more broken in the transition state and thereby making reaction of compound **7** a more E1-like reaction than a normal Hofmann (18,20). An alternate explanation has been offered (19) which predicts a more normal transition state. This assumes that compound **7** reacts in a twist form in which the only coplanar β-hydrogen is the *trans* hydrogen on the ring carbon bearing the methyl group. Thus, since coplanar elimination is favored, the product is expected to be 1-methylcyclohexene. This example will be discussed in greater detail later. The early work on 2-methylcyclohexyl models was generally regarded as more evidence that Hofmann eliminations tended to follow *anti* stereochemistry.

There are examples of *syn* Hofmann elimination in the early literature but they tend to be relatively rigid systems in which *anti* coplanar elimination is prohibited by bridged systems. One of these employing deuterium labeling is N,N,N-trimethyl-3-*exo*-d_1-bicyclo[2.2.1]heptyl-2-*exo*-ammonium hydroxide (**9**) which was observed to undergo complete loss of deuterium on elimination (21) (eq. 13). These results are only compatible with a *syn* elimination but the

$$\text{(13)}$$

question of whether it is an E2 or ylid mechanism was not investigated.

Even though compound **6** undergoes a *syn* elimination that is not an ylid mechanism, it bears a phenyl group at one of the reaction sites and so it was not clear at the time whether *syn* eliminations, if they occur in systems containing only alkyl or hydrogen substituents, would be *syn* ylid or *syn* E2 mechanisms. One attempt to answer this involved the use of a compound which, because of steric crowding, exists in a conformation that should only allow *syn* elimination. It was observed that N,N,N-trimethyl-2-t-butyl-3,3-dimethyl-2-d_1-butylammonium hydroxide (**10**) gives approximately 75% trimethylamine-d_1 by an ylid mechanism (22) (eq. 14).

$$t\text{—Bu}—\overset{\overset{\displaystyle D}{|}}{\underset{\underset{\displaystyle t\text{—Bu}}{|}}{C}}—CH_2—\overset{+}{N}Me_3\overset{-}{O}H \longrightarrow t\text{—Bu}—\overset{}{\underset{\underset{\displaystyle t\text{—Bu}}{|}}{C}}{=}CH_2 + CH_2D—NMe_2 + NMe_3 \quad (14)$$

$$\qquad\qquad\qquad \textbf{10} \qquad\qquad\qquad\qquad\qquad\quad 75\% \qquad\quad 25\%$$

From the type of data illustrated by the few preceding examples there emerged a general view that Hofmann eliminations would always show *anti* stereochemistry unless special factors were present, such as activating substituents like a phenyl group or a rigid system preventing *anti* coplanarity or no *trans* β-hydrogens. Even if *syn* elimination was observed, it was often assumed that it might be largely ylid rather than E2. This assumption of preferred *anti* stereochemistry became so widely held that it found its way into virtually every major organic chemistry textbook. The vast majority of the early work on cyclic systems involved use of cyclohexane compounds as models. This is understandable because in conformational analysis the cyclohexane ring is really the standard system and steric effects are best understood in six-membered rings. However, in light of very recent work it appears unfortunate that cyclohexane models were used first. The cyclohexane ring appears to be exceptional in Hofmann eliminations and even at present it is not well understood.

C. Methods and Limitations

1. Product Stereochemistry

In much of the early work on stereochemistry of Hofmann eliminations, alkyl or aryl substituents were placed about the reaction sites in such a way as to create asymmetric centers at each carbon bearing the leaving hydrogen and nitrogen. By observing whether elimination produced *cis* or *trans* olefins from a given stereoisomer of the starting material one could deduce whether *syn* or *anti* elimination was occurring. The examples in eqs. 7 and 8 are of this type. This method is, of course, completely rigorous for the particular model used, but the limitation is that the steric or electronic factors present in the model are unique to that molecule. Thus the results of such a model cannot be used directly to predict the results expected from some other compound. This approach may give interesting results but it does little to elucidate the mechanism in simple systems.

2. Rate Profile

One of the more recent attempts to examine Hofmann eliminations has employed what can be called a "rate profile" method (23–26). This approach essentially compares the rate profile or plot of log k vs. ring size for various kinds of eliminations in cyclic compounds. The basic assumption is that eliminations that have the same stereochemistry, *syn* or *anti*, will show similar rate profiles, and reactions that do not have the same stereochemistry will show different rate profiles. This presumes that steric factors in a given ring system will affect all *syn* mechanisms or all *anti* mechanisms in the same way. While this may be an intuitively attractive hypothesis, it remains unproven.

As an illustration of the rate profile method, it was shown that a plot of log k for *trans* cycloolefin formation vs. ring size showed very similar profiles for Hofmann elimination and N-oxide pyrolysis (23). Because the latter reaction is known to follow *syn* stereochemistry (3), it was concluded that the former must also. In the same manner it was found that a plot of log k for *cis* cyloolefin formation vs. ring size showed dissimilar profiles for Hofmann elimination and N-oxide pyrolysis. It was concluded that because the latter was a *syn* elimination, the former must not be, and it therefore had to be an *anti* elimination. It was this type of reasoning coupled with some comparison of *cis* and *trans* olefin formation in Hofmann elimination with other reactions and some deuterium tracer work that led proponents of the rate profile method to conclude that *cis* olefins are formed by *anti* mechanisms and *trans* olefins are formed by *syn* mechanisms in Hofmann eliminations (26). This conclusion has since turned out to be an oversimplification and should not be used in predicting stereochemistry in a given case. There are now many examples of both *cis* and *trans* olefins formed by varying degrees of *syn* and *anti* mechanisms, and it appears that even changing the solvent or conditions can change the balance between these two mechanistic paths.

It is not surprising that the rate profile method has flaws since even if the basic assumption of similar profiles for the same stereochemistry is granted, the method by its very nature is not sensitive enough to be very quantitative. In spite of all the disadvantages of the rate profile method, it should be given a great deal of credit because it was the first major challenge (23) to the prevailing assumption that Hofmann eliminations always followed *anti* stereochemistry.

3. Deuterium Labeling

Of all the recent methods for determining stereochemistry of Hofmann eliminations, the most ideal is deuterium labeling. It introduces no steric or electronic effect. With the common accessibility of mass spectral analysis, gas chromatography, and a number of stereospecific methods for introduction of deuterium, it has become the method of choice by nearly all workers in this field. One of the advantages in a deuterium label is that not only can the stereochemistry of reaction be determined, but frequently one can obtain other evidence about the mechanism such as the degree of bond breakage in the transition state.

Attractive as the deuterium label method is, it still has several pitfalls that must be avoided. For example, if the olefinic products from Hofmann elimination are separated and analyzed for deuterium, it must be known that both the olefins and their deuterium label are stable to the reaction conditions. It is known (27) that olefins sometimes isomerize rather rapidly under strongly basic conditions similar to those used in eliminations. It is essential to look at the primary products from elimination. In general, the safest and best way to rule out isomerization or loss of deuterium after elimination is to subject each product separately to the reaction conditions and observe the purity and deuterium content as a function of time. Frequently a reaction will have some intrinsic feature that allows product isomerization to be ruled out.

In order to interpret the results of elimination on a deuterium-labeled compound in terms of the parent hydrogen analog, it is necessary to know either the *syn* or *anti* kinetic isotope effect for hydrogen and deuterium depending on the particular mechanism by which they are lost. The literature on deuterium isotope effects in Hofmann eliminations must be used with caution. The early literature simply gives isotope effects based on an assumed *anti* mechanism so that both their magnitude and significance are questionable. Even in some recent literature the values listed are implied to be kinetic isotope effects (28) when in fact they are simply the effect on elimination of substituting deuterium into a given position. It is only in the event of a pure *syn* mechanism to give *trans* olefins and a pure *anti* mechanism to give *cis* olefins that these latter values (28) become kinetic isotope effects. Kinetic isotope effects in Hofmann eliminations can be determined in cases where a single stereochemistry is operating (29) or where a mixture of *syn* and *anti* mechanisms operate simultaneously (30).

One of the major considerations in using deuterium labeling is the stereospecificity of the method used to introduce the deuterium. One method for obtaining a *cis* deuterium-labeled amine is the reduction of an epoxide with lithium aluminum deuteride, either with or without aluminum chloride, to the corresponding *trans*-labeled alcohol (28,31). The alcohol is then converted to its tosylate and displaced with some appropriate reagent, leading to inversion and resulting in a *cis*-labeled derivative which by further modification can be used in a Hofmann study. The greatest uncertainty in this method usually comes in the epoxide reduction stage. It is known that a considerable portion of the product contains two deuteriums when lithium aluminum deuteride is used alone (32). This product presumably arises by way of a ketone formed by rearrangement of epoxide or hydride transfer from some alkoxide (32). The dideuterated material not only interferes with the deuterium analysis in the Hofmann product but, more seriously, the ketone intermediate makes that portion of the product nonstereospecifically labeled. In addition there is the possibility, to an unknown extent, of the small amount of ketone and the aluminum alkoxide salts which are specifically labeled entering into an Oppenauer-type equilibrium. This type of reaction would result in scrambling the deuterium label in a portion of the alcohol product with the exact extent being unknown. In spite of whether or not the above side reactions actually occur, it is extremely difficult to rule them out experimentally. It thus appears unwise to use the above sequence to prepare *cis*-labeled amines for Hofmann studies, especially since a much more rigorous and simpler method is available.

It appears that the best method for introducing a *cis* deuterium label is the reaction of a *cis* olefin with diborane-d_6 followed by either hydroxyl-amine-*O*-sulfonic acid or chloramine (33,34). It has been shown that the initial addition of the deuterium and boron (35) and the boron to nitrogen migration of the alkyl group are stereospecific for both the hydroxyl-amine-*O*-sulfonic acid method (36) and the chloramine method (29). It should be mentioned that in medium-to-large ring compounds where both *cis* and *trans* olefins are available this method leads to a *cis* label if the *cis* olefin is used or to a *trans* label if a *trans* olefin is used.

The best method for introducing a *trans* deuterium label is the reaction of a *cis* olefin with diborane-d_6 followed by hydrogen peroxide to give a *cis*-labeled alcohol (35). The alcohol is then converted to its tosylate and displaced with some reagent, preferably sodium azide, to give inversion to a *trans*-labeled azide which on reduction will give a *trans*-labeled amine, and this can be used in a Hofmann study (28).

One advantage of the deuterium labeling method is that from a *cis*-labeled compound and the *syn* kinetic isotope effect one can calculate the amount of *syn* and *anti* mechanisms that operate on the hydrogen analog. One can also calculate the results from the *trans*-labeled compound and the *anti* kinetic isotope effect. The results from both methods should be in agreement if no errors are made at any step of the study. These two approaches allow completely independent checks on the mechanistic conclusions.

4. Reaction Conditions

One of the more perplexing problems is to try to compare results found under one set of solvent and base conditions with those found under a different set of conditions. Preparative Hofmann eliminations are usually run by pyrolysis of quaternary hydroxides, but much of the recent work on stereochemistry has also been done using various bases in a number of solvents. There have not been enough models studied thoroughly under all sets of conditions to fully clarify the effect of solvent and base. Only one kind of system, the cyclopentyl ring, has been studied rigorously enough (37,38) to conclude that the stereochemistry of Hofmann elimination is very much a function of the conditions used. It would thus appear that when conclusions about a mechanism are reached the conditions should be stated, and that caution should be used in comparing results under different conditions, especially where even the substrate or model is also different.

III. RECENT WORK

A. Bicyclic Compounds

It would seem logical that if *syn* Hofmann eliminations are to be found in simple systems, then they should most likely occur in systems where a fairly rigid *syn* coplanar arrangement of the β-carbon–hydrogen and carbon–nitrogen bonds is present and where *anti* coplanarity is difficult because of steric constraint. This has been found to be true and the bicyclo[2.2.1]heptyl system in eq. 13 is a classic example. Compound 9 is known to undergo completely *syn* elimination (21), presumably because of the difficulty in achieving an *anti* coplanar conformation of the leaving groups.

Since compound 9 reacts exclusively by *syn* elimination the question arises as to whether its mechanism is *syn* E2 or *syn* ylid, as might have been expected from the results in eq. 14. This question is answered by the results shown in eq. 15.

$$+ \ CH_2D\!-\!NMe_2 \ + \ NMe_3$$
$$\quad\quad 6\% \quad\quad\quad 94\%$$

(15)

It has been found (29,39) that compound 9 undergoes elimination on pyrolysis with a maximum of 6% ylid mechanism and even this 6% may be the result of exchange (22). That the remaining 95% of the mechanism is *syn* E2 is shown by the fact that compound 9 shows a *syn* k_H/k_D kinetic isotope effect of 1.86. This was shown in a competitive elimination between compound **9** and its 3-hydrogen analog which had been labeled with deuterium in a remote position for mass spectral identification. It thus appears that *syn* stereochemistry is not a sufficient condition for an ylid mechanism and unless special factors are present it will not be observed. Models indicate that the ylid mechanism observed in compound **10** is caused not so much by a *syn* conformation as by steric shielding of the β-hydrogen by the *t*-butyl groups. The β-hydrogen is so sterically inaccessible to external attack that the reaction proceeds mostly by internal abstraction of the β-hydrogen.

There is an inherent feature about compound **9** that creates an uncertainty. The two hydrogens on carbon-3 are different, with the *cis* hydrogen being *exo* and the *trans* hydrogen being *endo*. The point of view could be taken, as in other elimination work (40), that *exo* hydrogens are more easily abstracted and that reaction at the *exo* position giving *syn* elimination is simply a result of greater reactivity of the *exo* hydrogen rather than the constrained geometry. The fact that the *endo* isomer of the hydrogen analog of **9** gives a very poor yield of olefin on Hofmann elimination (41) could be taken as support for this view.

A model similar to **9** clears up the ambiguity and this is N,N,N-trimethyl-3-*cis*-d_1-bicyclo[2.2.2]octyl-2-ammonium hydroxide (**11**) which is slightly more flexible than **9** but should still be constrained enough to prevent an *anti* coplanar elimination (42). Compound **11** has been found (29) to undergo a completely *syn* elimination on pyrolysis (eq. 16) with a maximum of 13% *syn* ylid mechanism and even this small amount may be the result of exchange

$$
\begin{array}{ccc}
 & + \ CH_2D-NMe_2 & + \ NMe_3 \\
 & 13\% & 87\%
\end{array}
$$

$$\text{(16)}$$

(22). There is no inherent structural difference between the two hydrogens on carbon-3 except that one is *cis* and one is *trans* to the trimethylammonium group, and it would thus appear that constrained geometry favoring a *syn* coplanar conformation is a sufficient condition for *syn* E2 Hofmann eliminations.

At the time the above work was done it appeared that there was not as much of an energy difference between *syn* and *anti* E2 Hofmann eliminations as had been generally thought and that *syn* E2 mechanisms probably were reasonably common. The reason they had not been detected except in special cases was that models used either favored the *anti* mechanisms for steric reasons or were incapable of detecting the *syn* mechanism.

B. Five-Membered Rings

In principle, any *syn* elimination occurring in small ring compounds can be detected using a *cis* deuterium label on the β-carbon. In simple parent ring systems one must know the *syn* k_H/k_D kinetic isotope effect in order to calculate what happens in the hydrogen analog from what is observed with the deuterated analog. These kinetic isotope effects are usually difficult or impossible to obtain directly. One method used to circumvent this difficulty is to use a model to obtain the kinetic isotope effect and then use this model kinetic isotope effect for the parent system. The fundamental assumption is that the kinetic isotope effect found for the model is valid for the parent compound. In the examples where this method has been used there are enough independent checks so that the method appears valid. One of the first examples of this method in Hofmann elimination is a tetramethyl-cyclodecane system (28). The complexity of the method is greatly increased by using a ten-membered ring because for each isomeric olefin produced there are two stereoisomers formed, *cis* and *trans*. This makes it difficult to separate all of the olefinic products by gas chromatography although if this can be accomplished the approach is still valid. It is simpler to use a model with a ring structure that restricts the products to only *cis* isomers. This has been done for a dimethylcyclopentane ring (30,37,38).

Since the method used to obtain the kinetic isotope effect for Hofmann elimination of *N,N,N*-trimethyl-3,3-dimethylcyclopentylammonium hydroxide (30,37) is a perfectly general method which has also been applied to several other kinds of eliminations, and since it illustrates how this type of model is handled, it seems appropriate to discuss the derivation of the general

(17)

(18)

equations required to analyze the data. All that is needed is a functional group in a molecule of the type that will give positional isomers of the olefin by elimination to each side of the functional group. Both the hydrogen analog and the stereospecifically monodeuterated analog must be available. Compounds **12** and **13** represent examples of this type. These compounds were prepared from 4,4-dimethylcyclopentene using diborane or diborane-d_6 followed by chloramine and subsequent quaternization (37), and eqs. 17 and 18 show the products of pyrolysis on each compound. It must be assumed that reactions involving all the positions in compounds **12** and **13** are of the same kinetic order and that the deuterium in compound **13** has only a primary isotope effect and gives no secondary isotope effect of any consequence on any other position. If k_A, k_B, $k_{A'}$, and $k_{B'}$ represent the rate constants for formation of olefins A, B, A', and B', respectively, and if k_i represents the rate constant for removal of a given hydrogen H_i while k_D represents the rate constant for removal of deuterium, then the following relationships can be written:

$$k_A = k_3 + k_4; \qquad k_B = k_1 + k_2; \qquad k_{A'} = k_D + k_4; \qquad k_{B'} = k_1 + k_2$$

The per cent of each olefin formed is proportional to its rate constant for formation, so that if Z and Q are proportionality factors, the following relationships can be written:

$$\%A = Zk_A; \qquad \%B = Zk_B; \qquad \%A' = Qk_{A'}; \qquad \%B' = Qk_{B'}$$

It is therefore true that

$$\frac{\%A}{\%B} = \frac{Zk_A}{Zk_B} = \frac{k_3 + k_4}{k_1 + k_2}$$

and

$$\frac{\%A'}{\%B'} = \frac{Qk_{A'}}{Qk_{B'}} = \frac{k_D + k_4}{k_1 + k_2}$$

If these two relationships are now divided and simplified, the result is

$$\left(k_3 + k_4\right)\left(\frac{\%A'}{\%B'}\right) = (k_D + k_4)\left(\frac{\%A}{\%B}\right)$$

and if this is divided by k_D, it can be written:

$$\left(\frac{k_3}{k_D} + \frac{k_4}{k_D}\right)\left(\frac{\%A'}{\%B'}\right) = \left(1 + \frac{k_4}{k_D}\right)\left(\frac{\%A}{\%B}\right)$$

This relationship can be simplified because the per cent of olefin A' that contains no deuterium is proportional to k_D while the per cent of olefin A'

that contains deuterium is proportional to k_4, and if P is the proportionality constant, then the following relationships can be written:

$$\%A'd_0 = Pk_D; \qquad \%A'd_1 = Pk_4$$

and

$$\frac{\%A'd_1}{\%A'd_0} = \frac{Pk_4}{Pk_D} = \frac{k_4}{k_D}$$

If this is now substituted into the above relationship, and if it is taken into account that $k_3/k_D = syn\ k_H/k_D$ kinetic isotope effect, the following final expression can be written (30):

$$syn\ \frac{k_H}{k_D} = \frac{k_3}{k_D} = \left[\frac{\left(\frac{\%A}{\%B}\right) + \left(\frac{\%A'd_1}{\%A'd_0}\right)\left(\frac{\%A}{\%B}\right)}{\left(\frac{\%A'}{\%B'}\right)} \right] - \left(\frac{\%A'd_1}{\%A'd_0}\right)$$

There is now no term on the right-hand side of this expression that cannot be obtained experimentally. All that is needed is the per cent of each olefin A and B in eq. 17 and A' and B' in eq. 18 and the deuterium analysis of olefin A'. Using this data, a value of $syn\ k_H/k_D = 1.71$ can be calculated for elimination to give the symmetrical olefin from **12** and **13**. This kinetic isotope effect and the deuterium analysis on A' can be used to calculate the per cent syn elimination leading to olefin A because

$$\frac{(\%A'd_0)\left(syn\ \frac{k_H}{k_D}\right)(100)}{\%A'd_1 + (\%A'd_0)\left(syn\ \frac{k_H}{k_D}\right)} = \%\ syn \text{ elimination to } A$$

It is calculated that compound **12** yields olefin A by a 76% syn mechanism.

If the above conclusions are correct, then the same results should be achieved using the *trans* deuterium-labeled model **14** and this has been tested (43). The synthesis of **14** started with 4,4-dimethylcyclopentene and used diborane-d_6 and hydrogen peroxide to prepare the *cis*-labeled alcohol (35). The alcohol was converted to the tosylate and subsequent azide displacement,

$$\begin{array}{ccccc}
\text{Me} \quad \text{Me} & & \text{Me} \quad \text{Me} & & \text{Me} \quad \text{Me} \\
& \longrightarrow & & + & \\
D \qquad \overset{+}{N}Me_3\ \bar{O}H & & D(H) & & D \\
\mathbf{14} & & 43\% & & 57\% \\
& & (90.7\%\ d_1) & & (100\%\ d_1) \\
& & (9.3\%\ d_0) & &
\end{array} \qquad (19)$$

reduction, and quaternization led to **14**. The results of pyrolysis of **14** are shown in eq. 19. Using the data from eqs. 19 and 17 and solving the previous expressions in such a way as to obtain the *anti* kinetic isotope effect, a value of *anti* $k_H/k_D = 5.57$ can be calculated for elimination of **12** and **14** to give symmetrical olefin. From this kinetic isotope effect and the deuterium analysis of the 4,4-dimethylcyclopentene obtained from **14**, it is calculated that compound **12** yields olefin *A* by a 64% *syn* mechanism.

Each of the calculated mechanisms for Hofmann elimination of **12** to give 4,4-dimethylcyclopentene, 76% *syn* from the *cis* deuterium model **13** and 64% *syn* from the *trans* deuterium model **14**, is within experimental error of the other. The probable error in terms of the per cent *syn* mechanism by the deuterium tracer method is in the order of ± 5 to 10%. This comes primarily from the error in mass spectral analysis for deuterium.

Using the preceding approach it has been shown (38) that the conditions used for elimination of trimethylammonium compounds very definitely has an influence on the ratio of *syn* to *anti* mechanism. The *syn* kinetic isotope effects were obtained using compounds **12** and **13** with a variety of bases in place of hydroxide ion and carrying out the eliminations in several different solvents. The mechanism for elimination of **12** to give 4,4-dimethylcyclopentene was calculated for each set of conditions and the results are shown in Table 1.

TABLE 1[a]

Elimination to give 4,4-Dimethylcyclopentene and Cyclopentene

Base	Solvent	°C	*syn* k_H/k_D	Dimethylcyclopentyl, % *syn*	Cyclopentyl, % *syn*
NaOH	H_2O	190	[b]	10[c]	4[d]
NaOH	H_2O–DMSO	130	1.62	52	1[d]
t-BuOK	*t*-BuOH	70	1.85	63	17
t-BuOK	*t*-BuOH–DMSO	70	1.92	72	45
Hofmann	pyrolysis	110[e]	1.71[e]	76[e] (64[f])	46[e,g]

[a] Based on Ref. 38.
[b] Could not be determined accurately due to small loss of deuterium.
[c] Calculated using *syn* $k_H/k_D = 1.17$ which is probably too low. By using a more realistic value of 1.7, this would be 14% *syn*.
[d] The experimental error would make this value indistinguishable from zero.
[e] Data from Refs. 29, 30, and 44.
[f] Data from the *trans* deuterium model 14, Ref. 43.
[g] Calculated using an average isotope effect of 1.84.

If it can be assumed that the *syn* kinetic isotope effects found under various conditions from compounds **12** and **13** are valid for the parent cyclopentyl ring, then the per cent *syn* mechanism for the parent ring can be determined simply by observing the elimination results of *N,N,N*-trimethyl-*cis*-2-d_1-cyclopentylammonium ion (**15**) under any set of conditions for which

15

the kinetic isotope effect has been determined. The fact that all the *syn* kinetic isotope effects observed in Hofmann eliminations in a wide variety of structures under several conditions are extremely similar seems to make the above assumption valid. If the relative rate constants for reaction of **15** by both *syn* and *anti* elimination are set on a scale so that their sum is 100, then the following expressions can be written for elimination of **15**:

$$\frac{k_{syn}}{syn\,\dfrac{k_H}{k_D}} + k_{\varepsilon yn} + 2k_{anti} = 100$$

and

$$\frac{k_{syn}}{syn\,\dfrac{k_H}{k_D}} = \%\ \text{olefin-}d_0$$

The per cent olefin-d_0 can be measured by mass spectral analysis of the cyclopentene from **15** and, if the *syn* k_H/k_D is assumed from a model such as **12** and **13**, then the relative *syn* and *anti* rate constants for loss of hydrogen from **15** can be calculated. From these the per cent *syn* mechanism can be calculated for the parent hydrogen analog of **15**. This has been done for Hofmann pyrolysis (37,44) and for eliminations in solution (38). The results are shown in Table 1.

Several features about the data in Table 1 are rather interesting. The fact that the cyclopentyl system shows mechanisms all the way from completely *anti* to nearly equal amounts of *syn* and *anti* and the 3,3-dimethylcyclopentyl system shows mechanisms from predominantly *anti* to predominantly *syn* makes it abundantly clear that the present understanding of the stereochemistry of Hofmann eliminations is inadequate. Whether these trends are due to a change of base strength leading to a change in transition state for each mechanism (38) is not clear because the *syn* kinetic isotope effects found for all the conditions are virtually within experimental error of each other.

It has been suggested (38) that the constraint provided by the five-membered ring against achieving an *anti* coplanar conformation is responsible for the observation of a reasonable balance between *syn* and *anti* mechanisms in these five-membered rings. In this situation changes in base and solvent can show more effect on the mechanisms than in other systems in which one mechanism is heavily favored. The fact that the 3,3-dimethylcyclopentyl system gives more *syn* elimination than the cyclopentyl system under all conditions is probably due to steric factors in the former (37). In order to achieve an *anti* coplanar conformation leading to *anti* elimination of **12** to give 4,4-dimethylcyclopentene, rather severe steric interactions between the bulky trimethylammonium group and the *cis* methyl group on carbon-3 must be introduced. There is also severe eclipsing of the 3,3-dimethyl group with one of its adjacent methylene groups. Both these effects are absent in the cyclopentyl case.

It has been found that in the elimination of the 3,3-dimethylcyclopentyl system **12**, the ratio of olefins produced is very sensitive to the base and solvent. The ratio of 4,4-dimethylcyclopentene to 3,3-dimethylcyclopentene varies from 1.07 for Hofmann pyrolysis (37) to 7.4 for elimination with potassium *t*-butoxide in *t*-butanol (38). The *syn* kinetic isotope effects and per cent *syn* mechanisms leading to 4,4-dimethylcyclopentene are, within experimental error, identical under the two sets of conditions. The mechanism for formation of 3,3-dimethylcyclopentene from **12** with the *t*-butoxide base is not known but under Hofmann pyrolysis of **12** the mechanism is completely *syn* (43) (discussed below). It would seem attractive to propose that the larger base, *t*-butoxide, encounters more steric hindrance from the ring methyls in the transition state leading to 3,3-dimethylcyclopentene by what is probably a *syn* mechanism than does the smaller hydroxide ion (45).

By using the per cent *syn* mechanism for elimination of the hydrogen analog of **15** with potassium *t*-butoxide in *t*-butanol, it has been calculated (38) that for elimination of *N,N,N*-trimethyl-*trans*-2-d_1-cyclopentylammonium ion (**16**) under the same conditions (46) there is an *anti* k_H/k_D kinetic isotope

16

effect of 4.75. This can be compared to two *anti* isotope effects determined directly on **16** for mechanisms thought to be *anti*. These are in aqueous sodium hydroxide and in ethanol with sodium ethoxide for **16** and are 3.99 and 4.66, respectively (46). Whether the difference between these and the value of 5.57 obtained directly (43) from **14** (discussed above) is simply experimental error or due to different conditions is not clear. The important thing is that when these *anti* isotope effects are compared to the *syn* isotope effects in Table 1 a striking difference between the *syn* and *anti* mechanisms becomes apparent. This has been interpreted (38) as support for the view that in the *syn* mechanism the β-carbon has a very high electron density, very carbanionic, with little double bond character developed in the transition state while the *anti* mechanism involves less electron density on the β-carbon and more double bond character developed in the transition state. The proton transfer from the β-carbon is pictured as being much further progressed in the *syn* mechanism.

The difference between the conclusions from the deuterium work on the cyclopentyl system (37,38) and the conclusions reached from rate profile work (23–25) is more apparent than real. The rate profile work was done under conditions that probably do give largely *anti* elimination leading to cyclopentene, so the conclusion of a largely or exclusive *anti* mechanism is not in error. It is clear from the deuterium work that widely different results can be obtained depending on the conditions used for elimination. It is also clear that the rate profile has some rather serious limitations with regard to how quantitative it is.

The fact that compound **12** undergoes a higher per cent *syn* mechanism to give 4,4-dimethylcyclopentene than does the parent cyclopentyl system to give cyclopentene has been attributed to a steric effect (37). There is an even more dramatic illustration of a steric effect in the elimination of **12** to give 3,3-dimethylcyclopentene (43). In order to examine this reaction the 3,3-dimethylcyclopentyl system must be labeled with deuterium in the 2-position. This presents a greater problem than the synthesis of **13** or **14** which utilized the symmetrical olefin 4,4-dimethylcyclopentene as a starting material. To place the deuterium in the 2-position, 3,3-dimethylcyclopentene must be used as a starting material and, being an unsymmetrical olefin, leads to two isomeric products (**17** and **18**) on reaction (eq. 20). These must be separated

$$\text{(20)}$$

by gas chromatography. Compound **17** can be isolated and converted to its quaternary hydroxide **19** which gives a mixture of olefins on Hofmann pyrolysis (eq. 21). Using the data from elimination of compounds **12** and **19** and developing the same kind of expression outlined for compounds **12** and **13**, the elimination to give 3,3-dimethylcyclopentene from compounds **12** and **19** was found to show a *syn* k_H/k_D kinetic isotope effect of 1.81. This value is within experimental error of the *syn* isotope effect of 1.71 found previously for elimination to give 4,4-dimethylcyclopentene (30). By using

$$
\begin{array}{ccc}
\textbf{19} & \substack{65.4\% \\ (100\%\ d_1)} & \substack{34.6\% \\ (5\%\ d_1) \\ (95\%\ d_0)}
\end{array}
\qquad (21)
$$

the *syn* isotope effect of 1.81 and the deuterium analysis of 3,3-dimethylcyclopentene from **19**, it can be calculated that compound **12** yields 3,3-dimethylcyclopentene by a 97% *syn* mechanism. This is in contrast to the formation of 4,4-dimethylcyclopentene by a 76% *syn* mechanism under identical conditions.

The nearly total *syn* mechanism to give 3,3-dimethylcyclopentene from **12** is easily understood from models. In order to achieve an *anti* coplanar conformation for *anti* elimination leading to 3,3-dimethylcyclopentene, the *cis* methyl group on carbon-3 and the bulky trimethylammonium must both be placed in *cis* pseudoaxial positions. No such steric interaction is encountered in a *syn* coplanar conformation leading to 3,3-dimethylcyclopentene. It can be concluded from the formation of equal amounts of the two olefins from **12** and the analysis of the stereochemistry of mechanisms leading to each olefin that in compound **12** *syn* elimination leading to 3,3-dimethylcyclopentene is slightly faster than *syn* elimination to give 4,4-dimethylcyclopentene in Hofmann pyrolysis. The reason for this is not clear unless in the transition state for *syn* elimination leading to 3,3-dimethylcyclopentene the proton transfer to hydroxide has progressed far enough to flatten carbon-2 and thereby relieve the eclipsing strain with the 3,3-dimethyl group. It might also be due to a difference in the electronic effect of carbon-3 which is quaternary versus carbon-4 which is secondary.

C. Four-Membered Rings

The results of stereochemical studies on five-membered rings would seem to indicate that slight steric constraint by ring structures working against an *anti* coplanar conformation of leaving groups is enough to make a *syn* mechanism competitive with an *anti* mechanism. If this is correct, then decreasing the ring size to a cyclobutane structure should favor *syn* elimination even more than in the cyclopentane ring. Hofmann elimination has been studied in the cyclobutyl system by deuterium labeling (37,44). By using cyclobutene as a starting material and diborane-d_6 to introduce the deuterium

(33), N,N,N-trimethyl-*cis*-2-d_1-cyclobutylammonium hydroxide (**20**) was synthesized. Pyrolysis indicates a 90% *syn* mechanism for the hydrogen analog (37). The isotope effect used here is the average of *syn* isotope effects found for three other *syn* eliminations and may or may not be exactly correct for the cyclobutyl ring. This does not affect the conclusions because

20

even if the *syn* kinetic isotope effect for the cyclobutyl ring were 1 (no isotope effect) the mechanism would have to be 62% *syn*. The results indicate that the *syn* isotope effect could not be greater than 2.2. It seems safe to assume that 1.84 is a reasonable *syn* isotope effect for the cyclobutane ring because this is within experimental error of the *syn* isotope effect found for a wide variety of substrates under several conditions.

It is clear that the parent cyclobutyl system undergoes a much greater degree of *syn* elimination than does the cyclopentyl system. The reason for this is clear from models. In order to achieve an *anti* coplanar conformation leading to *anti* elimination the cyclobutane ring must be very severely distorted or puckered. Not only does this lead to a great deal of bending strain but the bulky trimethylammonium group is brought into close proximity to the *cis* hydrogen on carbon-3 in the ring. The *syn* mechanism encounters none of these steric difficulties and thus predominates.

D. Seven-Membered Rings

Because of its size the seven-membered ring is much more difficult to understand than the five-membered ring. It is reasonably flexible but in no conformation can all adjacent carbons be perfectly staggered. This leads to an uncertainty as to the exact conformational preference of groups or the conformational energy of a given transition state. Nevertheless, the cycloheptyl system has been investigated by deuterium labeling and by the rate profile method. The results are different but not necessarily in disagreement.

In the deuterium work (37,44) cycloheptene was the starting material and diborane-d_6 was used to introduce the label (33) in the synthesis of N,N,N-trimethyl-*cis*-2-d_1-cycloheptylammonium hydroxide (**21**). Hofmann pyrolysis of compound **21** was found to give cycloheptene with loss of about 10% deuterium depending on the exact conditions. It can be calculated that

290 JAMES L. COKE

the hydrogen analog of **21** eliminates by approximately a 34% *syn* mechanism when a *syn* isotope effect of 1.84 (an average for three systems) is used. It is

21

difficult to decide exactly what the conformation looks like for each mechanism, but it appears from models that neither *syn* nor *anti* elimination has any great advantage. It is not known exactly what shape to assume for the two carbons undergoing loss of groups, but if the transition states resemble the ground states to any reasonable extent, and they probably do, at least for the parts of the rings not actually undergoing reaction, then the *anti* coplanar elimination still involves considerable eclipsing strain adjacent to the trimethylammonium group but on the side away from the carbon losing the proton. The *syn* coplanar elimination appears to involve rather serious steric interactions between *cis* hydrogens across the ring as well as eclipsing strain around the ring. The point to be made is that in the seven-membered ring the *anti* mechanism is being worked against to a considerable extent just as it is in the five-membered ring but for a somewhat different reason. The *anti* mechanism in the seven-membered ring, looking only at the β-carbon and its attached groups and their interaction with the trimethylammonium group, looks nearly identical to aliphatic systems. However, unlike aliphatic systems the ring size causes eclipsing strain. There is a striking similarity in the rate constants for *cis* olefin formation for the seven- and five-membered rings and aliphatic systems for Hofmann elimination in solutions (23–25). It appears to be the six-membered ring that is exceptionally slow.

Rate profile work on the cycloheptyl ring (23–25) leads to the conclusion that the elimination follows a predominate or exclusive *anti* mechanism. It is important to remember that the rate work was done in solution using alkoxides as the base. There is reason to believe that electrostatic interaction between the base and the positive trimethylammonium group would be more important under hydroxide pyrolysis conditions and thus *syn* elimination might be expected to be more favored in normal Hofmann pyrolysis than it is in solution eliminations. This appears to be the case in the five-membered rings. There may be solution conditions that give reasonable amounts of *syn* elimination in the cycloheptyl system. In fact, some *syn* elimination may be occurring under the solution conditions studied, but the rate profile method probably is not sensitive enough to detect it. At present there are no results available from deuterium work with solution conditions on the seven-membered ring.

E. Six-Membered Rings

Hofmann elimination has been very extensively studied in six-membered rings. In spite of the fact that both deuterium labeling and rate profile methods have been used with rather general agreement of *anti* elimination in most cases, there still remains considerable uncertainty as to what conformation best represents the transition state.

The earliest work on the parent cyclohexyl system (31) used lithium aluminum deuteride reduction of cyclohexene epoxide to introduce a *trans* deuterium with subsequent displacement of the tosylate (inversion) to prepare *N,N,N*-trimethyl-*cis*-2-d_1-cyclohexylammonium hydroxide (22). Hofmann

pyrolysis of 22, prepared in this way, was found to give cyclohexene containing about 5% dideuterated cyclohexene as well as the monodeuterated olefin (31). The dideuterated olefin evidently comes from formation of monodeuterated cyclohexanone during the reduction (32) and the subsequent reduction to dideuterated cyclohexanol of mixed stereochemistry. As pointed out earlier in the discussion of this method of synthesis, the formation of cyclohexanone in the presence of aluminum alkoxide salts lends some uncertainty as to the exact stereochemical purity of the monodeuterated alcohol. However, the above results have been checked starting with cyclohexene and using diborane-d_6 followed by chloramine and quaternization to prepare 22. Compound 22 prepared by this method was found to undergo Hofmann pyrolysis with loss of 1% deuterium (37,44). By using a *syn* kinetic isotope effect of 1.84 (an average for three systems), it can be calculated that the hydrogen analog of 22 undergoes elimination by a 4% *syn* mechanism. This is within experimental error of being completely *anti*.

If it can be assumed for the cyclohexyl system that Hofmann pyrolysis will give at least as much *syn* elimination as other conditions, then the *anti* k_H/k_D kinetic isotope effect can be determined directly by examining the elimination of *N,N,N*-trimethyl-*trans*-2-d_1-cyclohexylammonium hydroxide (23). Compound 23 was prepared (46) from cyclohexene by hydroboration with diborane-d_6 to obtain the *cis*-labeled alcohol which was converted to the tosylate and subsequently displaced with inversion. If an *anti* mechanism is correct for elimination of 23, the ratio of olefin-d_1 to olefin-d_0 is the *anti* k_H/k_D kinetic isotope effect. This was observed to be 4.33 for aqueous sodium hydroxide conditions (46). This can be compared to the values already discussed for the cyclopentyl (38,46) and 3,3-dimethylcyclopentyl (43) systems and that range from 3.99 to 5.57 depending on the conditions. It appears reasonable that in the transition state for *anti* elimination in the cyclohexyl ring, proton transfer has progressed much less than for *syn* mechanisms in other systems (38).

A number of substituted cyclohexyl systems have been investigated using deuterium labeling (47). By starting with the appropriate olefin and using diborane-d_6 followed by hydroxylamine-O-sulfonic acid for *cis* labeling or hydroboration with diborane-d_6 to give the *cis*-labeled alcohol with subsequent tosylate formation, azide displacement, and reduction for *trans* labeling, compounds **24** through **31** were prepared. The products from

$\overset{+}{N}Me_3 \ \bar{O}H$

Me₃C   D

24

$\overset{+}{N}Me_3 \ \bar{O}H$

Me₃C   D

$\overset{+}{N}Me_3 \ \bar{O}H$

25

$\overset{+}{N}Me_3 \ \bar{O}H$

D

26

$\overset{+}{N}Me_3 \ \bar{O}H$

Me₃C

D

27

Me₃C $\overset{+}{N}Me_3 \ \bar{O}H$

D

28

D

Me₃C $\overset{+}{N}Me_3 \ \bar{O}H$

29

$\overset{+}{N}Me_3 \ \bar{O}H$

D

30

Me₃C $\overset{+}{N}Me_3 \ \bar{O}H$

D

31

Hofmann pyrolysis of these hydroxides were examined by infrared (ir) spectroscopy to determine whether the deuterium in the olefins was allylic or vinylic. Compounds **24** through **27** gave high yields of olefins but compounds **28** through **31** gave very poor yields of olefins with most of the product being dimethylalkylamines. The latter result from nucleophilic substitution on the methyls attached to nitrogen. The olefins from **24**, **25**, and **26** were found to contain both allylic and vinylic deuterium while the olefin from **27** was found to contain only allylic deuterium. These findings were interpreted as proof that compounds **24** through **27** eliminate by an *anti* mechanism. The olefins from **28**, **29**, and **31** were found to contain some vinylic deuterium and thus they appear to undergo some *syn* and some *anti* elimination. The *syn* elimination was discussed (47) in terms of an ylid mechanism although the trimethylamine was not examined. In light of other work (29) the *syn* mechanism is likely to be *syn* E2.

The above observations were thought to be the result of compounds **24** through **27** reacting by *anti* elimination from a chair conformation with the trimethylammonium group in an axial position. It was pointed out that the results from compounds **28** through **31** indicate some elimination can take place even when the trimethylammonium group prefers to exist in the equatorial chair conformation in the starting material, but the mechanism of this elimination is not stereospecific. It is uncertain what conformation should be drawn for the transition state for elimination in **28** through **31**.

No limits of error were discussed (47) with regard to how little vinylic deuterium could be detected by the ir analysis used on the elimination products of compounds **24** through **27**. It is worthwhile discussing just how sensitive the ir method might be since this has a bearing on what upper limits of *syn* mechanism could take place on **24** through **27** without being detected. It must be emphasized that the ir method used on the elimination products detects the presence of vinylic and allylic deuterium and is insensitive to that portion of the product from which deuterium has been completely removed. It also must be emphasized that, if deuterium is lost by a *syn* or *anti* mechanism, then it is lost to a lesser extent than a hydrogen would be from the same position by whatever factor the kinetic isotope effect predicts for that stereochemistry.

The point to be emphasized is that in compounds **24**, **25**, and **26** any amount of *syn* mechanism that might occur on the hydrogen analog would result in a relatively small loss of deuterium from these *cis*-deuterated models. As an example, if it were assumed that the hydrogen analog of **24** would eliminate by a 50% *syn* mechanism, and if the *syn* k_H/k_D isotope effect for **24** were 1.8, then compound **24** would give 4-*t*-butylcyclohexene which would show 56% allylic deuterium, 28% vinylic deuterium, and 16% no deuterium. It is thus clear that for compounds **24**, **25**, and **26** it is not a question of whether the olefins contain vinylic deuterium or not but of what the ratio of vinylic to allylic deuterium is. Infrared spectroscopy is not particularly accurate in answering this question. Compound **26** is more sensitive than **24** or **25** because it gives two isomeric olefins that can be separated, but it still suffers the same kind of disadvantage. If the olefins had been analyzed by mass spectral or combustion analysis, then the above uncertainty might have been cleared up.

Although the results from compounds **24** through **26** simply indicate that at least some *anti* mechanism must take place without necessarily determining the amount with much accuracy, compound **27** is a different situation. Compound **27** now has the *anti* k_H/k_D kinetic isotope effect working in favor of the accuracy of the ir method. This can best be illustrated by an example. If it is assumed that the hydrogen analog of **27** would eliminate by a 5% *syn* mechanism, and if the *anti* k_H/k_D kinetic isotope effect for **27** were 5.0, then compound **27** would give 4-*t*-butylcyclohexene which would show 4% vinylic deuterium, 81% allylic deuterium, and 16% no deuterium. While the sensitivity of the ir method is not known for sure, it seems likely that it would be able to detect 4% of vinylic deuterium. The results from compound **27** appear to demonstrate convincingly an *anti* mechanism within experimental error. There is no reason to think that compounds **24** through **26** are different, and thus it appears that all these compounds show the same *anti* stereochemistry for elimination that is exhibited by the parent cyclohexyl ring (31,37).

One interesting aspect of the work on the substituted cyclohexyl systems (47) is that compounds **25** and **26** show a predominance in elimination away from the *t*-butyl group or the adjacent ring. This apparently is a steric effect but the magnitude of the effect seems strange in view of the lack of steric effect to approach of hydroxide ion in the elimination of **12** (eq. 17) in which nearly equal amounts of the two olefins are formed (both by predominately *syn* mechanism). Whether the difference noted for these examples results from the fact that **12** eliminates by a *syn* mechanism and **25** and **26** by an *anti* mechanism or whether some other factor is responsible is unknown.

Compounds **24** through **31** were also examined (47) under Wittig elimination conditions (48) using butyllithium as the base. All the compounds were found to undergo *syn* elimination in agreement with earlier work (41). Elimination brought about by bases like butyllithium are similar to the ylid mechanism shown in eq. 5 except that under Wittig conditions (butyllithium) proton transfer in the first step is an irreversible reaction. In Hofmann eliminations with hydroxide or alkoxides as the base, this proton transfer step is reversible.

Hofmann elimination of the parent cyclohexyl system has also been examined by the rate profile method (23–25) using a variety of alkoxides as the base and several alcohols as solvents. The conclusion reached is that the elimination follows a predominately or exclusively *anti* mechanism. This is an agreement with the results from deuterium labeling. One interesting result of this work is that it allows comparison of the rates of elimination in various rings. The rates indicate that the cyclopentyl and cycloheptyl systems eliminate from five to twenty-six times as fast as the cyclohexyl system depending on the particular base and solvent used. Under one set of conditions, potassium *t*-butoxide in *t*-butanol, in which it is known that the cyclopentyl system shows 17% *syn* elimination (38), one can calculate that the rate of *anti* elimination in the cyclopentyl system is twenty-two times as fast as *anti* elimination in the cyclohexyl system (24). This assumes a completely *anti* mechanism in the cyclohexyl system, but if this assumption is incorrect, then the rate difference may be slightly greater. The slowness of the cyclohexyl system has generally been regarded as being caused by having the trimethyl-ammonium group placed in an axial position (46). The relative rate difference for *anti* elimination in the cyclopentyl and cyclohexyl systems indicates a difference in activation energy of about 2 kcal with the cyclohexyl compound having the larger activation energy. This is a relatively small amount of energy difference and in view of the uncertainties in describing each transition state it seems unwise to speculate on its exact origin.

Even though most normal Hofmann eliminations in cyclohexyl systems show *anti* stereochemistry, it is possible, by locking the ring into an inflexible boatlike conformation, to observe completely *syn* elimination. Compound **11** illustrates this (29). The *anti* mechanism in cyclohexyl systems has generally been thought of as taking place from a conformation having the bulky trimethylammonium axial on a chair conformation. Because the energy required to do this should be about the same as is required for the ring to exist as a twist conformation, it has been proposed that some substituted cyclohexyl systems undergo elimination from a twist conformation (19). It was proposed (18) that the Saytzeff-type product (eq. 11) from compound **7** results from a more carbonium-ion-like, or more productlike, transition state due to steric crowding of the trimethylammonium group. Because little difference in product composition was found when the hyperconjugative ability of the *cis*-2-alkyl group changes, this idea was rejected in favor of elimination from a twist form (19). It was proposed that compounds like **7** eliminate from a twist form that has only the *trans* hydrogen on carbon-2 *anti* and coplanar with the trimethylammonium group. The hydrogens on carbon-6 are not *anti* or *syn* coplanar and thus the observed product is to be expected. It was further argued that in compounds like **8** a twist form is even more likely to be responsible for elimination. A chair conformation of **8** with the trimethylammonium group axial would also require the methyl group to be axial and thus would be extremely unfavorable. The twist form of **8** is probably lower in energy, and it has been proposed (19) that in this twist conformation only the *trans* hydrogen on carbon-6 is *anti* and coplanar and thus the observed product is expected.

Models indicate that in some cases a twist form may be more likely and in some cases an axial trimethylammonium group on a chair conformation may be more likely. It is not possible to resolve the problem of what picture is best for the transition state in simple cyclohexyl systems at the present time. However, one factor that should never be ignored is the effect of placing the trimethylammonium group *gauche* to two equal size groups on the β-carbon in the transition state. This tends to hold the trimethylammonium group *anti* and coplanar to the third group on the β-carbon. In the case of compound **7** the methyl group and carbon-3 of the ring appear sterically the same to the trimethylammonium group and the third group on carbon-2 is the hydrogen that is lost. The other β-carbon involves placing the trimethylammonium group *gauche* to both carbon-5 in the ring and the *cis* hydrogen on carbon-6. Since the hydrogen and ring carbon are sterically different size groups, the trimethylammonium group will lie at a smaller than 60° dihedral angle to a *cis* hydrogen on carbon-6. The effect of this is to rotate the *trans* hydrogen on carbon-6 out of coplanarity. The above arguments apply whether one assumes a chair or a twist conformation for the transition state. One of the difficulties with any of these arguments is that it is not clear to what extent the transition state resembles the ground state conformations. Thus the magnitude of steric effects is unclear.

F. Eight-Membered and Larger Rings

The question of stereochemistry of elimination in the medium ring compounds becomes more complex than in smaller rings because both *cis* and *trans* olefin isomers are formed. Because the *cis* and *trans* isomers of cycloalkenes are usually separable, it is possible to look at the mechanistic stereochemistry leading to each olefin independently. Both deuterium labeling and rate profile methods have been used to examine medium ring compounds.

The smallest ring size that gives both *cis* and *trans* cycloalkenes by elimination is the eight-membered ring, and this has been examined by deuterium labeling (44,49). Using both *cis*- and *trans*-cyclooctene as starting materials and diborane-d_6 followed by chloramine (33) as the method for introducing the deuterium label, the *cis* and *trans* isomers of N,N,N-trimethyl-2-d_1-cyclooctylammonium hydroxide (**32** and **33**) were prepared. Hofmann

$$34\% \ \textit{cis-cyclooctene} \ (85\% \ d_1)$$
$$\text{and} \tag{22}$$
$$66\% \ \textit{trans-cyclooctene} \ (100\% \ d_1)$$

32

$$38\% \ \textit{cis-cyclooctene} \ (89\% \ d_1)$$
$$\text{and} \tag{23}$$
$$62\% \ \textit{trans-cyclooctene} \ (66\% \ d_1)$$

33

pyrolysis of **32** (eq. 22) and **33** (eq. 23) gave mixtures of *cis* and *trans* olefins which were separated into the pure isomers and analyzed separately by mass spectrometry to determine deuterium content. It can be noted that since *trans*-cyclooctene from **32** retains all the deuterium, it must be formed by an exclusively *syn* mechanism (see eq. 2). Therefore deuterium analysis of *trans*-cyclooctene from **33** reveals the *syn* k_H/k_D kinetic isotope effect for this reaction since this olefin is formed by direct competition between *syn* loss of hydrogen (66%) and *syn* loss of deuterium (34%). The *syn* kinetic isotope effect is found to be 1.94 for *trans*-cyclooctene formation. This is within the experimental error of *syn* isotope effects found in two other systems (29,37). If this *syn* isotope effect is valid for the *syn* elimination of **32** to give *cis*-cyclooctene, then the data from eq. 22 can be interpreted. The loss of deuterium (15%) from **32** in giving *cis*-cyclooctene must take place by a *syn* mechanism (see eq. 1), and if the above isotope effect of 1.94 is assumed for this reaction, then it can be calculated that formation of *cis*-cyclooctene from the hydrogen analog of **32** takes place by a 51% *syn* elimination.

The above relative rates for *syn* and *anti* mechanism to give *cis*-cyclooctene were used to calculate an *anti* isotope effect of 2.64 for formation of *cis*-cyclooctene from **33**. This seems excessively low in light of more recent work (38,43,46) (see eq. 19), and because of the roundabout way in which it was obtained it probably has a large enough experimental error that it should be disregarded.

Elimination in the cyclooctyl system has been examined by the rate profile method using a variety of bases and solvents (23–26). The conclusion was reached that *trans*-cyclooctene is formed by a *syn* mechanism and *cis*-cyclooctene is formed by an *anti* mechanism with the qualifying condition "either largely or exclusively." Whether or not the rate profile conclusions are in conflict with the deuterium labeling results on the cyclooctyl system is unknown since the reaction conditions are quite different. It seems appropriate to examine the rate profile method with regard to how precise it is and how rigorously its basic assumptions have been proven. Simply because it is a comparative method, it would be expected to work better for *syn* mechanisms because the *syn* Hofmann eliminations can be compared to a rigorously proven *syn* mechanism, namely the pyrolysis of *N*-oxides (3). There is no such good analogy for the *anti* mechanisms, even though *anti* elimination of 1,2-dibromides by iodide ion was used to try to correlate suspected *anti* Hofmann mechanisms (23). The unreliability of the rate profile method for proving the mechanism of *cis* olefin formation is revealed dramatically by examining the rate profile plots for *cis* olefins produced under different conditions (25,26). These plots show beyond doubt that there is no such thing as a typical rate profile for formation of *cis* olefins in cyclic systems by Hofmann elimination. This leads to one of two inescapable conclusions:

either different conditions can change the mechanism of *cis* olefin formation (mixed *syn* and *anti* mechanisms) or different rate profiles can predict the same mechanism. If the latter conclusion is correct, then the whole rate profile method must be discredited. If the former conclusion is correct, then *cis* olefins are formed by a variety of mechanisms depending on the ring size or the conditions used and this variety of mechanisms must be some combination of *syn* and *anti* mechanisms. In light of recent work on the cyclopentyl ring (38), it seems much more likely that a varying combination of *syn* and *anti* mechanisms leading to *cis* olefins is correct.

Even the rate profile plots for *trans* olefin formation (25,26) show some slight variations depending on the conditions used for elimination. The similarity of these rate profiles to that for *trans* olefin formation by *N*-oxide pyrolysis was taken as evidence that Hofmann elimination in cyclic systems to give *trans* olefins takes place by a *syn* mechanism. The validity of this conclusion depends directly on the proof that the assumption of similar rate profiles for similar elimination stereochemistry is correct.

Because of the widespread acceptance of conclusions from the rate profile method and frequent reference to this method as a proof of mechanism, an examination of how the method developed is justified. The first application of this method to formation of *trans* olefins by Hofmann elimination (23) contained the statement "Provided we accept the tenet that processes possessing mechanisms similar with respect to their salient steric features will exhibit rate curves of similar type . . ." which was in part a reference to the similarity in rate profiles for *N*-oxide and Hofmann elimination to give *trans* olefins. The conclusion, if one accepts this tenet, is that both reactions follow *syn* stereochemistry because this is the stereochemistry that has been rigorously proven for the *N*-oxide pyrolysis. Thus the fundamental principle of the rate profile method started out as a hypothesis.

$$(24)$$

$$(25)$$

In order to test this hypothesis a deuterium labeling experiment was carried out on a cyclodecyl system (28). Compounds **34** and **35** were prepared and Hofmann elimination of each (eq. 24 and 25) was found to give a mixture of symmetrical and unsymmetrical olefin, each as its *cis* and *trans* isomer. The olefinic mixture from each of the compounds **34** and **35** was analyzed by gas chromatography and the composition was compared to the mixture obtained from the hydrogen analog of **34**. The principle is that the deuterium in **34** and **35** should not influence the rate of formation of the unsymmetrical olefin. If the deuterium has an influence on the rate of formation of either the *cis* or *trans* symmetrical olefin, then it must be lost in the rate determining step for their formation. From the stereochemistry of the deuterium label and the stereochemistry of the symmetrical olefin formed, the mechanism that must operate can be deduced depending on whether hydrogen or deuterium is lost (see eqs. 1–4). The effect of the deuterium label on the rates of symmetrical olefin formation were calculated as isotope effects. As an example, the ratio of per cent *trans* symmetrical olefin to per cent *trans* unsymmetrical olefin from the hydrogen analog was divided by the ratio of per cent *trans* symmetrical olefin to per cent *trans* unsymmetrical olefin from compound **35**. This was called the kinetic isotope effect for formation of *trans* symmetrical olefin from **35**. It should be noted that these are not kinetic isotope effects. They are simply the effect of an isotope in a given position. In order to be a kinetic isotope effect the *trans* olefin would have to be formed by an exclusive *syn* mechanism and the *cis* olefin would have to be formed by an exclusive *anti* mechanism. This was the question under investigation in the first place.

The results from **34** indicate almost no effect from the deuterium in formation of either *trans* or *cis* symmetrical olefin (1.1 and 1.0, respectively). The results from **35** indicate a much larger effect from the deuterium in the formation of either *trans* or *cis* symmetrical olefin (2.3 and 2.8, respectively). It was argued that the above results are only compatible with a *syn* mechanism leading to *trans* symmetrical olefin and an *anti* mechanism leading to *cis* symmetrical olefin. The uncertainty in this interpretation is the magnitude of the experimental error in the isotope effects and how much of a minor stereochemical mechanism these experimental errors would allow to go undetected. From the magnitude of the isotope effects from **35**, 2.3 for what should be a *syn* mechanism and 2.8 for what should be an *anti* mechanism, it is extremely probable that either the experimental error in the isotope effects is fairly large, making them poor tests for stereochemistry, or the mechanism leading to each olefin is not pure *syn* or *anti*. The olefins from Hofmann elimination of **34** and **35** were not analyzed individually for deuterium.

There is a potentially serious error made in using the hydrogen analog of **34** and **35** as a model for determining the stereochemistry of elimination in the parent cyclodecyl system. The reactions of this tetramethyl analog are already known to be different from the parent unsubstituted analog (50). A striking illustration of this situation is the Hofmann elimination on compound **12** when compared to the parent unsubstituted cyclopentyl system. Under the same pyrolysis conditions, the parent cyclopentyl ring gives 46% *syn* elimination to *cis* olefin (37), while the dimethyl derivative **12** gives approximately 70% *syn* elimination to the *cis* olefin 4,4-dimethylcyclopentene (37,43) and 97% *syn* elimination to give the *cis* olefin, 33-dimethylcyclopentene (43). Thus compound **12** is simply not a valid model for testing the stereochemistry of elimination in the parent ring system. Compound **12** may provide reliable isotope effects for the parent system, but the stereochemistry of elimination is very much a function of conformational effects. Thus the conclusions from compounds **34** and **35** as to stereochemistry of elimination cannot be used directly to determine the stereochemistry of the formation of *trans*-cyclodecene in the parent ring system.

Therefore, the basic assumption of the rate profile method with regard to *trans* olefin formation, that similar mechanisms should show similar rate profiles, still remains a hypothesis and nothing more at this point. The next usage of the method appears in a series of investigations of Hofmann eliminations under a variety of conditions. In the first of these (24) the rate profile method is clearly described as a hypothesis with the tenet of similar mechanisms showing similar rate profiles being required for validity. In the second (25) it is clearly stated that "There appears to exist a rate profile characteristic for *syn* elimination mechanisms and a rate profile characteristic for *anti* elimination elimination mechanisms." There is only passing reference to the fact that this conclusion requires acceptance of the basic tenet given earlier. In the third investigation (26) no mention is made of the basic tenet and it is simply stated that it has been shown by the rate profile method that *cis* olefins are formed by *anti* elimination and *trans* olefins are formed by *syn* elimination. It is also stated once again that clear evidence exists that two distinct mechanisms operate, *syn* and *anti*, and that each shows its own distinct rate profile. Thus, what started as a hypothesis and has no further rigorous unequivocal proof has now become a method for proving a mechanism.

In spite of claims to the contrary, the rate profile method as it applies to Hofmann eliminations remains a hypothesis and not a proof. There is very definitely a similarity in the rate profiles for *trans* olefin formation by *N*-oxide pyrolysis and Hofmann elimination in solution. This is only indicative that the Hofmann elimination to give *trans* olefins is a *syn* mechanism in those medium ring compounds where the similar rate profiles are observed and under those conditions used. It is not clear to what degree an *anti* mechanism could compete in *trans* olefin formation without being detected by the rate profile method. It is clear that the rate profile method is neither as versatile nor as accurate as the deuterium labeling method.

G. Aliphatic Compounds

Investigations on the stereochemistry of Hofmann elimination in aliphatic compounds encounter the same difficulties as those for medium ring compounds because both *cis* and *trans* isomers are produced. In addition, if the aliphatic ammonium ion is unsymmetrical with regards to the location of the nitrogen, then usually two different olefins will be produced, each as a mixture of *cis* and *trans* isomers. There have been several deuterium labeling studies carried out with somewhat different conclusions being reached.

One of the first investigations of aliphatic systems determined the influence of a deuterium label on the olefin compositions (51). Using two methods for introducing the deuterium label, hydroboration with diborane-d_6 and displacement of the tosylate or reduction of the epoxide with lithium aluminum deuteride–aluminum chloride and displacement of the tosylate, the *threo* and *erythro* ammonium ions **36** and **37** were prepared. Hofmann elimination of **36** (eq. 26) and **37** (eq. 27) with a variety of base–solvent pairs

$$n\text{-Bu} \overset{H}{\underset{D}{\diagup}} \overset{\overset{+}{N}Me_3}{\underset{CH_2\text{-}t\text{-Bu}}{\diagup}} H \longrightarrow
\begin{array}{c}\text{2,2-dimethyl-3-nonene (}cis\text{ and }trans\text{)} \\ \text{and} \\ \text{2,2-dimethyl-4-nonene (}cis\text{ and }trans\text{)}\end{array} \quad (26)$$

36 (*threo*)

$$\overset{n\text{-Bu}}{\underset{D}{\overset{H}{\diagup}}} \overset{\overset{+}{N}Me_3}{\underset{CH_2\text{-}t\text{-Bu}}{\diagup}} H \longrightarrow
\begin{array}{c}\text{2,2-dimethyl-3-nonene (}cis\text{ and }trans\text{)} \\ \text{and} \\ \text{2,2-dimethyl-4-nonene (}cis\text{ and }trans\text{)}\end{array} \quad (27)$$

37 (*erythro*)

was carried out, and by comparison of the olefin composition from the deuterium-labeled compounds with the olefin composition from the hydrogen analog the isotope effect was calculated (28). As was pointed out for compounds **34** and **35**, this method does not give kinetic isotope effects unless only a single mechanism, *syn* or *anti*, operates to give a single olefin, *cis* or *trans*. The isotope effect calculated by this method simply gives the effect of an isotope in a given position on the composition of the olefins in the product. From the stereochemistry of the deuterium in the starting material and the stereochemistry of the 2,2-dimethyl-4-nonene, the mechanism for elimination can be deduced depending on whether hydrogen or deuterium is lost (see eqs. 1–4). If the deuterium label does not change the olefin composition, then it cannot be involved in elimination and hydrogen must be the group that is lost in giving 2,2-dimethyl-4-nonene.

The results of elimination on the *erythro* isomer **37** indicate little effect by deuterium on the rate of formation of either *cis*- or *trans*-2,2-dimethyl-4-nonene. The effects range from 0.7 to 1.2 with an experimental error of ± 0.3. The results from the *threo* isomer **36** indicate a larger effect by deuterium on the rate of formation of both *cis*- and *trans*-2,2-dimethyl-4-nonene. The effects range from 2.9 to 4.7 with an experimental error of ± 0.8. The above observations were interpreted as being in agreement with the view that *cis*-2,2-dimethyl-4-nonene arises by an *anti* elimination and *trans*-2,2-dimethyl-4-nonene arises by a *syn* elimination, either largely or exclusively.

The uncertainty in the above data is apparent if the limits of error on the isotope effects are considered. The magnitude of the error frequently allows the isotope effect found from both **36** and **37** for a given olefin formation to be almost identical. This allows the possibility of mixed *syn* and *anti* mechanisms leading to a given olefin rather than constituting proof that only a single mechanism leads to a single olefin as has been claimed (51). Considering the magnitude of the error in isotope effect, the results are still compatible with a reasonable amount of *syn* mechanism leading to *cis*-2,2-dimethyl-4-nonene and a reasonable amount of *anti* mechanism leading to *trans*-2,2-dimethyl-4-nonene.

One interesting feature of the above data is that the effect of deuterium in the *threo* isomer **36** in general tends to be smaller for the formation of *trans*-2,2-dimethyl-4-nonene than for formation of *cis*-2,2-dimethyl-4-nonene. The effect comes from *syn* loss of deuterium in the former and *anti* loss of deuterium in the latter. One exception is the value of 4.2 for formation of *trans* olefin from **36** with potassium *t*-butoxide in dimethylsulfoxide which seems to be so excessively large when compared to other *syn* isotope effects that its magnitude should be reexamined. This trend of larger isotope effects for *anti* mechanisms and smaller values for *syn* mechanisms has been noted in cyclic compounds also (37,38).

A recent study of aliphatic systems utilized both olefin composition and deuterium loss for each olefin as a means of analyzing the stereochemistry of elimination (52). By using lithium aluminum deuteride–aluminum chloride reduction of an epoxide and subsequent displacement of the tosylate as a way to introduce the deuterium label, the *threo* ammonium ion **38** was prepared. Elimination of **38** with a variety of base–solvent pairs was carried

$$
\begin{array}{c}
\text{H} \quad \overset{+}{\text{NMe}_3} \\
n\text{-Bu} \diagdown\!\!\diagup \quad \diagdown\text{-H} \\
\text{D} \quad n\text{-Bu} \\
\textbf{38} \; (\textit{threo})
\end{array}
\quad\longrightarrow\quad
\begin{array}{c}
\text{4-decene (\textit{cis} and \textit{trans})} \\
\text{and} \\
\text{5-decene (\textit{cis} and \textit{trans})}
\end{array}
\qquad (28)
$$

out (eq. 28). Complete experimental details were not given but the general method for analysis of the products consisted of separating the mixture of cis-4-decene and cis-5-decene from the mixture of trans-4-decene and trans-5-decene by preparative gas chromatography. Each pair of cis isomers and trans isomers was then ozonized and the products were reduced with lithium aluminum hydride. Separation of the alcohols was then possible and deuterium analysis of the alcohols determined the deuterium content of each original olefin. The 5-decenes each give 2 moles of 1-pentanol while the 4-decenes each give 1 mole of 1-butanol and 1 mole of 1-hexanol. The assumption was made that for the hydrogen analog of 38 the rates of formation of cis-4-decene and cis-5-decene are equal and the rates of formation of trans-4-decene and trans-5-decene are equal.

The mechanisms leading to the isomers of 5-decene were calculated using the data from the above treatment. The results indicate that in the formation of trans-5-decene conditions of pyrolysis or potassium t-butoxide in the solvents dimethylsulfoxide, benzene, or t-butanol give 87 to 95% syn mechanism but potassium methoxide in methanol gives only 32% syn mechanism. In the formation of cis-5-decene conditions of pyrolysis, potassium t-butoxide in dimethyl sulfoxide or t-butanol, or potassium methoxide in methanol give from 92 to 95% anti mechanism but potassium t-butoxide in t-butanol gives only 75% anti mechanism. These results strongly indicate that the reaction conditions can change the mechanism of elimination and that until the effects of solvent and base on the stereochemistry are better understood it is hazardous to make generalizations that cis or trans olefins are formed by one mechanism or another.

In a study of a system quite analogous to compound 38 the erythro and threo trimethylammonium ions 39 and 40 were prepared using diborane-d_6

39 (erythro)   40 (threo)

followed by hydroxylamine-O-sulfonic acid (53). Hofmann eliminations of 39 and 40 were carried out in five different base–solvent pairs and the olefin mixture was analyzed for each reaction. By using the olefin composition from 39 and 40 and the olefin composition from the hydrogen analog, an isotope effect was calculated for the formation of cis- and trans-3-hexene under each condition. The method used to calculate the isotope effect (51) does not give kinetic isotope effects but simply reflects the effect of deuterium in a given position on the olefin composition. However, if deuterium is not lost from a given reactant in forming an olefin, then the effect of the isotope will be 1.

If it is lost, then the effect of the isotope will be some fraction of the kinetic isotope effect which is dependent on the extent to which loss of deuterium contributes to the formation of that particular olefin.

In the formation of *cis*-3-hexene the *erythro* isomer **39** gives isotope effects of about 1.1, but the *threo* isomer **40** gives isotope effects ranging from 2.8 to 4.7. These results are strongly indicative that the *cis*-3-hexene is formed by a predominantly *anti* mechanism (see eqs. 1 and 3) with probably a small amount of *syn* mechanism. In the formation of *trans*-3-hexene the *erythro* isomer **39** gives isotope effects of 1.2 to 3.2 and the *threo* isomer **40** gives isotope effects of 1.2 to 2.2. These results indicate that the *trans*-3-hexene is formed by both *syn* and *anti* mechanisms under the conditions used, with both mechanisms making a significant contribution.

The above conclusion of mixed *syn* and *anti* mechanisms for the formation of *trans*-3-hexene using alkoxides in alcohols is similar to the results found earlier (52) in the formation of *trans*-5-decene from the hydrogen analog of **38** with alkoxides in alcohols.

A deuterium labeling study has appeared on the 2-butyl system (54). The *threo* isomer of N,N,N-trimethyl-3-d_1-2-butylammonium ethoxide (**41**) was prepared. Hofmann elimination of **41** gave a mixture of olefins (eq. 29).

$$
\begin{array}{c}
\text{Me} \quad \overset{+}{\text{N}}\text{Me}_3 \ \bar{\text{O}}\text{Et} \\
\text{H} \\
\diagup \diagdown \text{-Me} \\
\text{D} \quad \text{H} \\
\textbf{41} \ (threo)
\end{array}
\qquad \longrightarrow \qquad
\begin{array}{c}
\text{1-butene} \\
\text{and} \\
\text{2-butene} \ (cis \ \text{and} \ trans)
\end{array}
\qquad (29)
$$

This was separated and each olefin was analyzed for deuterium separately. The *trans*-2-butene was reported to form with loss of 11% deuterium but there was an uncertainty of 9% in this analysis. The *cis*-2-butene was reported to contain 23% deuterium and it was claimed that all this deuterium content was the result of rearrangement of 0.1% of 1-butene-d_1 to *cis*-2-butene. The precision required to make this claim on the basis of control experiments can be appreciated by pointing out that only 0.5% of *cis*-2-butene was formed from **41**. It is doubtful that the precision and accuracy of any analytical technique or control method is good enough to support the contention that exactly 0.1% of the *cis*-2-butene is formed from 1-butene. However, even if the 23% deuterium in the *cis*-2-butene from **39** is the result of direct elimination, the conclusion would be that *cis*-2-butene formed from the hydrogen analog of **41** would arise by at least a 90 to 95% *anti* mechanism. The data for *trans*-2-butene from **41** is not as clear in its implication. The 9% uncertainty in deuterium analysis would allow from 0 to probably 30% *syn* elimination leading to *trans*-2-butene in the hydrogen analog of **41**. The experimental uncertainties are large in dealing with Hofmann elimination of

41 because of the extremely small quantities of 2-butenes formed. The relationship of any conclusions from this work to those reached in other studies (51–53) is uncertain because of different substrate structure and reaction conditions.

IV. SUMMARY

If the number of possible variables are considered, it can be safely said that elimination mechanisms are an experimentalist's dream and a theoretician's nightmare. There have been several attempts to develop theoretical methods for examining the steroechemistry of elimination reactions. These vary from molecular orbital treatments (55) to application of the principle of least motion (11,56). It is unrealistic to expect theoretical approaches of this type to predict the results of eliminations on complex molecules. Their value is in being able to analyze differences in mechanisms as variables are changed.

In an attempt to rationalize the stereochemistry of Hofmann eliminations an empirical approach based on the proposition that there is an inherent difference in the reactivity of the two hydrogens on the β-carbon has been suggested (26,57). In many simple systems these hydrogens are diastereomeric, and it was proposed that regardless of whether a *cis* or *trans* olefin is the product, the same hydrogen reacts to yield either olefin. The reactive one was thought to be the one required to be lost if *syn* elimination leads to *trans* olefins and *anti* elimination leads to *cis* olefins. One difficulty with this proposal is the fact that the assumed mechanisms, *syn* leading to *trans* olefins and *anti* leading to *cis* olefins, frequently are not valid. One of the examples cited in arriving at the above empirical approach is the selective exchange of only one of a pair of diastereomeric hydrogens adjacent to a sulfoxide group (58,59). This empirical approach deserves further scrutiny.

The stereochemistry of Hofmann eliminations is best understood and analyzed by considering trends in mechanisms within a series of reasonably similar molecules. It is useful to summarize a few of the various factors that have been shown to influence the stereochemistry of elimination. The *anti* mechanism for compound **5** and the *syn* mechanism for compound **6**, as compared to the results from compounds **7** and **8**, demonstrate that electronic factors, presumably caused by the phenyl group, can control the stereochemistry. The differences in degree of *syn* elimination at the two positions of compound **12** as compared to the parent cyclopentyl ring clearly show that the stereochemistry can be controlled by steric effects. The *syn* mechanism in compound **11** and the *anti* mechanism in the parent cyclohexyl ring demonstrate a reversal of stereochemistry by changing the dihedral angle between leaving groups. The change in amount of *syn* mechanism for compound **12** as the reaction conditions are changed (Table 1) shows how large an effect base and solvent have on the stereochemistry. Finally, the difference in the *syn* and *anti* kinetic isotope effect found for compounds **13** and **14** demonstrates that there is a fundamental difference in the bond breakage for the *syn* and *anti* transition states.

When one considers that the above effects are probably only some of the factors controlling the stereochemistry of Hofmann eliminations, it is not surprising that a clear over-all picture of this reaction has failed to emerge.

References

1. C. K. Ingold, *Structure and Mechanism in Organic Chemistry*, Cornell Univ. Press, Ithaca, New York, 1953, chap. VIII.
2. D. J. Cram, in M. S. Newman, Ed., *Steric Effects in Organic Chemistry*, Wiley, New York, 1956, p. 304.
3. A. C. Cope and E. R. Trumbull, *Org. Reactions*, **11**, 317 (1959).
4. J. F. Bunnett, *Angew. Chem. Int. Ed. Eng.*, **1**, 225 (1962).
5. D. V. Banthorpe, *Elimination Reactions*, Elsevier, Amsterdam, 1963.
6. W. H. Saunders, Jr., in S. Patai, Ed., *The Chemistry of Alkenes*, Wiley-Interscience, New York, 1964, p. 149.
7. D. V. Banthorpe, in J. H. Ridd, Ed., *Studies on Chemical Structure and Reactivity*, Methuen, London, 1966, p. 33.
8. J. F. Bunnett, in A. F. Scott, Ed., *Survey of Progress in Chemistry*, Vol. 5, Academic Press, New York, 1969, p. 53.
9. C. Ingold, *Proc. Chem. Soc.*, **1962**, 265.
10. C. H. DePuy, R. D. Thurn, and G. F. Morris, *J. Amer. Chem. Soc.*, **84**, 1314 (1962).
11. J. Hine, *J. Amer. Chem. Soc.*, **88**, 5525 (1966).
12. D. J. Cram, F. D. Greene, and C. H. DePuy, *J. Amer. Chem. Soc.*, **78**, 790 (1956).
13. R. T. Arnold and P. N. Richardson, *J. Amer. Chem. Soc.*, **76**, 3649 (1954).
14. J. Weinstock and F. G. Bordwell, *J. Amer. Chem. Soc.*, **77**, 6706 (1955).
15. A. C. Cope, G. A. Berchtold, and D. L. Ross, *J. Amer. Chem. Soc.*, **83**, 3859 (1961).
16. G. Ayrey, E. Buncel, and A. N. Bourns, *Proc. Chem. Soc.*, **1961**, 458.
17. S. J. Cristol and F. Stermitz, *J. Amer. Chem. Soc.*, **82**, 4692 (1960).
18. T. H. Brownlee and W. H. Saunders, Jr., *Proc. Chem. Soc.*, **1961**, 314.
19. H. Booth, N. C. Franklin, and G. C. Gidley, *J. Chem. Soc.*, C, **1968**, 1891.
20. E. D. Hughes and J. Wilby, *J. Chem. Soc.*, **1960**, 4094.
21. C. W. Bird, R. C. Cookson, J. Hudec, and R. O. Williams, *J. Chem. Soc.*, **1963**, 410.
22. A. C. Cope and A. S. Mehta, *J. Amer. Chem. Soc.*, **85**, 1949 (1963).
23. J. Sicher, J. Zavada, and J. Krupicka, *Tetrahedron Lett.*, **1966**, 1619.
24. J. Sicher and J. Zavada, *Coll. Czech. Chem. Commun.*, **32**, 2122 (1967).
25. J. Zavada and J. Sicher, *Coll. Czech. Chem. Commun.*, **32**, 3701 (1967).
26. J. Sicher and J. Zavada, *Coll. Czech. Chem. Commun.*, **33**, 1278 (1968).
27. S. Bank, C. A. Rowe, Jr., A. Schriesheim, and L. A. Naslund, *J. Amer. Chem. Soc.* **89**, 6897 (1967).
28. J. Zavada, M. Svoboda, and J. Sicher, *Tetrahedron Lett.*, **1966**, 1627.
29. J. L. Coke and M. P. Cooke, Jr., *J. Amer. Chem. Soc.*, **89**, 6701 (1967).
30. J. L. Coke and M. P. Cooke, Jr., *Tetrahedron Lett.*, **1968**, 2253.
31. K. T. Finley and W. H. Saunders, Jr., *J. Amer. Chem. Soc.*, **89**, 898 (1967).

32. B. Rickborn and J. Quartucci, *J. Org. Chem.*, **29**, 3185 (1964).
33. H. C. Brown, W. R. Heydkamp, E. Breuer, and W. S. Murphy, *J. Amer. Chem. Soc.*, **86**, 3565 (1964).
34. M. W. Rathke, N. Inoue, K. R. Varma, and H. C. Brown, *J. Amer. Chem. Soc.*, **88**, 2870 (1966).
35. H. C. Brown, *Hydroboration*, Benjamin, New York, 1962.
36. L. Verbit and P. J. Heffron, *J. Org. Chem.*, **32**, 3199 (1967).
37. M. P. Cooke, Jr., and J. L. Coke, *J. Amer. Chem. Soc.*, **90**, 5556 (1968).
38. K. C. Brown and W. H. Saunders, Jr., *J. Amer. Chem. Soc.*, **92**, 4292 (1970).
39. J. L. Coke and M. P. Cooke, Jr., *J. Amer. Chem. Soc.*, **89**, 2779 (1967).
40. J. K. Stille, F. M. Sonnenberg, and T. H. Kinstle, *J. Amer. Chem. Soc.*, **88**, 4922 (1966).
41. A. C. Cope, E. Ciganek, and N. A. LeBel, *J. Amer. Chem. Soc.*, **81**, 2799 (1959).
42. A. F. Cameron, G. Ferguson, and D. G. Morris, *J. Chem. Soc.*, *B*, **1968**, 1247.
43. G. H. Britton and J. L. Coke, Results to be published.
44. J. L. Coke, M. P. Cooke, Jr., and M. C. Mourning, *Tetrahedron Lett.*, **1968**, 2247.
45. H. C. Brown, I. Moritani, and Y. Okamoto, *J. Amer. Chem. Soc.*, **78**, 2193 (1956).
46. W. H. Saunders, Jr., and T. A. Ashe, *J. Amer. Chem. Soc.*, **91**, 4473 (1969).
47. G. Lamaty, C. Tapiero, and R. Wylde, *Bull. Soc. Chim. Fr.*, **1968**, 2039.
48. G. Wittig and R. Polster, *Justus Liebigs Ann.*, *Chem.*, **599**, 13 (1956).
49. J. L. Coke and M. C. Mourning, *J. Amer. Chem. Soc.*, **90**, 5561 (1968).
50. J. Sicher, M. Svoboda, and V. A. Vaver, *Chem. Commun.*, **1965**, 12.
51. M. Pankova, J. Sicher, and J. Zavada, *Chem. Commun.*, **1967**, 394.
52. M. Pankova, J. Zavada, and J. Sicher, *Chem. Commun.*, **1968**, 1142.
53. D. S. Bailey and W. H. Saunders, Jr., *Chem. Commun.*, **1968**, 1598.
54. H. D. Froemsdorf and H. R. Pinnick, Jr., *Chem. Commun.*, **1968**, 1600.
55. N. T. Anh, *Chem.*, *Commun.*, **1968**, 1089.
56. O. S. Tee, *J. Amer. Chem. Soc.*, **91**, 7144 (1969).
57. J. Sicher, J. Zavada, and M. Pankova, *Chem. Commun.*, **1968**, 1147.
58. A. Rauk, E. Buncel, R. Y. Moir, and S. Wolfe, *J. Amer. Chem. Soc.*, **87**, 5498 (1965).
59. S. Wolfe and A. Rauk, *Chem. Commun.*, **1966**, 778.

Regioselectivity in the Reductive Cleavage of Cyclopropane Rings by Dissolving Metals

STUART W. STALEY

Department of Chemistry, University of Maryland,
College Park, Maryland

I. INTRODUCTION

Most of the research on the reductive cleavage of conjugated cyclopropyl rings by dissolving metals, and certainly that involving the regioselectivity of these reactions, has been published in the past 5 or 6 years. This chapter, therefore, represents more of a progress report on an expanding area of research rather than a review of a long-established field. Emphasis will accordingly be placed on a discussion of the various factors thought to influence the regioselectivity of these reactions. Those areas where current knowledge is most deficient will also be pointed out.

The *over-all* reaction involves the addition of two electrons by a metal (usually lithium or sodium) and two protons by some donor (often ammonia) with concomitant opening of one of the two cyclopropyl bonds conjugated with an adjacent π center. Generally the two conjugated cyclopropyl bonds are nonequivalent and herein lies the origin of regioselectivity, or the directional preference of bond cleavage (eq. 1).

$$(1)$$

An early precursor to reductive cleavages by dissolving metals is an example of Baeyer (1) who reported in 1895 the "suspended metal" reduction of carone (1) to carvomenthol (2) by sodium in moist ether, although the structures of 1 and 2 were not known at the time. The same transformation

$$(2)$$

was later demonstrated by several groups (2,3) with lithium in liquid ammonia (eq. 2). In 1949 the reduction of methyl cyclopropyl ketone (3) to a mixture of a 2-pentanone (4) and 2-pentanol (5) with sodium and ammonium sulfate in liquid ammonia was reported by Boord and co-workers (4) (eq. 3). In contrast, reductive cleavage of the cyclopropyl ring in 2-cyclopropyl-1-pentene (6) by sodium in liquid ammonia (with or without methanol) was

$$(3)$$

$$(4)$$

not observed; instead the double bond was reduced in the presence of methanol (5) (eq. 4). However, it is possible that small quantities of ring-opened products may have escaped detection since the product mixture was analyzed by infrared spectroscopy. Subsequently, several other examples of double bond reduction without concomitant cleavage of a conjugated cyclopropyl ring were reported [7, by lithium and ethanol in ammonia (6), and 8, by lithium in ammonia (7)]. Cleavage of the cyclopropyl ring in 7 was

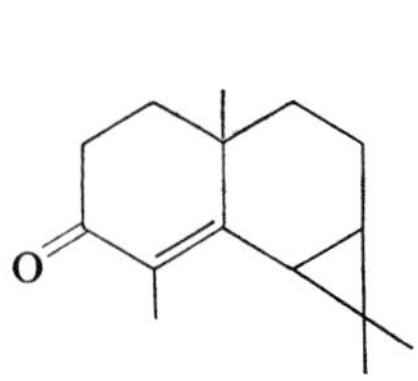

7 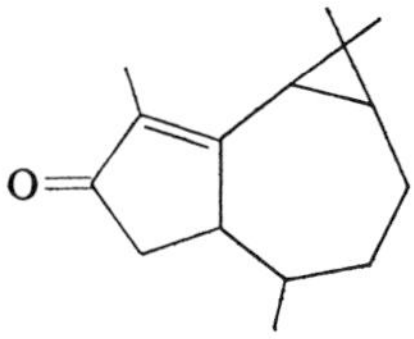 8

achieved with zinc and acetic acid, but not with acetic acid alone (6). In contrast, opening of more highly conjugated vinylcyclopropanes [**9** (8), **10** (9), **11** (10), and **12** (11)] can be readily achieved (eqs. 5–8).

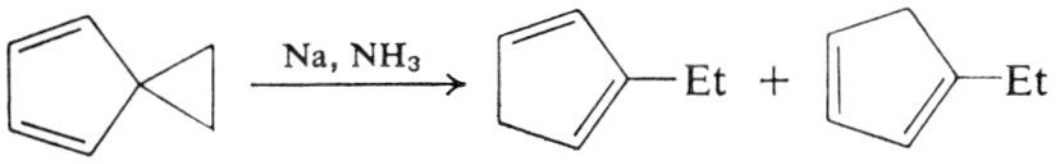

$$\xrightarrow{\text{Na, NH}_3} \quad -\text{Et} \quad + \quad -\text{Et} \tag{5}$$

9

$$\xrightarrow[-75°]{\text{Na, NH}_3} \tag{6}$$

10

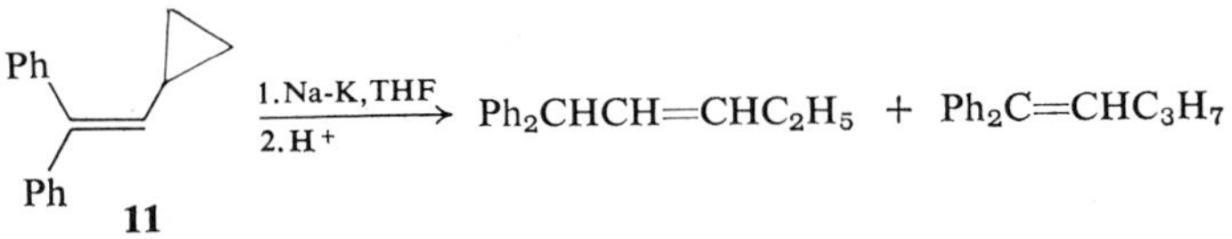

Ph
  \
   $\diagup$\triangle$ $\xrightarrow[\text{2.H}^+]{\text{1.Na-K,THF}}$ $Ph_2CHCH{=}CHC_2H_5 \; + \; Ph_2C{=}CHC_3H_7$ (7)
  /
Ph **11**

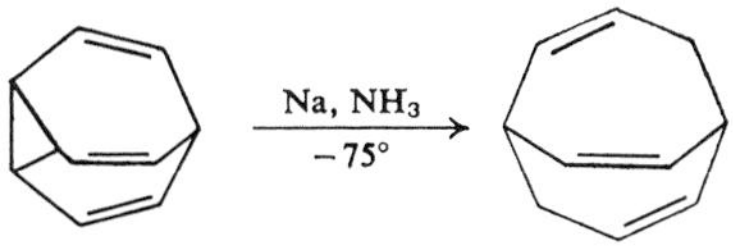

$$\underset{Me}{\overset{}{}}\underset{CN}{\overset{COOEt}{C{=}C}} \xrightarrow[\text{H}^+]{\text{electrolysis}} \underset{Me}{\overset{n\text{-Pr}}{C{=}C}}\underset{CN}{\overset{COOEt}{}} \; + \; \underset{Me}{\overset{n\text{-Pr}}{CHCH}}\underset{CN}{\overset{COOEt}{}} \tag{8}$$

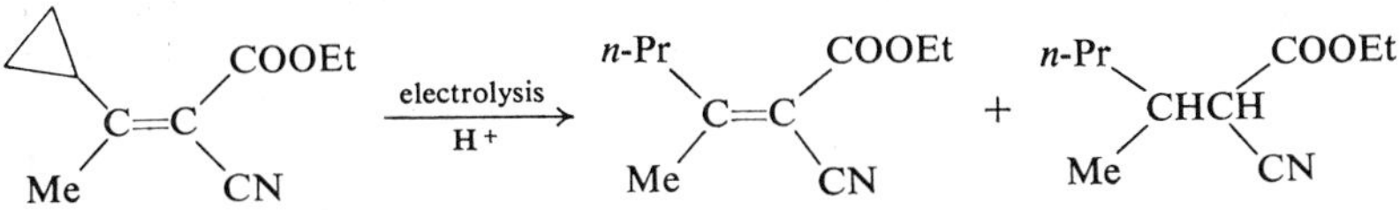

12   + other products

An indication of the scope of this reaction and the conditions under which it can be effected is provided by the following observations. Spiro[2.4]-hepta-4,6-diene (**9**) (8) and its 1-methyl derivative (**12**) are cleaved instantaneously and exothermically in sodium–liquid ammonia at ca. $-33°$ with or without an added proton donor such as ethanol (13). On the other hand, phenylcyclopropane (**13**), which is cleaved much more slowly by lithium (or sodium) in liquid ammonia (14,15), undergoes normal Birch reduction in the presence of added methanol (16) (eq. 9). Reductive opening of phenylcyclo-butane (**14**) has also been observed (eq. 10), although the rate is about 10^3–10^4 times slower than that of **13** at $25°$ (14); however, only double bond

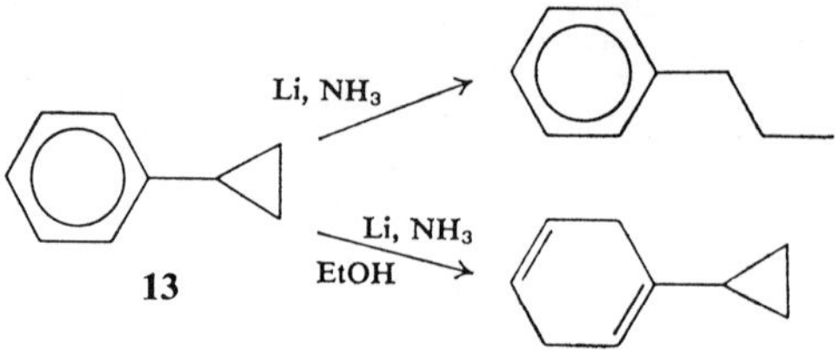

Li, NH$_3$

(9)

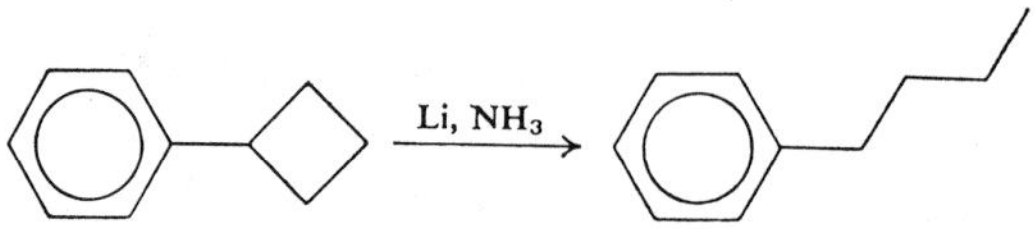

13

Li, NH$_3$
EtOH

Li, NH$_3$

(10)

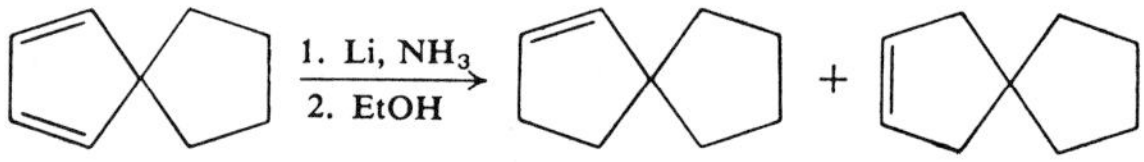

14

1. Li, NH$_3$
2. EtOH

+

(11)

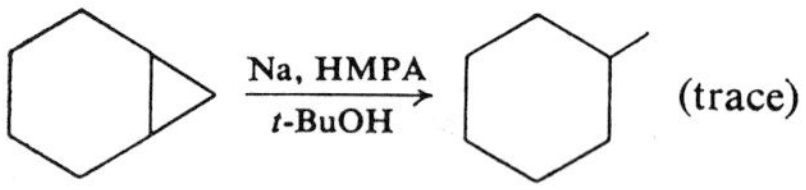

15

Na, HMPA
t-BuOH

(trace)

(12)

16

reduction has been observed upon treatment of **15** with lithium in ammonia or ethylamine, followed by ethanol (17) (eq. 11). These results implicate a competition between double bond reduction and reductive ring cleavage with the former being promoted by good proton donors and the latter by ring strain and product stability (e.g., an aromatic cyclopentadienide anion is initially formed upon cleavage of **9**). Finally, sodium and t-butyl alcohol in hexamethylphosphoramide (HMPA), which is capable of reducing non-conjugated tetrasubstituted olefins, also effects cleavage of norcarane (**16**) at room temperature, but only at an exceedingly low rate (18) (eq. 12).

Turning now to synthetic applications, Jeger and co-workers have reported several examples where this type of reaction has been successfully employed for the introduction of angular methyl groups into steroids [**17** (19) (eq. 13) and **18** (20) (eq. 14)]. Williams and Djerassi (21) extended this reaction to prepare a deuterated C_{19} in **19** (eq. 15). Several examples in conformationally mobile steroids were provided by Burn, Davies, and Petrow (22) who reported that the lithium in liquid ammonia reduction of 16α,17α-cyclomethylenepregnan-20-ones (**20**) effects cleavage of the cyclopropane ring to produce 16α-methylpregnan-20-ones (**21**) (eq. 16); the cyclopropane ring of the corresponding 20-ols is stable under these conditions. In 1963 Norin (23), in connection with studies on the configuration of thujopsene,

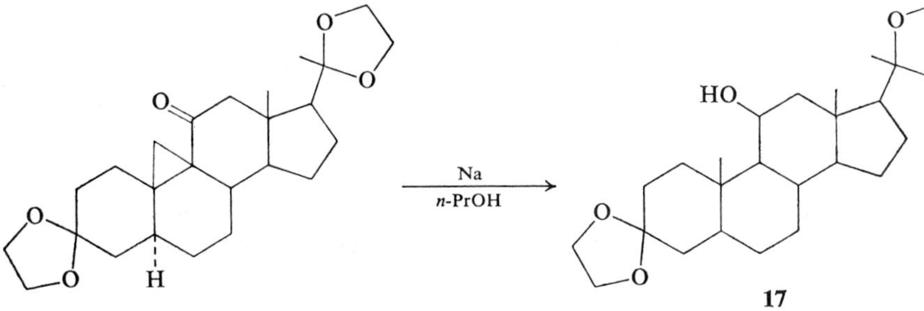

$$\text{(13)}$$

17

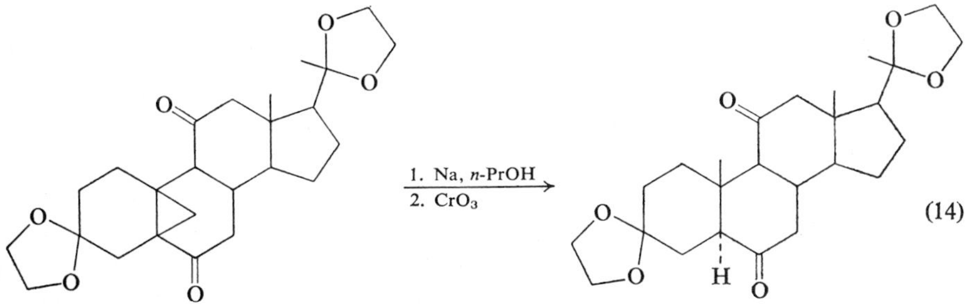

$$\text{(14)}$$

18

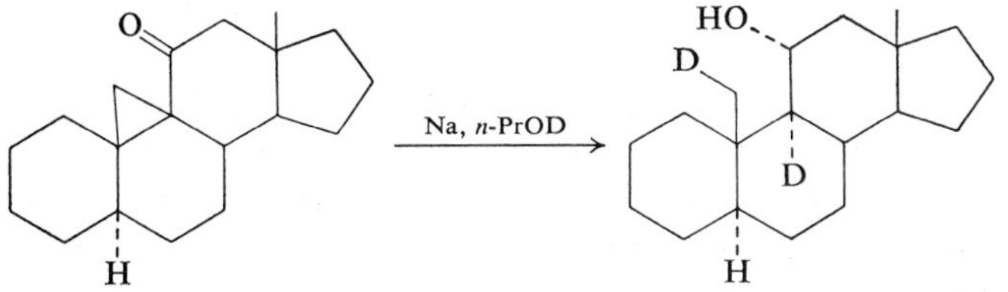

$$\text{(15)}$$

19

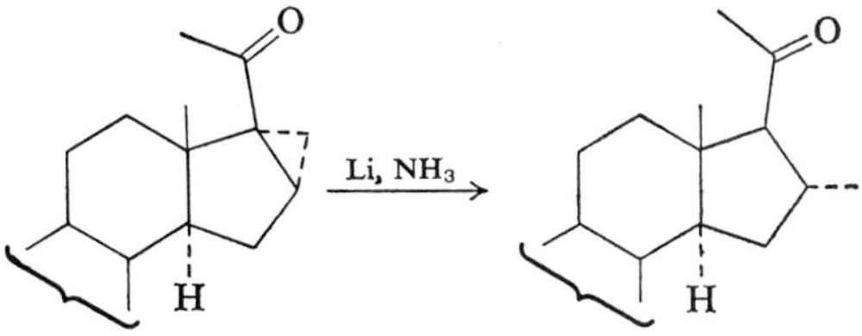

$$\text{(16)}$$

20   **21**

reported an almost quantitative yield in the reduction of cyclopropyl ketone **22** to **23** (eq. 17). He subsequently recognized the generality of the regiospecific opening of a three-membered ring of a conjugated cyclopropyl ketone with lithium in liquid ammonia and made the important observation that, in the reductions investigated to that time, "the cyclopropane bond which is cleaved is the one possessing maximum overlap with the π-orbital of the carbonyl groups" (2). In the past 5 years many additional examples of this

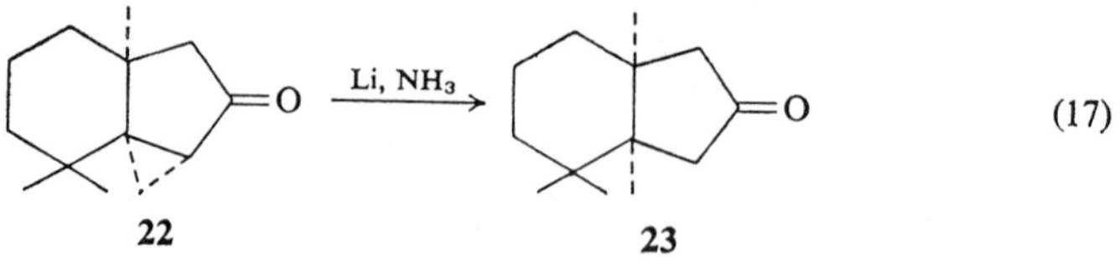

$$\text{(17)}$$

22   **23**

effect have been reported and will be considered in detail. In addition, other effects have been recognized and are also discussed in subsequent sections.

II. MECHANISM

The important mechanistic question concerning regioselectivity in reductive cleavage reactions centers around whether the ring-opening step, which is assumed to be irreversible in most cases, occurs as a radical anion or as a dianion (either with or without simultaneous transfer of a proton). Although dianion cleavages have been implied in several cases (3,24), most workers have assumed a radical-anion opening. Different compounds, of course, may open by different mechanisms [or, as in certain Birch-type reductions (25), both may actually be operative in the same compound], but little evidence has been reported in support of either mechanism.

We have recently considered this problem in connection with the reductive cleavage of phenylcyclopropane (14). Various postulated kinetic steps (analogous to those generally considered for the Birch reduction) are given in Scheme 1. The observed pseudo-first-order rate constant for reduction of 0.02 M phenylcyclopropane to ethylbenzene at 25° in 0.43 M lithium in liquid ammonia was found to be 1.1×10^{-2} sec^{-1}. Variation of the concentration of lithium [the Oswald isolation method (26)] showed the reaction to be

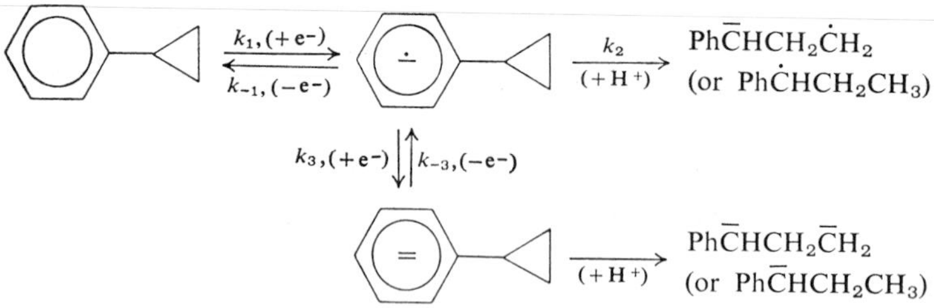

SCHEME 1

approximately first order in lithium. If one makes the assumption that ring opening is the rate-determining step [i.e., $k_2 \ll k_1$ or k_{-1} (Scheme 1)], then the dependence of the rate of reduction (which, by application of the steady-state approximation, can be shown to be equal to $(k_1 k_2/k_{-1})$[phenylcyclopropane][Li]) on lithium concentration is consistent only with a radical-anion cleavage. Evidence that ring opening is the slow step is obtained from the effect of substituents on the rate (14). For example, opening of Bond a in **24** under the same conditions as for phenylcyclopropane (**13**) is 16 times

24

slower than cleavage of the parent compound (statistically corrected) whereas opening of Bond b in **24** is 3.5 times faster. This is clearly not consistent with $k_1 < k_2$ but can be explained by steric and electronic effects of the methyl group on ring opening (see Sections III.B and III.C).

Additional evidence for a radical-anion mechanism is provided by the isolation of a small amount of two ring-opened isomers (**27**, mixture of 1- and 2-methylallyl isomers) in the reaction of **25** with sodium in liquid ammonia

| 25 | 26 | 27 | (18) |

28

(12) (eq. 18). This is more in accord with radical-anion intermediate **26**, than with dianion intermediate **28**, which would be expected to undergo protonation. It is, of course, possible that a small amount of **28** was oxidized to **26** following ring opening. Evidence of this sort must therefore be regarded with caution.

In the case of cyclopropyl ketones, which possess a highly electronegative oxygen atom, a dianion cleavage mechanism is somewhat more likely. In addition, a more extensive π system, such as in cyclopropylnaphthalene, would also be expected to make a dianion cleavage mechanism more competitive, as is expected on the basis of studies of Birch-type reductions (25,27).

It is evident that more work on the mechanism of the reductive cleavage of cyclopropyl ketones and related compounds is necessary. In any case, it is clear that the transition state for ring opening is *negatively charged*, regardless of whether it occurs after addition of one or two electrons.

III. FACTORS AFFECTING REGIOSELECTIVITY

Almost all studies of reductive cleavage reactions have been formulated in terms of the regioselectivity of the reactions, i.e., the relative rates of cleavage of two or three nonequivalent bonds *within the same molecule.* Because all competing processes originate from the same state (which may be a rigid molecule or an equilibrium mixture of rapidly interconverting conformers), the relative rates of cleavage are determined *solely* by the relative Gibbs energies of formation of the corresponding activated complexes ($\Delta\Delta G_f^{\ddagger}$) (the Curtin-Hammett principle) (28). Accordingly, the analysis of regioselectivity becomes one of analyses of the various factors which influence $\Delta\Delta G_f^{\ddagger}$ for competing transition states. We shall initially subdivide these factors into internal and external effects. Internal effects are those which apply only to the molecule undergoing cleavage and include effects commonly treated by classical mechanical methods such as bond angle and bond length deformations, torsional or eclipsing strain, and nonbonded interactions, and those generally treated by a quantum mechanical approach, such as delocalization effects in the π and σ systems (including inductive effects and orbital symmetry considerations). The classical and quantum mechanical treatments, of course, often deal with the same "effects." External effects include ion pairing and solvation. Most of the discussion which follows will be concerned with internal energy effects (although it should be recognized that entropy effects, especially those related to ion pairing and solvation, may be quite significant).

A. Geometric Factors in Rigid and Semirigid Systems

As already mentioned, the initial studies of Norin with several bi- and tricyclic cyclopropyl ketones led him to conclude that, for the cases investigated, the cyclopropane bond cleaved is the one better disposed to overlap with the carbonyl π orbital (maximum overlap criterion) (2). An identical conclusion was subsequently reached by Dauben and Deviny (3) and additional examples, which were studied in order to test this point, have been provided by Fraisse-Jullien and co-workers (29,30), Bellamy and Whitham (31), and Monti and co-workers (32). A number of the reactions of rigid and semirigid systems which seem to be in accord with the maximum overlap criterion are listed in Table 1. It should be noted that some of these reactions were performed for synthetic purposes and quantitative data are not available; conclusions in these cases should accordingly be drawn with caution. By convention, in Table 1 and subsequently, the more substituted cyclopropyl bond will be referred to as Bond *a* and the less substituted as Bond *b*, if such a distinction is applicable; the regioselectivity (i.e., the ratio of products from Bond a cleavage to those from Bond b cleavage, k_a/k_b) is assumed to apply at $-33°$ unless indicated otherwise.

It is my view that, in many cases in which the maximum overlap criterion leads to the prediction of the observed regioselectivity, other factors can also be considered to be important; indeed, in several cases, this criterion leads to the wrong prediction. It will be seen that direct consideration of the competing activated complexes (or approximations thereof) affords a more meaningful rationale for regioselectivity.

If we first consider bicyclo[3.1.0]hexan-2-one and its derivatives, it can be seen in Table 1 that the criterion of maximum overlap often leads to prediction of the correct product, despite the fact that [at least in the case of the parent compound (29)] the observed cleavage leads to what is probably the less stable product (33). However, it should also be noted that in almost every example in Table 1 in which the cyclopropyl bonds adjacent to the carbonyl group differ in substitution, the major pathway involves cleavage of the less substituted bond. Since alkyl substituents on the cyclopropyl ring have been shown to retard the rate of cleavage of the proximate cyclopropyl bonds (see Section III.C), this factor probably also plays a role in the observed regioselectivity.

I do not mean to imply that overlap considerations are not important, but I suggest it is better to estimate overlap in structures which approximate the transition states rather than the ground state. A similar point has been made by Monti and co-workers (32). There are certainly many examples where consideration of overlap in the ground state leads to prediction of the correct product. In certain cases this may be due to the fact that the activated complex resembles the ground state closely enough so that the same steric factors (angle and torsional strain, nonbonded interactions) are operative in both. However, consideration of other factors often leads to prediction of the same result. More importantly, in certain cases considerations of maximum overlap sometimes lead to predictions exactly opposite to those observed.

Several cases where the latter result is obtained are found in the work of Fraisse-Jullien and co-workers (29,30). In contrast to the examples cited in Table 1, and to the reduction of the closely related ketone **29** (30) (eq. 19), reduction of **31** (29) effects opening of the more poorly overlapping Bond a to afford **32** (eq. 20). There are several possible effects that may contribute to this difference in behavior. First, the phenyl group at C_1 in **31** may promote opening of Bond a more than of Bond b by providing an additional π system for delocalization of charge in the transition state. (This, in part, would circumvent the problem that Bond a overlaps less well with the carbonyl π orbitals). Second, cleavage of Bond a in **31** *reduces* a C_4 methyl–C_6 hydrogen interaction, whereas opening of Bond b *creates* a *cis*-3,4-dimethyl eclipsed

TABLE 1
Reductive Cleavage Reactions of Some Bicyclo[3.1.0]hexyl Systems

	Ref.

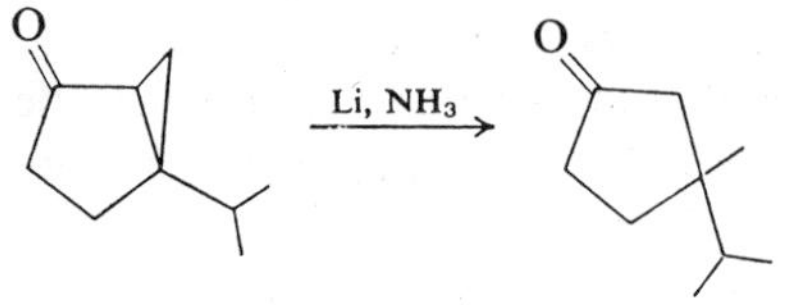

23

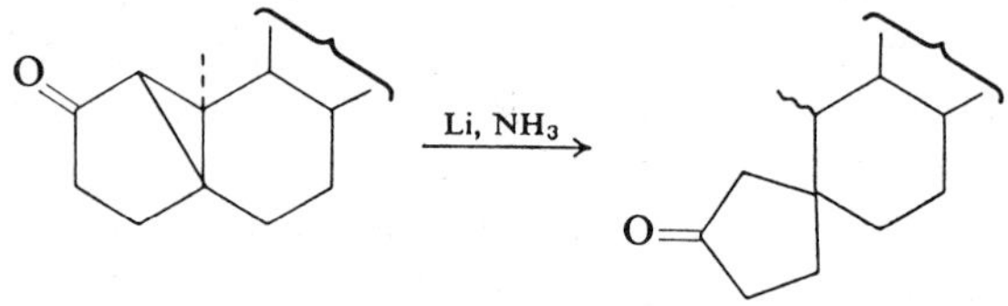

3

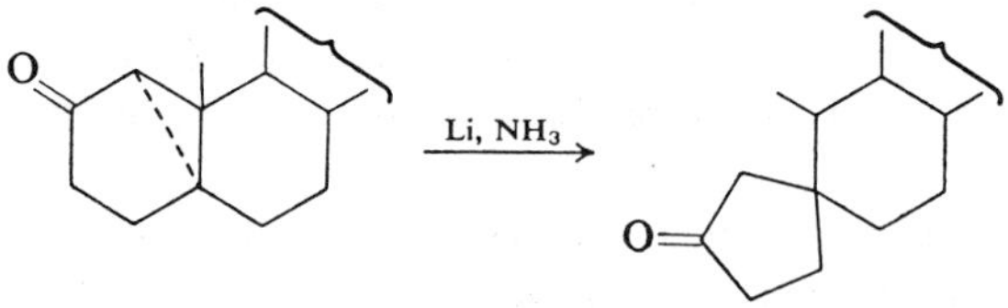

34

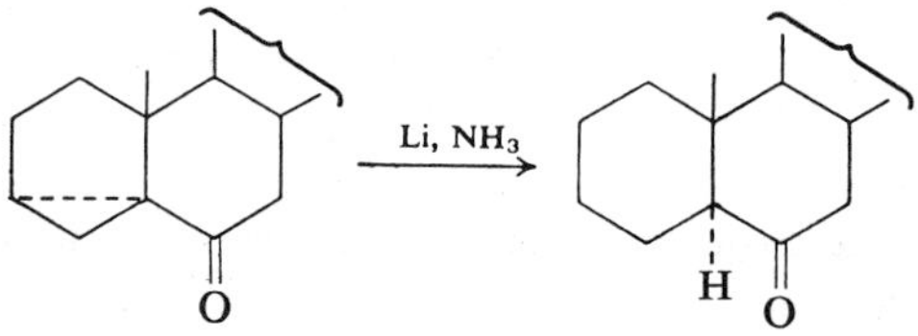

3

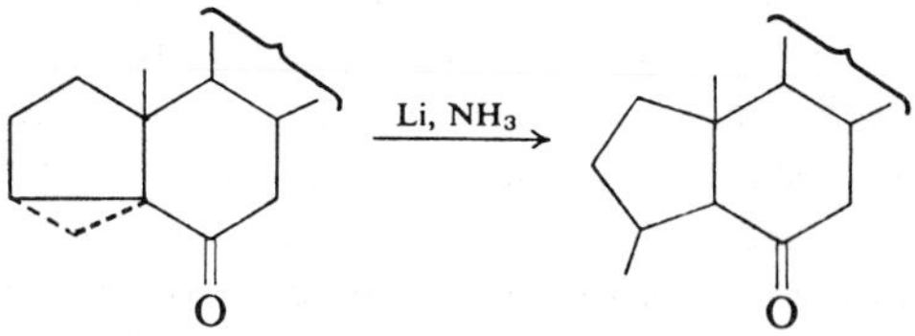

3

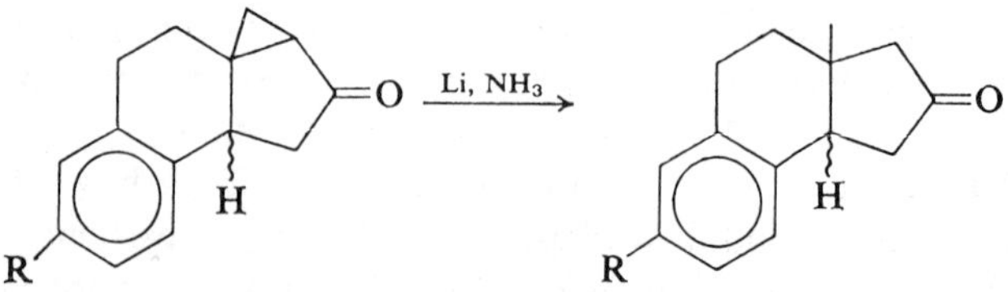

30

R = H or OCH₃

R = H or OCH$_3$

		Ref.

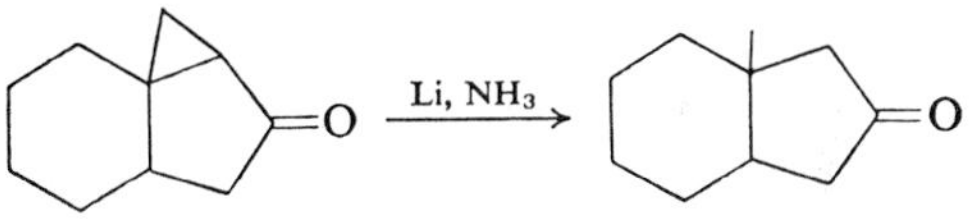

Li, NH₃ → ... 35

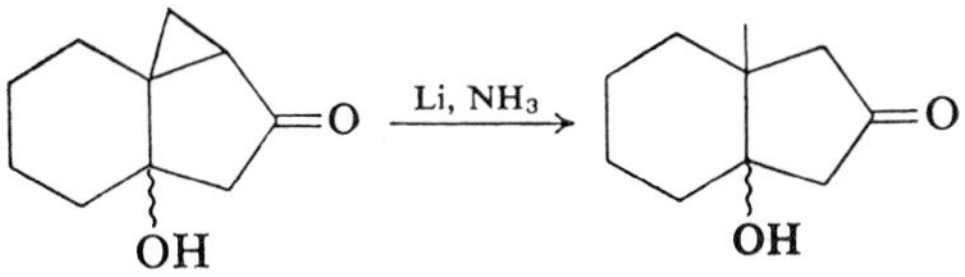

Li, NH₃ → ... 35

OH ... OH

Li, NH₃ → 29

H∼∼COOMe ... CH₂CH₂OH

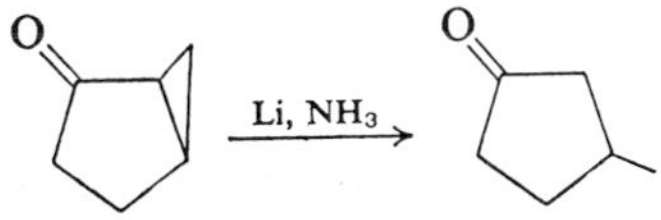

Li, NH₃ / t-BuOH → 36

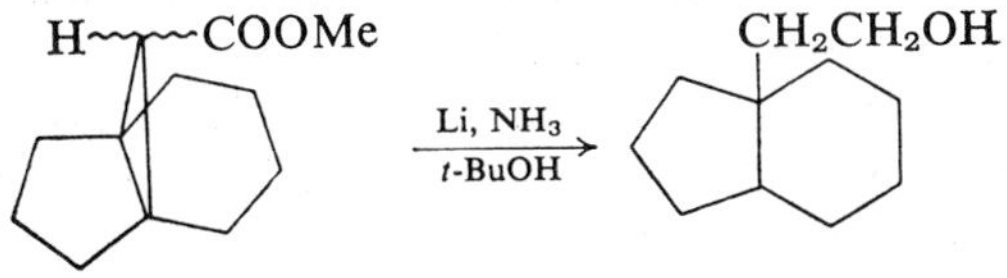

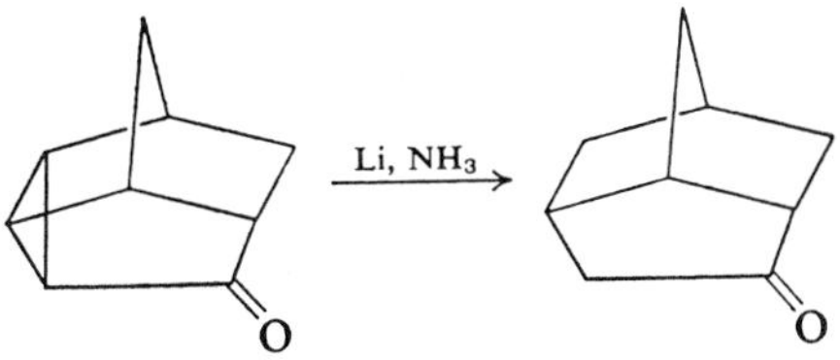

Li, NH₃ → 37

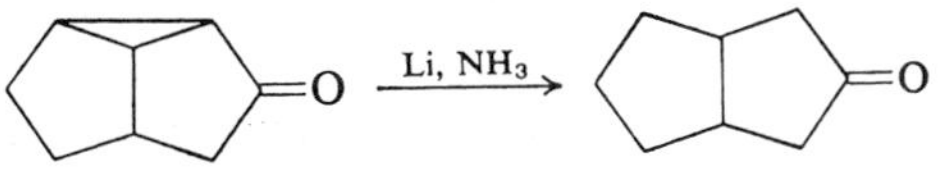

Li, NH₃ → 32

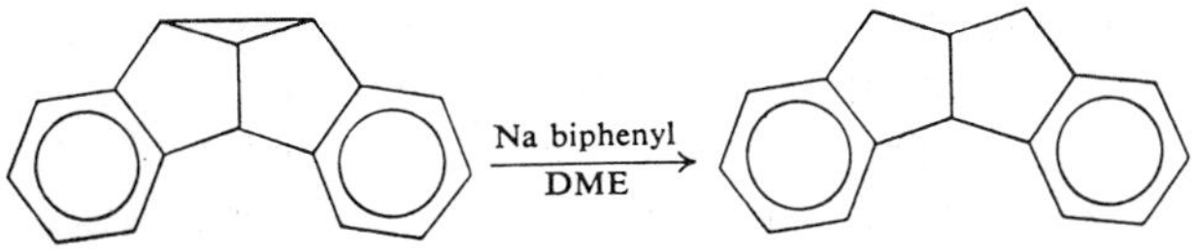

Na biphenyl / DME → 24

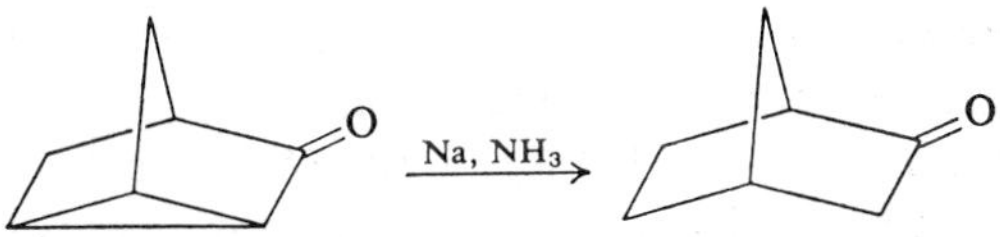

Na, NH₃ → 38

interaction in **33**. Based on nmr studies of ring flipping in 9,10-disubstituted-9,10-dihydrophenanthrenes (**39**), the analogous interactions in **30** should be

$$\text{(19)} \qquad \text{R = H or Me}$$

29 → 30

$$\text{(20)}$$

31   32   33

somewhat less. This explanation could be tested in part by examining the regioselectivity of the cleavage of the *exo*-4-methyl isomer of **31** which should have a greater propensity than **31** to undergo Bond *b* cleavage.

A second example in which the criterion of maximum overlap leads to the wrong prediction is the case of **34** (eq. 21), which suffers opening of Bond *a* exclusively (29); this process, of course, permits more stabilization of the

$$\text{(21)}$$

34

activated complex by the benzo group via delocalization and inductive/field effects than does opening of Bond *b*.

Further evidence concerning the role of maximum overlap in these reactions is provided by studies of tricyclic derivatives of bicyclo[3.1.0]hexan-2-one by House et al. (40). Overlap considerations lead to the prediction that reductive cleavage of **35** should afford **37** as the major product (eq. 22), whereas it is actually the minor product on the basis of reported yields. The difference in relative rates of the two competing cleavage processes is actually quite small and appears to be better correlated with the stabilities of the products, which should be about equal on the basis of data for the corresponding hydrocarbons (41). In an even more striking example, it was found that the closely related tetracyclic derivative of **35** (**38**) affords **39** as essentially the only product (40) (eq. 23). Again, consideration of product stabilities (or, more rigorously, the cleaved but unprotonated intermediates) leads to a reasonable rationale for this high selectivity. From studies of the minimum

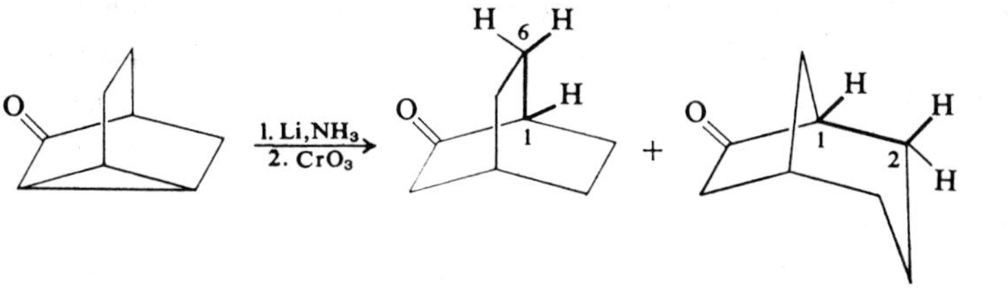

$$(22)$$

35   **36 (42%)**   **37 (37%)**

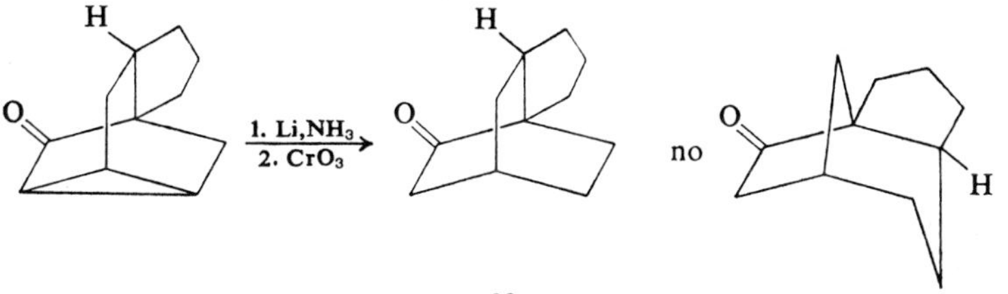

$$(23)$$

39   **40**

energy conformation of bicyclo[2.2.2]octyl (**42**) and bicyclo[3.2.1]octyl (**43**) ring systems, it appears that the HCCH dihedral angle subtended by the three heavy bonds in **36** is ca. 55° whereas that for the three bonds indicated in **37** is ca. 70°. In view of the fact that the largest dihedral angle for the

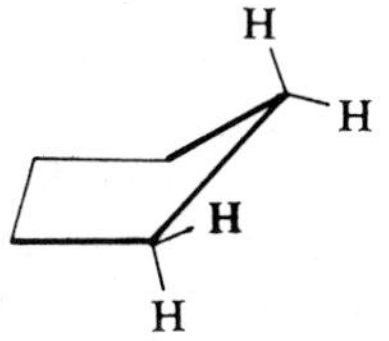

41 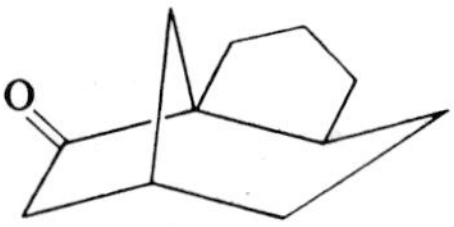   **40a**

C—C—C—C bonds in cyclopentane (**41**) is ca. 42° (44), introduction of a cyclopentane ring at C_1 and C_6 in **36** and at C_1 and C_2 in **37** will cause distortion of these two bicyclic systems. Calculations clearly indicate that the bicyclo[2.2.2]octyl system can accommodate the cyclopentane ring by twisting (**42**) whereas the bicyclo[3.2.1]octyl system, which is rigid, can only reduce the dihedral angle under discussion by flipping into the much less stable boat conformation (**40a**) (43). If one assumes that 1% of **40** might have gone undetected, it can readily be calculated from the relationship

$$\Delta\Delta G^{\ddagger} = -RT \ln (k_a/k_b)$$

where $R = 1.987$ cal deg^{-1} mole^{-1} and k_a/k_b is the regioselectivity, that the smaller Gibbs energy of activation ($\Delta\Delta G^{\ddagger}_{-33°}$) for cleavage of **38** to **39** can be accounted for by only 2 kcal/mole extra torsional and angle strain in the

activated complex for cleavage to **40**. It should therefore be kept in mind
that relatively small energy differences between the activated complexes for
competing processes can lead to high regioselectivities; e.g., 4 kcal/mole of
strain energy could change the regioselectivity from one measurable extreme
to the other.

Another example of high regioselectivity in a tricyclic system for which
the criterion of maximum overlap is unsatisfactory is found in the reduction
of (+)-pericyclocamphanone (**42**) with lithium in liquid ammonia, which was
reported to give a nearly quantitative yield of (+)-camphor (**43**) (2) (eq. 24).

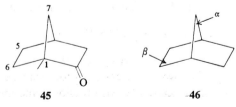

This is a remarkable result in view of the fact that Dreiding models indicate
that the C_3—C_4 and C_3—C_5 bonds overlap the carbonyl orbitals equally.
Norin rationalized this high regioselectivity by suggesting that nonbonded
interactions between the *gem*-dimethyl group and both the carbonyl group and
a C_6 hydrogen slightly distort the ring system, thereby causing better overlap
of the carbonyl with the C_3—C_5 bond than is indicated by models (2). To
the extent that the activated complexes leading to **43** and **44** resemble the
corresponding products, it is perhaps more correct to rationalize the course
of the reaction by assuming that **43** is more stable than **44**. Such an ordering
of relative stabilities could result from a number of factors of which two are
discussed here. Addition of a *gem*-dimethyl group to C_6 in **45** (to give **44**)
would be expected to cause a greater increase in torsional (eclipsing) strain
(due to interaction with the substituents at C_1 and C_5) than would addition
of the same group to C_7 (to give **43**); note that the substituents at the latter
position are essentially completely staggered. Second, it has been postulated
that the increase in the binding energy of C—C σ bonds with increasing s
character of the overlapping orbitals is greater (at least in the sp^2-sp^3 hybridization range) than the corresponding increase for C—H bonds (45). Since
$\angle C_1C_7C_4$ in norbonane (α in **46**) is 96° whereas $\angle C_1C_6C_5$ (β in **46**) is 104°
(46), there will be more s character in the bonds to substituents at C_7 than

7

5

6   1

O

45

α

β

46

to those at C_6. It is therefore possible that substitution of a *gem*-dimethyl group at C_7 in **45** will lead to a greater increase in σ binding energy than will substitution at C_6. However, there is some evidence which indicates that change of σ binding energy with hybridization is more a function of bond type (i.e., digonal, trigonal, and tetrahedral) than simply of changes in s character (47). Finally, solvation and ion-pairing factors may also play a role; these will be discussed in a subsequent section.

A number of examples of reductive cleavage of bicyclo[4.1.0]heptan-2-one and its derivatives are given in Table 2. In all cases the selectivity of the reaction appears to be in accord with predictions based on considerations of maximum overlap. Two cases are of particular interest, that of carone (**2**) (2,3) in which only the peripheral or *a* Bond is cleaved despite the fact that this bond is the more heavily substituted with alkyl groups (eq. 25), and that of 3-methylcar-4-en-2-one (**48**)(31) in which Bond *a* cleavage is also greatly favored even though Bond *b* cleavage should be favored not only by the substitution factor just mentioned but also because of stabilization of the activated complex through interaction of the reacting orbitals with the double bond at C_4—C_5 (eq. 26).

The high regioselectivity of these two reactions can also be explained by considering the factors that stabilize or destabilize the activated complexes, or, as a close approximation, the ring opened products which immediately follow. It can be seen from inspection of models that **47** possesses large nonbonded interactions between the quasi-axial hydrogen on C_2 and the quasi-axial methyl on C_6 whereas **49** appears to possess an even larger 1,4-methyl-methyl interaction. Thus it appears that good overlap for Bond *b* cleavage

$$(25)$$

$$(26)$$

cannot be achieved in the activated complexes primarily because of the nonbonded interactions discussed above. This result is more readily seen by analysis of the conformations of the products rather than of those of the starting materials. It would be of interest to determine the regioselectivity in

TABLE 2
Reductive Cleavage Reactions of Some Bicyclo[4.1.0]heptyl Systems

	Ref.
Na, EtOH	31
Li, NH₃	3
Li, NH₃	3
Li, NH₃	30,48
Li, NH₃	49
Li, NH₃	49
Li, NH₃	50

the cleavage of **50** (eq. 27) and **53** (eq. 28), in which the two cyclopropyl bonds are more or less equivalently substituted and in which the steric influence of the methyl group should be relatively minor. Since it seems

$$\text{50} \xrightarrow{\text{Li, NH}_3} \text{51} + \text{52} \qquad (27)$$

$$\text{53} \xrightarrow{\text{Li, NH}_3} \text{54} + \text{55} \qquad (28)$$

likely that the stability of **52** is greater than that of **51** whereas **54** is probably more stable than **55** (51), such a study would allow one to determine how well rates of ring opening correlate with product stabilities in these bicyclic systems. This is particularly interesting in the case of **50** where product stabilities and maximum overlap considerations lead to opposite predictions.

It is interesting to note that, in a study of the reduction of a number of bicyclo[1.1.0]butanes with lithium in ethylamine or ethylenediamine, Moore and co-workers (52) found only reduction of the bridge bond to afford cyclobutanes. Thus, what is probably the most strained bond (53) in each case is the one that undergoes opening. This might be taken to imply that the rate of reduction is dependent on release of angle strain in going to the transition state. However, an alternate interpretation is that the transition state looks very much like the initially formed radical anion and, due to the

56

hybridization in **56** (i.e., the C_1—C_3 bond has a high degree of p character and therefore has a relatively low-lying antibonding orbital), the extra electron will be located primarily in the C_1—C_3 bond, thereby leading to its rupture. There is very little information concerning the extent to which the rates of reduction of double bonds and cyclopropyl rings can be correlated with the corresponding heats of hydrogenation (54) and energies of the lowest unoccupied molecular orbitals (55). It is probable that both factors will be found to play a role and that the former will be more important for reductions in which there is substantial proton transfer from ammonia in the transition states. This is clearly an area in which further investigation will prove profitable.

B. Substituent Effects

This section deals with the directive effects of substituents on reductive cleavages which arise from causes other than intramolecular steric inter-actions. Most substituents influence the course of these reactions by delocal-ization of charge. The case of alkyl groups, which appears to be more com-plex, will be considered last.

Reductive cleavage of *trans*-2-phenylcyclopropyl methyl ketone (**57**) with lithium in liquid ammonia affords a single product, 5-phenylpentan-2-one (**58**), in high yield (29,56) (eq. 29). A similar result was obtained by Walsh and Ross (57) in a study of the reductive cleavage of triptycene derivatives, in which **59** was cleaved exclusively to **60** by potassium in tetrahydrofuran (THF) (eq. 30). Evidence was also presented which suggests that the opening occurs via a radical anion rather than a dianion.

$$\text{(29)}$$

$$\text{(30)}$$

It may reasonably be assumed that the high regioselectivity in the above two reactions is caused by resonance and inductive/field effects of phenyl in the transition state. Qualitative evidence that this is indeed the case is provided by data of Zimmerman and co-workers (58–60) on the reductive opening of *cis*- and *trans*-5,6-diarylbicyclo[3.1.0]hexan-2-ones. As can be seen in Table 3, there is a large change in the product distribution with sub-stituents in the *para* position of the 6-phenyl group. The remote position of these substituents precludes their having any steric influence so that the effect on relative rates must be electronic in nature. If we make the common assumption that cleavage occurs at a radical-anion stage, then two basic sets of canonical resonance forms may be used to describe the transition state for opening of a cyclopropyl bond. Set **A** has the negative charge delocalized in the enolate system whereas Set **B** has it delocalized in the benzyl

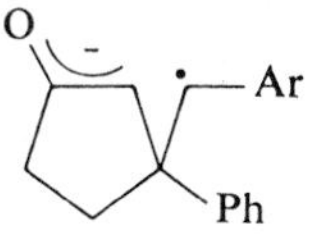

A 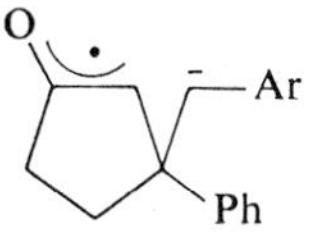   B

system. It might be anticipated that Set **A** has a greater weight due to the electronegative oxygen atom but the relative importance of the two sets will also depend on the nature of the *para* substituent in Ar. In view of the fact that (*1*) in several cases the total yields are low, and (*2*) the relative rates of cleavage for bonds in different molecules are not known and cannot readily be estimated from the experimental data, one must exercise caution in drawing conclusions from these data.

Considering the *trans* series first (**61–63**), if one makes the reasonable

TABLE 3

Isolated Yields of Products Resulting from the Reductive Cleavage of Various Cyclopropyl Bonds in **61–66** by Lithium in Liquid Ammonia (58,60).

---

trans Series

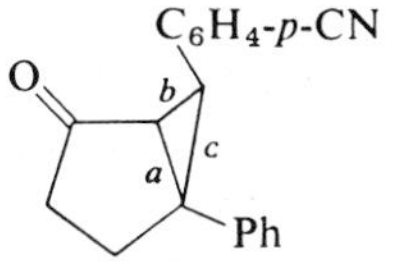

61	62	63
a	*a* 20%	*a* 39%
b 28%	*b* 1%	*b*
c	*c* 64%	*c*

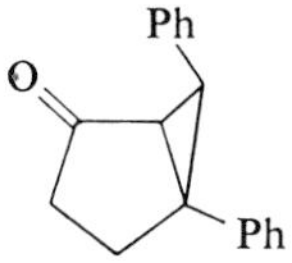

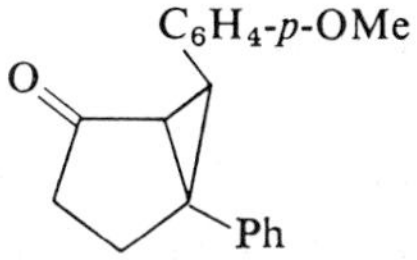

cis Series

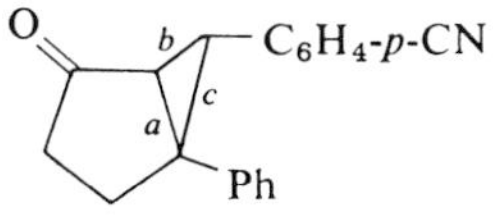

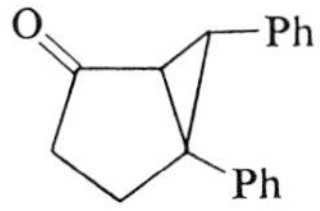

64	65	66
a	*a*	*a*
b 21%	*b* 46%	*b* 83%
c ca. 15%[a]	*c* 54%	*c*

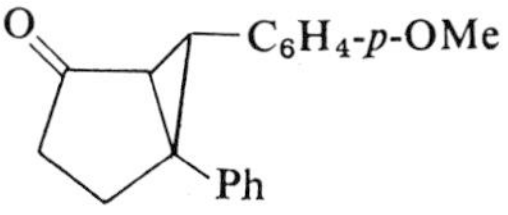

---

[a] This product was identified only as a cyclopentanone (by infrared spectroscopy).

assumption that the rate of cleavage of Bond a is independent of the *para* substituent (R), one can then conclude that the rate of Bond c cleavage decreases on going from R = H to R = OMe whereas the rate of Bond b cleavage is greater for R = CN than for R = H. These results are consistent with generally accepted notions that p-OMe is electron donating by resonance and would tend to destabilize an electron-rich activated complex whereas p-CN is electron withdrawing and has the opposite effect. The data are not sufficient to determine whether $k_{61c} > k_{62c}$ or $k_{62b} > k_{63b}$, but both of these relationships seem to be reasonable. Furthermore, if one assumes that the rate of Bond b opening in **67** is approximately the same as that in **65**, then the finding that

67

reduction of the former affords only a single product (Bond a cleavage) in 21% yield whereas the latter shows no Bond a opening (Table 3) provides additional evidence for the rate-enhancing effect of a p-cyano group.

It can be seen in Table 3 that no products of Bond a cleavage (whose rate is assumed to be constant) were detected in the *cis* series so that one cannot tell whether the rates of cleavage of the b or c Bonds increases or decreases upon changing the substituent. However, it is obvious that the rate of Bond b opening is increased *relative to Bond* c *opening* on going from R = H to R = OMe. This is particularly interesting because the p-anisyl group participates in both of these cleavages. This result can be rationalized on the basis of resonance, Form **B** being less important for Bond b cleavage (eq. 31) than **B′** is for Bond c cleavage (due to the greater electronegativity of a

$$(31)$$

$$(32)$$

carbonyl group relative to a phenyl group) (eq. 32), and thus the p-methoxy group should retard the rate of Bond b cleavage less than that of Bond c cleavage. Alternatively, one might argue that canonical Form **A** for Bond b opening is relatively more important than **A′** is for Bond c cleavage (again, due to the greater electronegativity of the carbonyl) and that a p-methoxy group stabilizes radicals. There is some evidence for the latter effect in the literature (61).

Actually, both of these effects probably contribute to the stability of the activated complexes, but the problem is too complex to be analyzed by simple valence bond representations. It would be most useful to have quantitative rate data for cleavage reactions of compounds such as 61–67 so that the distribution of charge in the activated complexes may be more fully evaluated.

It can be seen in Table 3 that the amount of Bond b cleavage is greatly decreased on going from the *cis* (65 and 66) to the *trans* (62 and 63) series. As Zimmerman (60) has pointed out, there is substantial steric hindrance in the transition state for cleavage of Bond b in the *trans* series (68) due to interaction of an *ortho* hydrogen on the C_6 phenyl ring and the *endo* methylene hydrogen on C_4 when the phenyl group is rotated so as to overlap favorably with the developing p orbital on C_6 (68). Such interaction is not present in the corresponding activated complexes for the *cis* series. Similar steric interactions

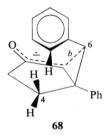

68

are discussed in Section III.C.

When an *endo* phenyl group is introduced at C_4 in 62 (to give 69), reductive cleavage occurs predominantly at Bond a (59). This result parallels that found for 31 and can be related to relief of the $C_4Ph—C_6Ph$ interaction on opening of Bond a. Surprisingly, there appears to be a complete reversal in the amount of b and c Bond cleavage products in going from 62 to 70 (59). Because the latter ketone simply has an added *endo* phenyl at C_4 (which

69; a 78%
    b
    c

70; a 27%
    b 70%
    c

should have only a slight steric influence), this result is difficult to explain and should probably receive closer scrutiny.

Turning now to other substituents, 2,2-diphenylcyclopropanecarboxylic acid (**71**) was found to suffer reductive cleavage of Bond *a* exclusively (eq. 33) whereas reduction of **72** afforded ca. a 5:1 ratio of **73** and **74** (eq. 34). Although

$$\text{(33)}$$

$$\text{(34)}$$

$$\text{(35)}$$

steric factors also play a role in these product ratios (see Section III.C), the increase in the relative amount of Bond *a* cleavage on going from **72** to **71** may be attributed to the stabilization of an electron-rich system by the carboxylate anion relative to methyl. This is consistent with the much greater rate of Birch reduction of benzoic acid compared with toluene (as well as with the 1,4-dihydro product from the former compared with the 2,5-dihydro product from the latter) (64). House (36) also recognized the stabilization provided by the carboxylate anion when he commented on the high regioselectivity of the opening of **75** (eq. 35). In the latter case, however, the influence of "maximum overlap" was also considered to be of possible significance.

A particularly interesting example of high directional control by a substituent was provided by LeBel and Liesemer (65) who observed that on lithium–ammonia reduction of *anti*-8-bromo ketone **76**, no starting material was recovered and only **78** was obtained as product (eq. 36). This is in striking contrast to the reduction of tetracyclic ketone **79** which proceeds more slowly to give mainly product resulting from cleavage of the *b* Bonds (see Section III.D). These data are consistent with a mechanism involving cleavage of Bond *a* in **77** in concert with expulsion of the bromide ion (65).

The nature of the methyl substituent effect in electron-rich systems

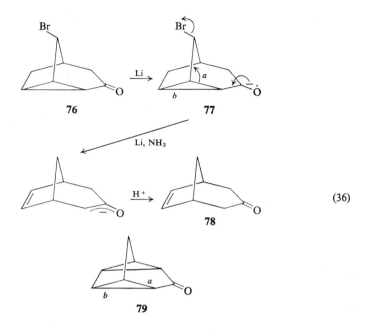

(36)

(carbanions and radical anions) in solution is a matter of current interest and will undoubtedly continue to be a subject of concern to many investigators. The basic facts are clear; methyl (or alkyl) groups tend to decrease (relative to hydrogen) the rates of reactions in which the transition state is more electron rich than the ground state when the methyl group is located at an electron-rich site. There is ample evidence to support this in the case of reductive cleavage reactions. Following the previously mentioned work by Burn and coworkers (22), several groups reported that reduction of *trans*-2-methylcyclopropyl methyl ketone (**80**) (29,56), in which the methyl group exerts only a slight intramolecular steric effect, leads primarily to opening of Bond b ($k_a/k_b = 0.06$) (eq. 37). An even greater regioselectivity ($k_a/k_b = 0.018$ at 25°; a smaller but less precise ratio was obtained at $-33°$) was obtained by us for *trans*-1-methyl-2-phenylcyclopropane (**81**) (12,14) (eq. 38). These examples provide additional evidence that the rate-retarding effect of methyl is proportional to the amount of negative charge on the methyl-substituted carbon because a larger k_a/k_b would be anticipated for **80** relative to **81** due to the already mentioned greater electronegativity of enolate system as opposed to the benzyl system. This view is further supported by the finding that for **82** and **83**, in which the rigid structures provide no con-formational advantage to either Pathway a or b, k_a/k_b is substantially larger than for **81** (12,14) (eq. 39). That is, the higher electronegativity of the incipient cyclopentadienyl ring relative to benzyl is expected to cause the

$$\text{80} \quad \xrightarrow{\text{Li, NH}_3} \quad + \quad \quad (37)$$

$$\text{81} \quad \xrightarrow{\text{Li, NH}_3} \quad \text{Ph} \quad + \quad \text{Ph} \quad (38)$$

$$\xrightarrow{\text{Na, NH}_3} \quad + \quad (39)$$

82; R = H, $k_a/k_b = 0.21$
83; R = Me, $k_a/k_b = 0.1$

withdrawal of more electron density from the cyclopropyl β carbons in the transition state. This is consistent with the much greater rate of reduction of **82** compared with **81** and is also supported by the results of HMO calculations employing the ω technique for a number of assumed transition state geometries (14). It is conceivable that the rate of cleavage of Bond a in **82** or **83** is enhanced slightly due to release of a nonbonded interaction between the C_1-methyl and a C_4-hydrogen, but it seems very unlikely that k_a/k_b for **82** would be as small as that in **81** in the absence of this factor.

Additional evidence for the rate-retarding effect of a methyl group, in this case one that is remote from the site of bond cleavage, is found in the

$$\text{84} \quad \xrightarrow{\text{K, THF}} \quad (40)$$

85   86

opening of 2-methyltriptycene (**84**) which affords **85** and **86** in a 2:1 ratio (57) (eq. 40). Examination of the resonance structures for the activated complex for opening of Bond b in **84** shows that negative charge will be located on the methyl-substituted carbon whereas it will not be in the opening of Bond a, the major pathway.

In view of the ample evidence for the rate-retarding effect of methyl, it is pertinent to inquire as to the origin of this effect. The traditional rationale in organic chemistry is that methyl is electron donating (or less electron withdrawing) relative to hydrogen and therefore methyl substitution would tend to destabilize carbanions and radical anions relative to the unsubstituted homologs. However, this view has certainly not been universally held, at least in the case of saturated anionic systems. Schubert (66), in particular, has advanced the view that alkyl groups can be either electron donating or electron withdrawing relative to hydrogen, depending on the environment, due to the greater polarizability of the former substituents. This proposal has received impressive experimental support from studies of gas-phase acidities by ion cyclotron resonance spectroscopy. By this technique Brauman and Blair (67) have shown that the acidity of aliphatic alcohols increases in the order $H_2O < CH_3OH < (CH_3)_2CHOH < (CH_3)_3COH < (CH_3)_3CCH_2OH$, and $CH_3CH_2OH < CH_3(CH_2)_2OH < CH_3(CH_2)_3OH \approx CH_3(CH_2)_4OH \approx (CH_3)_3COH$. This can be taken as indicating that, in alkoxide ions (RO^-), electron-withdrawing ability decreases in the order $R_3C > R_2CH > RCH_2 > CH_3 > H$. Similar conclusions have been suggested by a number of recent calculations (68).

If, then, alkyl groups withdraw electron density from negative centers, why do these groups serve as destabilizing influences? One proposal is that alkyl groups sterically hinder solvation of ions in solution (66a,67b,69). In view of the high energy of solvation possessed by many charged species, it is quite reasonable to expect that alkyl substituents can stabilize ions by intramolecular delocalization of charge but destabilize them by hindering intermolecular delocalization (by solvation) and thereby effect a net destabilization. In considering the role of alkyl groups it is important to separate out intramolecular steric effects; these will be discussed in the next section.

C. Nonbonded Interactions in Conformationally Mobile Systems

There have been a number of studies of regioselectivity in simple conformationally mobile cyclopropane derivatives in the past few years.

72

(41)

87   88

Walborsky and Pierce (63) reported that cleavage of Bond a in 1,1-diphenyl-2-methylcyclopropane (**72**) in sodium–liquid ammonia was 5.0–5.7 times faster than that of Bond b, and that **87** underwent opening (at ca. 25°) to **88** exclusively (eq. 41). These results were rationalized on the basis of an electronic argument, to be discussed later. At about the same time Fraisse-Jullien and Frejaville (29) reported that products from the opening of Bonds a and b in **89** were obtained in a 6:4 ratio (eq. 42). In this case the selectivity

(42)

89

was explained on the basis of maximum overlap considerations, i.e., in the minimum energy conformation of **89** the carbonyl π orbitals overlap better with Bond a than with Bond b.

Shortly thereafter a similarly small regioselectivity was reported for the reductive cleavage of **90**; again, opening of Bond a predominated (by a

(43)

90

91

factor of 2.3) (12) (eq. 43). In addition, a very high selectivity $k_a/k_b = 61$ in favor of opening of Bond a was found for the reduction of cis-1-methyl-2-phenylcyclopropane (**91**) (12,14). The explanation advanced was similar to that proposed for **89**; that is, the activated complex for cleavage of Bond b will possess substantial steric interaction between the methyl group and an ortho hydrogen on the phenyl ring and will therefore be less stable than that for Bond a in spite of the destabilizing substituent effect of the methyl in the latter case.

Recently Dauben and Wolf (56) confirmed the greater rate of cleavage of Bond a in **89** [$k_a/k_b = 3.2$; a similar result was obtained for 2,2-dimethyl-cyclopropyl butyl ketone (**92**)] and also reported that $k_a/k_b = 19$ for cis-2-

(44)

92; R = Me, R′ = n-Bu
93; R = H, R′ = Me

methylcyclopropyl methyl ketone (**93**) (eq. 44). This is in agreement with the results for **91** and an explanation equivalent to that given (12) for the latter compound was advanced. However, in assessing the selectivity of **93** it is important to know whether cleavage occurs via a transition state related to the *cisoid* or the *transoid* conformation. These conformations are represented by **94** and **95**, respectively, for the case of Bond *a* cleavage. By trapping the

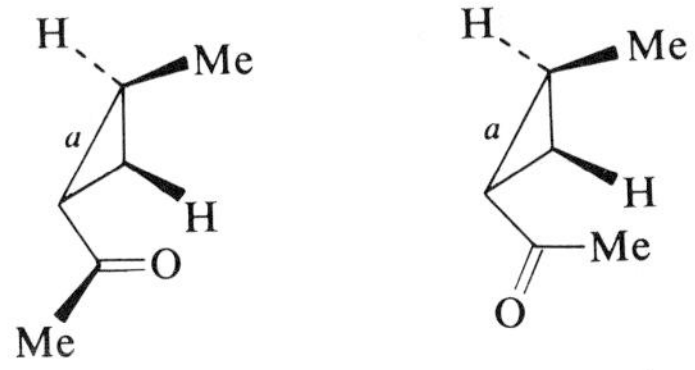

H .. Me   H .. Me

a   *a*

H   H

=O   Me

Me   O

cisoid   *transoid*

94   **95**

Bond *a* cleavage

intermediate enolate ions under conditions of kinetic control, Dauben and Wolf (70) showed that cleavage of Bond *a* in **80**, **89**, and **93** occurs predom-

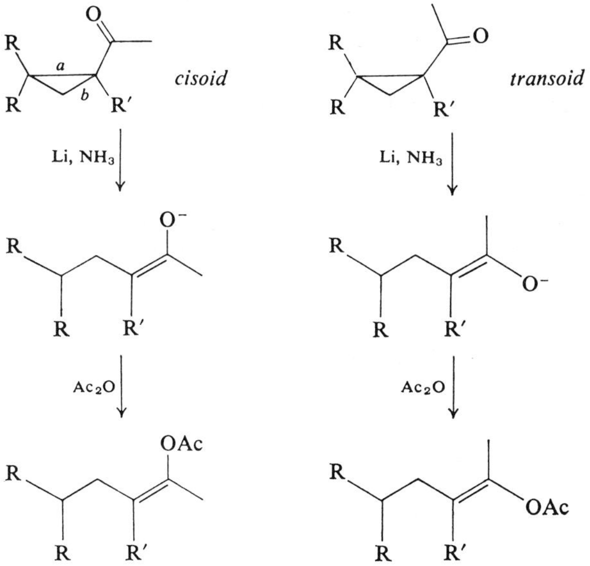

R   O   R   O

a   *cisoid*   *transoid*

R   *b*   R′   R   R′

Li, NH$_3$ Li, NH$_3$

R   O$^-$ R

R   R′   R   R′   O$^-$

Ac$_2$O   Ac$_2$O

R   OAc   R

R   R′   R   R′   OAc

SCHEME 2

inantly via the *cisoid* conformation (Scheme 2). These product ratios, which are consistent with a slightly larger steric size for methyl relative to a carbonyl group, also seem to parallel the ground state conformational populations for the starting materials. This suggests that there is little conformational change between the ground and transition states in these reactions (70).

In all of the studies of regioselectivity in reductive cleavages published to this time, only relative rates of opening of bonds within *the same compound* have been considered. However, in order to fully evaluate all of the factors that contribute to this problem, the relative Gibbs energies of activation corrected for strain in the ground state ($\Delta\Delta G^{\ddagger}_{corr}$) are required. This is necessary in order to tell whether rate differences for different compounds arise from effects in the ground or transition states, and requires a knowledge of the relative ground state strain energies and the relative intermolecular rates of reaction.

This problem can be illustrated by the reductive cleavage of *trans-* and *cis*-1-methyl-2-phenylcyclopropane (**81** and **91**) (14). On the basis of the small value of k_a/k_b in **81** and the large value in **91**, one might reasonably postulate that Pathway *a* is hindered by the substituent effect of the methyl group in **81** and that Pathway *b* is hindered by nonbonded interactions in **91**. However, this explanation is somewhat superficial. It tends to ignore, for example, the role of the methyl substituent effect in Pathway *a* of **91** and also the effect of the different steric environments of the methyl and phenyl groups in the two cases.

In order to sort out more completely the role of these various factors, the rates of cleavage of each bond in **81** and **91** were determined under pseudo-first-order conditions in 0.43 *M* lithium in liquid ammonia at 25°. From these

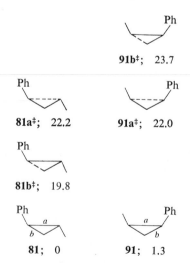

Figure 1. Relative Gibbs energies of formation (ΔG_f) (in kcal/mole) at 25° for **81, 91,** and their respective transition states for reductive cleavage in lithium–liquid ammonia (14).

data the following Gibbs energies of activation were calculated: $\Delta G^{\ddagger}_{81a}$ (Bond a of **81**) = 22.2 kcal/mole, $\Delta G^{\ddagger}_{81b}$ = 19.8 kcal/mole, $\Delta G^{\ddagger}_{91a}$ = 20.7 kcal/mole, and $\Delta G^{\ddagger}_{91b}$ = 22.4 kcal/mole. In addition, each isomer was equilibrated with 0.17 M potassium amide in liquid ammonia at 25° to a mixture consisting of 90.2% **81** and 9.8% **91**; this shows that the former is 1.3 kcal/mole more stable than the latter under these conditions. With these data we can now place the ground states and transition states for reductive cleavage of these isomers on a relative energy scale (Figure 1) which gives the relative Gibbs energies of formation for ground (ΔG_f) and transition ($\Delta G_f{}^{\ddagger}$) states. Inspection of this diagram reveals a significant fact, viz., that $\Delta G_f{}^{\ddagger}$ for **81a**‡ and **91a**‡ are essentially identical. This result can be accommodated by a transition state model in which the cyclopropyl bonds adjacent to the bond undergoing cleavage (C_1—C_3 and C_2—C_3 in **81a**‡ and **91a**‡) twist somewhat so as to minimize torsional and nonbonded interactions in the transition state. Inspection of models suggests that the direction of twisting

81a‡ 91a‡

is conrotatory in the case of **81a**‡ and disrotatory in the case of **91a**‡. It therefore appears that the destabilizing "methyl substituent effect" in transition state **91a**‡ is counterbalanced by the greater ground state strain in **91** relative to **81**. In addition, the importance of overlap between the phenyl ring and the developing p orbital on C_1 is indicated by the fact that transition state **81b**‡ is 4 kcal/mol more stable than **91b**‡ (see Figure 1) whereas **81** is only 1.3 kcal/mol more stable than **91**. This can be explained by the fact that the "maximum overlap" conformation in **91b**‡ possesses a substantial steric interaction between the methyl group and the *ortho* hydrogen on the phenyl ring.

H Me
91b‡

Information of this type can be used in the analysis of changes in regioselectivity in the following way (14). In Table 5 are listed four pairs of phenylcyclopropane derivatives along with the relative rates of a and b Bond cleavage. It can be seen that the first compound of each pair shows a lower k_a/k_b than the second (which is in each case the 1-methyl derivative of the first). The cause of this difference can be probed by comparing $\Delta G^{\ddagger}_{corr}$ for the various processes. From Table 4 it is clear that the increase in $\Delta G^{\ddagger}_{corr}$ for

TABLE 4

Regioselectivities (k_a/k_b) and Gibbs Energies of Activation Corrected for Estimated Strain in the Ground State (Relative to **13**) ($\Delta G^{\ddagger}_{corr}$) for the Reductive Cleavage of Phenylcyclopropanes in Lithium–Liquid Ammonia at 25° (14)

	k_a/k_b		$\Delta G^{\ddagger}_{corr}$(kcal/mole)
Ph **13**		**13**‡	20.5[a]
Ph **96**		**96**‡	22.5[a]
Ph **81**	0.018	**81a**‡	23.0
		81b‡	20.6
Ph **97**	0.14	**97a**‡	25.5
		97b‡	24.3
Ph **91**	20.5	**91a**‡	22.8
		91b‡	24.5
Ph **98**	80	**98a**‡	25.5
		98b‡	28.1
Ph **90**	1.9	**90a**‡	24.6
		90b‡	25.0
Ph **99**	> 200	**99a**‡	27.4
		99b‡	> 30.4

[a] Statistically corrected.

Bond *a* opening on introduction of a 1-methyl group falls in the relatively narrow range of 2.0–2.8 kcal/mole, but is somewhat larger in the case of Bond *b* opening and increases substantially in the case of **99b**‡. It therefore appears that there is an effect (probably steric in nature) of a 1-methyl group on Bond *b* cleavage in addition to the previously discussed "methyl substituent effect." Such an effect might be rationalized on the basis of (*1*) replacement of a partially eclipsed hydrogen-methyl interaction with a methyl-methyl interaction in going from **81b**‡ to **97b**‡, and (2) interaction of an *ortho* hydrogen with the 1-methyl group in addition to a *cis*-2-methyl group (as in **91b**‡) in the minimum energy conformation for cleavage of Bond *b* in

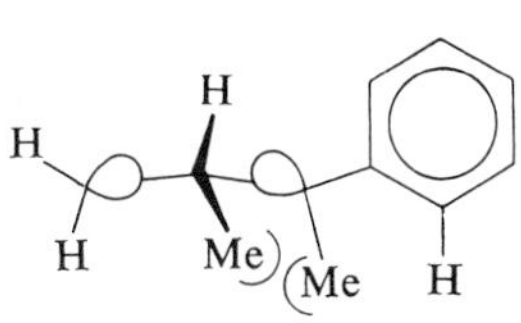

97b‡ 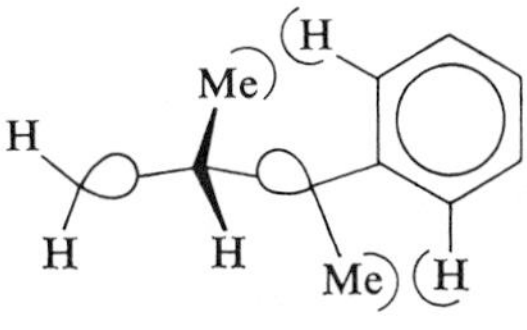   **98b‡**

98 (98b‡). A combination of these effects would be expected to lead to an even larger increase in $\Delta G^{\ddagger}_{corr}$ in going from **90b‡** to **99b‡**, as is observed (14).

A somewhat different approach to regioselectivity due to methyl substituents was employed by Walborsky and Pierce (63). These authors rationalized the greater rate of Bond a cleavage in **72** on the basis of canonical forms **100** and **102** being much more important than **101** and **103**, respectively, and, because 2° radicals are more stable than 1° radicals, the activation energy for cleavage of Bond a should be lower (eq. 45). It was subsequently pointed out that the regioselectivity in **72** is similar to that in **90**, and that a steric interpretation (*vide supra*) could also be invoked (12). In order to study this effect

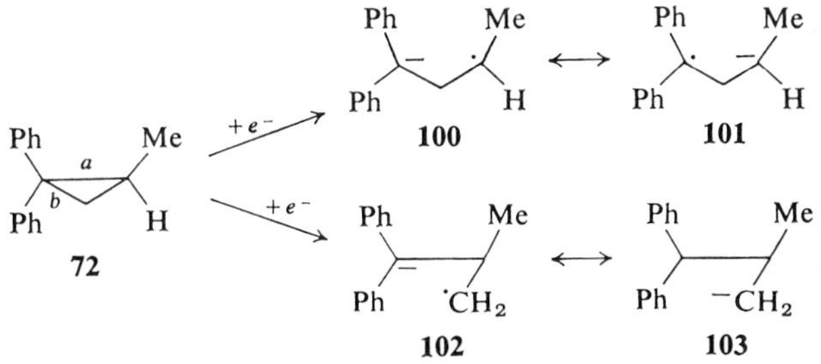

(45)

further, Walborsky and co-workers (71) have examined the reductive cleavage of **104**. It might be argued that the high regioselectivity in favor of Bond a cleavage ($k_a/k_b = 24$ in lithium–ammonia at $-78°$) supports an interpretation based on electronic considerations, as discussed for **72**. However, a strong argument can be made on the basis of steric considerations. One may regard **104** either as a derivative of 1-methylspiro[2.4]hepta-4,6-diene (**82**), in

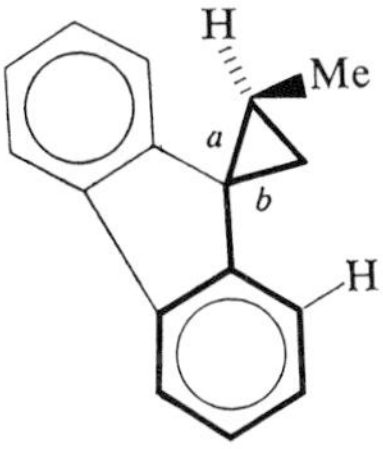

104

which case the regioselectivities are reversed, or as a derivative of *cis*-1-methyl-2-phenylcyclopropane (**91**, see heavy lines in **104**), in which case the regioselectivities are quite similar. In view of the fact that the rate of cleavage of **104** is much closer to that of **91** than to that of **82**, it is probable that the transition states of the former two are more closely related (with regard to bond lengths, charge distribution, etc.). From an examination of models it is also reasonable to conclude that steric effects are similar in these two cases. For example, as cleavage of Bond *b* in **104** takes place, Bond *a* will move towards the plane of the adjacent benzo rings, thereby causing strong steric interaction between the methyl group and the proximate *peri* hydrogen. This, of course, is similar to the steric argument made in the case of **82** (12).

This point can be checked by examining the reductive cleavages of **105**

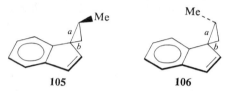

and **106**. On the basis of our interpretation (12,14) one would predict k_a to be less than k_b in the case of **105** and greater than k_b in **106**.

D. Orbital Symmetry

Reduction of tetracyclic ketone **79** with lithium in liquid ammonia afforded **108** in low yield as the major product (65) (eq. 46). It is interesting to note that two cyclopropane rings were cleaved although only 1 mole of hydrogen was added to the starting material, and that no other products were detected even though models indicate that the *a* Bonds overlap slightly better

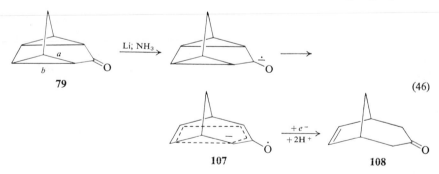

(46)

with the carbonyl group than do the *b* Bonds. LeBel and Liesemer viewed this reaction as an opening of both *b* Bonds in concert with protonation of a position adjacent to the carbonyl group (65). This regioselectivity could also be reasonably attributed to concerted opening to the bishomoaromatic intermediate **107**. The stabilization provided by delocalization similar to that in **107** has been shown to be significant in the case of several closely related hydrocarbon anions (**109** and **110**), both by nmr studies (72) and by kinetic studies of base-promoted deuterium exchange (73).

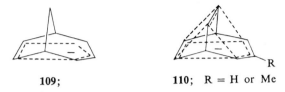

109;   **110;**   R = H or Me

Winstein and co-workers (74) and Katz and Talcott (75) have reported that *cis*-bicyclo[6.1.0]nona-2,4,6-triene (**111**) is converted to the monohomocyclo-octatetraene radical anion (**112**) upon reduction by various methods including treatment with alkali metals in ether solvents and electrolytic reduction in liquid ammonia (76). More recently Okamura and colleagues (77) have shown that **111** can be converted in good yield (*1*) to **113** by treatment with

$$\text{(47)}$$

111   **112**   **113**

sodium in HMPA-THF (eq. 47), and (*2*) to 1,3,6-cyclononatriene by treatment with sodium in liquid ammonia. The methylcyclooctatetraene radical anion has been observed during electrolytic (75) or alkali metal reduction (78) of **111** (cleavage of Bond *b*) in liquid ammonia but this presumably represents a relatively minor pathway.

In contrast to these results, Winstein and co-workers (79) have reported that no observable ring opening occurred upon treatment of the *trans* isomer of **111** (**114**) with potassium in dimethoxyethane (DME); the esr spectrum is clearly that of the radical anion of the starting material (**115**) (eq. 48). The

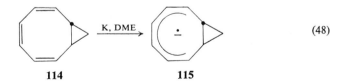

$$\text{(48)}$$

114   **115**

difference in behavior of these two isomers was explained on the basis of several factors. First, conservation of orbital symmetry would require that both **111** and **114** open in a disrotatory manner. In the former case this would lead to an unstrained homoconjugated system (**112**) whereas a rather strained *trans,cis*³-cyclononatetraene radical anion with poor overlap between adjacent π orbitals would be formed from the *trans* isomer. On the other hand, the unopened *trans* radical anion has excellent overlap between adjacent orbitals whereas there is much poorer overlap in the case of the unopened *cis* isomer. Thus a combination of stereoelectronic and quantum mechanical factors promote opening in a highly regioselective manner in one isomer and inertness in the other.

A reductive cleavage of particular interest has been reported by Miller and Jacoby (80). Reaction of dibenzonorcaradiene (**116**) with sodium in DME followed by quenching with water, deuterium oxide, or ammonium chloride gave varying amounts of three products (eq. 49). Arguments were

$$ (49) $$

presented in support of the interpretation that these products were formed by cleavage of the *a* or *b* Bond of the radical anion of **116**, followed by protonation, rather than by simultaneous ring opening and protonation. If the former mechanism is operative, then the formation of **117** is an apparent violation of the conservation of orbital symmetry if one makes the assumption that **116** is sterically constrained to open in a disrotatory manner.

A number of different mechanisms can be proposed to explain this behavior. However, since the product ratios vary significantly depending on the quenching agent, it is possible that the latter is involved in the cleavage step. It may therefore be premature to speculate on the mechanism until the exact conditions of this reaction are defined. Interestingly, essentially no **117** is formed when *t*-butyl alcohol is present throughout the whole reaction period (80).

That both solvent and orbital symmetry effects might influence the opening of **116** is suggested by studies of the somewhat related bicyclo-[5.1.0]octadienyl anion (**118**), the potassium salt of which was reported by Kloosterziel and Zwanenburg (81) to open in liquid ammonia exclusively to isomeric ion **120**. In parallel studies (Scheme 3) we have found that 1% of Path *a* product [possibly formed by a $6\pi + 2\sigma \rightarrow 8\pi$ "nonallowed" electrocyclic opening of **118**] is produced under these conditions at 25° and that in methyl- and dimethylamine (in which **119**, and not **120**, is the major product)

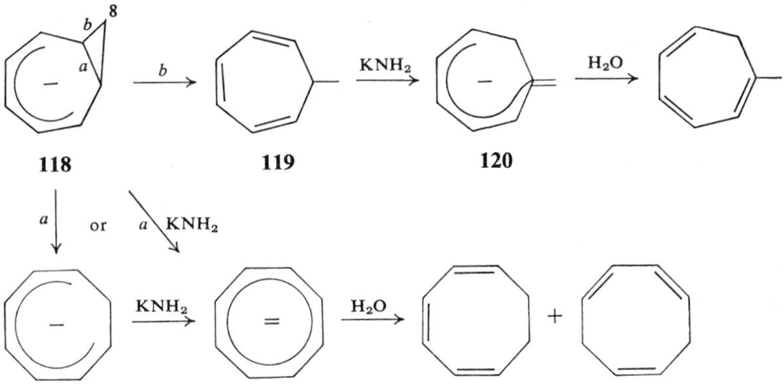

SCHEME 3

k_b/k_a is substantially reduced (to about 13 in the latter solvent) (82). Since the negative charge in both of the two hypothetical Pathway a transition states is highly delocalized whereas that in Pathway b is more localized on C_8, solvation (i.e., ion–dipole interactions or possibly hydrogen bonding) may stabilize the latter to a greater extent. Pathway b would therefore be more competitive in the better solvating solvents. In fact, strong hydrogen bonding at C_8 in the transition state for Pathway b is probably equivalent to concerted opening of Bond b and protonation at C_8.

E. Solvent and Counter Ion Effects

Although reductive cleavage reactions have been performed in a number of different solvents, including liquid ammonia, alkylamines, hexamethyl-phosphoramide, and various alcohols and ethers, very little is known concerning the effect of solvation on regioselectivity. As discussed in the previous section, the role of solvent (or an added proton donor) is probably quite important in those reactions in which the transition state involves ring opening with simultaneous proton transfer. Few data are available with regard to this point. The determination of solvent isotope effects and the study of the effect of added proton donors on the regioselectivity of these reactions should prove most informative.

In this connection it is interesting to note that k_a/k_b for **104**, which is 19 in sodium–liquid ammonia at $-78°$, is reduced to 5 in sodium naphthalide–DME at $-78°$, even though the reaction is slower in the latter case (71). Furthermore, the reaction rate and k_a/k_b for ammonia are not changed significantly by the addition of t-butyl alcohol to the reaction mixture. The nature of this solvent effect is not known at present, but it is interesting to note that the rate of cleavage of the more substituted bond (Bond a) is relatively slower in the more bulky solvent (DME). This is consistent with an explanation based on steric hindrance to solvation (or to proton transfer). It is hoped that additional data of this sort will become available in the near future.

The effect of the concentration of the dissolving metal has been studied by Walborsky and Pierce (63) who found that no reduction of **72** occurred at high (i.e., $>8\%$) concentrations of sodium in liquid ammonia and that at lower concentrations k_a/k_b varied slightly (from 5.0 to 5.7) as the metal concentration was reduced by a factor of 10^2. These authors suggested that this behavior is connected with the solubility of **72** in the sodium–ammonia solutions.

The data of Miller and Jacoby (80) also contain evidence for a change in regioselectivity in the cleavage of **116** with change in metal concentration. However, in view of the uncertainty as to the role of the quenching agent in these reactions, an interpretation based on concentration effects should be viewed with caution. In this regard it should be noted that the use of non-protic quenching agents, such as sodium benzoate (64) or acetone (14), will eliminate problems arising from proton transfer from the quenching agent.

The effect of the counter ion has also been investigated to a limited degree. Change of the metal from lithium to calcium, potassium, or sodium in the reduction of **89** (56), from lithium to potassium or cesium in the reduction of **90** (14), or from lithium to sodium in the reduction of **104** (71) (all in liquid ammonia) was, in each case, found to have little or no effect. A slight increase in the amount of Bond a cleavage in **89** on going from lithium to magnesium was suggested to be anomalous (56).

The regioselectivity (k_a/k_b) of the cleavage of **81** is significantly increased and that of **91** decreased on changing the metal from lithium to potassium or cesium. Furthermore, the value of k_a/k_b is essentially constant throughout each reaction in the case of lithium but is found to increase (decrease) during the reduction of **81 (91)** when potassium or cesium is employed (eq. 50). This has been found to result from interconversion of these two isomers by the potassium or cesium amide formed in the reductive cleavage reaction (14). Since **81** and **91** have opposite regioselectivities, this would tend to raise

major product   **81**   **91**   major product   (50)

k_a/k_b when starting with **81** and to lower it when starting with **91**.

The case of **97** appears to be one in which a small but real change in regioselectivity is observed with a change in counter ion; k_b/k_a $(25°) = 7.3$ for lithium, 6.1 for potassium, and 4.6 for cesium (14). Thus the more substituted bond (Bond a) is cleaved relatively faster in going from lithium to cesium.

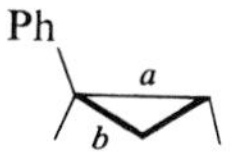

97

There are few data that indicate the nature of the ion pairing of carbanions or radical anions in liquid ammonia (83), but, based on studies in ether solvents (84), it is to be expected that, in cases where the negative charge is reasonably concentrated (as in **97a‡** and **97b‡**), the negatively-charged species will exist as contact ion pairs. If this is the case, then the activated complexes for lithium cleavage, in which coulombic attraction is greatest, should be the most susceptible to steric effects at the site of negative charge, as is observed. This explanation is somewhat speculative and additional data of this type should prove to be most useful.

IV. SUMMARY

The reductive cleavage of conjugated cyclopropyl rings may prove to have considerable synthetic utility. As Dauben (56) has pointed out, the over-all reaction in the *cis* series (**121**) (cyclopropanation followed by reductive cleavage) is equivalent to a chain-elongation and reduction reaction by

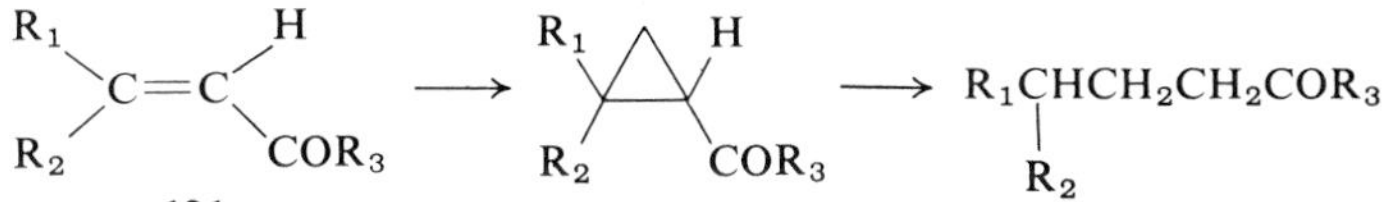

$$R_1CHCH_2CH_2COR_3 \quad (51)$$
$$\overset{|}{R_2}$$

121

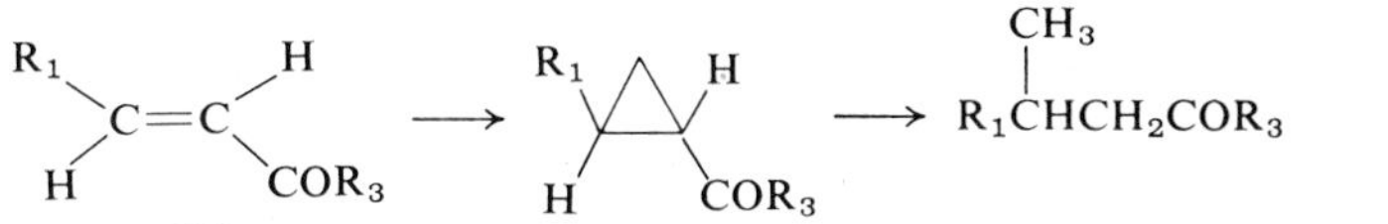

$$\overset{CH_3}{\underset{|}{R_1CHCH_2COR_3}} \quad (52)$$

122

insertion of the carbon between the original α and β carbons of the starting material (eq. 51). In the *trans* series (**122**), the process is equivalent to a 1,4 addition of methane to the unsaturated system to form a tertiary center β to the carbonyl group (eq. 52). The usefulness of this reaction in the formation of deuterated methyl groups, and of angular methyl groups in steroids and related ring systems, has already been mentioned. In addition, elucidation of the many factors that influence the course of these ring opening reactions will extend our knowledge of solvation, ion pairing, and charge distribution in the transition states and thereby allow the synthetic utility to be extended even further.

Acknowledgments

It is a pleasure to acknowledge the collaboration of Dr. Joseph J. Rocchio during much of our work in this field. Research support was provided by the National Science Foundation and the Petroleum Research Fund, administered by the American Chemical Society.

References

1. A. Baeyer, *Chem. Ber.*, **28**, 1586 (1895).
2. T. Norin, *Acta Chem. Scand.*, **19**, 1289 (1965).
3. W. G. Dauben and E. J. Deviny, *J. Org. Chem.*, **31**, 3794 (1966).
4. R. V. Volkenburgh, K. W. Greenlee, J. M. Derfer, and C. E. Boord, *J. Amer. Chem. Soc.*, **71**, 3595 (1949).
5. H. Greenfield, R. A. Friedel, and M. Orchin, *J. Amer. Chem. Soc.*, **76**, 1258 (1954).
6. R. B. Bates, G. Büchi, T. Matsuura, and R. R. Shaffer, *J. Amer. Chem. Soc.*, **82**, 2327 (1960).
7. R. E. Corbett and R. N. Speden, *J. Chem. Soc.*, **1958**, 3710.
8. K. Alder, H.-J. Ache, and F. H. Flock, Chem. Ber., **93**, 1888 (1960).
9. (a) G. Schröder, *Chem. Ber.*, **97**, 3140 (1964). (b) W. Hückel, S. Gupté, and M. Wartini, *Chem. Ber.*, **99**, 1388 (1966).
10. A. Maercker, *Justus Liebigs Ann. Chem.*, **732**, 151 (1970).
11. M. M. Baizer, J. J. Chruma, and P. A. Berger, *J. Org. Chem.*, **35**, 3569 (1970).
12. S. W. Staley and J. J. Rocchio, *J. Amer. Chem. Soc.*, **91**, 1565 (1969).
13. S. W. Staley and T. Copeland, Unpublished results.
14. S. W. Staley and J. J. Rocchio; J. J. Rocchio, Ph.D. Thesis, University of Maryland, 1970.
15. O. M. Nefedov, N. N. Novitskaya, and A. D. Petrov, *Dokl. Akad. Nauk SSSR*, **152**, 629 (1963).
16. R. Ya. Levina, V. N. Kostin, P. A. Gembitskii, and E. G. Treshchova, *Zh. Obshchei Khim.*, **31**, 829 (1961).
17. S. W. Staley and R. A. McCanner, Unpublished results.
18. G. M. Whitesides and W. J. Ehmann, *J. Org. Chem.*, **35**, 3565 (1970).
19. H. Wehrli, M. S. Heller, K. Schaffner, and O. Jeger, *Helv. Chim. Acta*, **44**, 2162 (1961).
20. M. S. Heller, H. Wehrli, K. Schaffner, and O. Jeger, *Helv. Chim. Acta*, **45**, 1261 (1962).
21. D. H. Williams and C. Djerassi, *Steroids*, **3**, 259 (1964).
22. D. Burn, M. T. Davies, and V. Petrow, *Steroids*, **3**, 583 (1964).
23. (a) T. Norin, *Acta Chem. Scand.*, **17**, 738 (1963). (b) See also P. L. Anderson, Ph.D. Thesis, University of Michigan, 1966; *Dissertation Abstr. B*, **28**, 91 (1967).

24. S. J. Cristol, P. R. Whittle, and A. R. Dahl, *J. Org. Chem.*, **35**, 3172 (1970).
25. W. A. Remers, G. J. Gibs, C. Pidacks, and M. J. Weiss, *J. Amer. Chem. Soc.*, **89**, 5513 (1967).
26. R. Livingston, in S. L. Friess and A. Weissberger, Eds., *Technique of Organic Chemistry*, Vol. VIII, Interscience, New York, 1959, p. 185.
27. H. Smith, *Organic Reactions in Liquid Ammonia*, Part 2, Wiley, New York, 1963, p. 237ff.
28. E. L. Eliel, *Stereochemistry of Carbon Compounds*, McGraw-Hill, New York, 1962, pp. 151, 152, 237–239.
29. R. Fraisse-Jullien and C. Frejaville, *Bull. Soc. Chim. Fr.*, **1968**, 4449.
30. G. Cueille, R. Fraisse-Jullien, and A. Hunziker, *Tetrahedron Lett.*, **1969**, 749.
31. A. J. Bellamy and G. H. Whitham, *Tetrahedron*, **24**, 247 (1968).
32. S. A. Monti, D. J. Bucheck, and J. C. Shepard, *J. Org. Chem.*, **34**, 3080 (1969).
33. Cf. thermochemical data compiled in J. D. Cox and G. Pilcher, *Thermochemistry of Organic and Organometallic Compounds*, Academic Press, New York, 1970, pp. 210–211.
34. S. B. Laing and P. J. Sykes, *J. Chem. Soc.*, *C*, **1968**, 937.
35. R. Fraisse-Jullien, C. Frejaville, and V. Toure, *Bull. Soc. Chim. Fr.*, **1966**, 3725.
36. H. O. House and C. J. Blankley, *J. Org. Chem.*, **33**, 47 (1968).
37. A. Nickon, H. Kwasnik, T. Swartz, R. O. Williams, and J. B. DiGiorgio, *J. Amer. Chem. Soc.*, **87**, 1615 (1965).
38. B. A. Loving, *U. S. Govt. Res. Develop. Rep.*, **69**, 66 (1969); *Chem. Abstr.*, **71**, 80782 (1969).
39. Cf. P. W. Rabideau, R. G. Harvey, and J. B. Stothers, *Chem. Commun.*, **1969**, 1005.
40. H. O. House, S. G. Boots, and V. K. Jones, *J. Org. Chem.*, **30**, 2519 (1965).
41. P. v. R. Schleyer, K. R. Blanchard, and C. D. Woody, *J. Amer. Chem. Soc.*, **85**, 1358 (1963).
42. O. Ermer and J. D. Dunitz, *Helv. Chim. Acta*, **52**, 1861 (1969).
43. (a) C. W. Jefford, D. T. Hill, and K. C. Ramey, *Helv. Chim. Acta*, **53**, 1184 (1970). (b) C. W. Jefford, D. T. Hill, and J. Gunsher, *J. Amer. Chem. Soc.*, **89**, 6881 (1967).
44. W. J. Adams, H. J. Geise, and L. S. Bartell, *J. Amer. Chem. Soc.*, **92**, 5013 (1970).
45. M. J. S. Dewar, *Hyperconjugation*, Ronald, New York, 1962.
46. J. F. Chiang, C. F. Wilcox, Jr., and S. H. Bauer, *J. Amer. Chem. Soc.*, **90**, 3149 (1968).
47. K. B. Wiberg and R. A. Fenoglio, *J. Amer. Chem. Soc.*, **90**, 3395 (1968).
48. R. K. Hill and J. W. Morgan, *J. Org. Chem.*, **33**, 927 (1968).
49. E. Piers, W. de Waal, and R. W. Britton, *Can. J. Chem.*, **47**, 4299 (1969).
50. E. Piers, R. W. Britton, and W. de Waal, *Can. J. Chem.*, **47**, 4307 (1969).
51. Cf. thermochemical data compiled in Ref. 33, pp. 208–209.
52. W. R. Moore, S. S. Hall, and C. Largman, *Tetrahedron Lett.*, **1969**, 4353.
53. R. B. Turner, P. Goebel, B. J. Mallon, W. v. E. Doering, J. F. Coburn, Jr., and M. Pomerantz, *J. Amer. Chem. Soc.*, **90**, 4315 (1968).
54. A. P. Krapcho and M. E. Nadel, *J. Amer. Chem. Soc.*, **86**, 1096 (1964).
55. B. R. Ortiz de Montellano, B. A. Loving, T. C. Shields, and P. D. Gardner, *J. Amer. Chem. Soc.*, **89**, 3365 (1967).
56. W. G. Dauben and R. E. Wolf, *J. Org. Chem.*, **35**, 374 (1970).
57. T. D. Walsh and R. T. Ross, *Tetrahedron Lett.*, **1968**, 3123.
58. H. E. Zimmerman, R. D. Rieke, and J. R. Scheffer, *J. Amer. Chem. Soc.*, **89**, 2033 (1967).
59. H. E. Zimmerman and R. L. Morse, *J. Amer. Chem. Soc.*, **90**, 954 (1968).

60. H. E. Zimmerman, K. G. Hancock, and G. C. Licke, *J. Amer. Chem. Soc.*, **90**, 4892 (1968).
61. (a) C. Rüchardt and R. Hecht, *Chem. Ber.*, **98**, 2471 (1965), and references cited. (b) S. T. Bowden, *J. Chem. Soc.*, **1957**, 4235, and references cited. (c) G. H. Williams, *Homolytic Aromatic Substitution*, Pergamon, New York, 1960.
62. A. Streitwieser, Jr., *Molecular Orbital Theory for Organic Chemists*, Wiley, New York, p. 135.
63. H. M. Walborsky and J. B. Pierce, *J. Org. Chem.*, **33**, 4102 (1968).
64. A. P. Krapcho and A. A. Bothner-By, *J. Amer. Chem. Soc.*, **81**, 3658 (1959).
65. N. A. LeBel and R. N. Liesemer, *J. Amer. Chem. Soc.*, **87**, 4301 (1965).
66. (a) W. M. Schubert, R. B. Murphy, and J. Robins, *Tetrahedron*, **17**, 199 (1962); for a review see J. E. Huheey, *J. Org. Chem.*, **36**, 204 (1971).
67. J. I. Brauman and L. K. Blair, *J. Amer. Chem. Soc.*, (a) **90**, 6561 (1968); (b) **92**, 5986 (1970).
68. (a) N. C. Baird and M. A. Whitehead, *Theor. Chim. Acta*, **6**, 167 (1966). (b) N. C. Baird, *Can. J. Chem.*, **47**, 2306 (1969). (c) T. P. Lewis, *Tetrahedron*, **25**, 4117 (1969). (d) W. Grundler, *Tetrahedron*, **26**, 2291 (1970). (e) P. H. Owens, R. A. Wolf, and A. Streitwieser, Jr., *Tetrahedron Lett.*, **1970**, 3385. (f) W. J. Hehre and J. A. Pople, *Tetrahedron Lett.*, **1970**, 2959.
69. F. E. Condon, *J. Amer. Chem. Soc.*, **87**, 4481, 4485, 4491, 4494 (1965).
70. W. G. Dauben and R. E. Wolf, *J. Org. Chem.*, **35**, 2361 (1970).
71. H. M. Walborsky, M. S. Aronoff, and M. F. Schulman, *J. Org. Chem.*, **36**, in press. (1971).
72. (a) S. Winstein, M. Ogliaruso, M. Sakai, and J. M. Nicholson, *J. Amer. Chem. Soc.*, **89**, 3656 (1967). (b) J. M. Brown, *Chem. Commun.*, **1967**, 638. (c) J. B. Grutzner and S. Winstein, *J. Amer. Chem. Soc.*, **90**, 6562 (1968).
73. (a) J. M. Brown and J. L. Occolowitz, *Chem. Commun.*, **1965**, 376; *J. Chem. Soc.*, *B*, **1968**, 411. (b) S. W. Staley and D. W. Reichard, *J. Amer. Chem. Soc.*, **91**, 3998 (1968). (c) J. W. Rosenthal and S. Winstein, *Tetrahedron Lett.*, **1970**, 2683.
74. R. Rieke, M. Ogliaruso, R. McClung, and S. Winstein, *J. Amer. Chem. Soc.*, **88**, 4729 (1966).
75. T. J. Katz and C. Talcott, *J. Amer. Chem. Soc.*, **88**, 4732 (1966).
76. Cf. F. J. Smentowski, R. M. Owens, and B. D. Faubion, *J. Amer. Chem. Soc.*, **90**, 1537 (1968).
77. W. H. Okamura, T. I. Ito, and P. M. Kellett, Personal communication.
78. R. M. Owens; cited in Ref. 76, footnote 14.
79. G. Moshuk, G. Petrowski, and S. Winstein, *J. Amer. Chem. Soc.*, **90**, 2179 (1968).
80. L. L. Miller and L. J. Jacoby, *J. Amer. Chem. Soc.*, **91**, 1130 (1969).
81. H. Kloosterziel and E. Zwanenburg, *Rec. Trav. Chim. Pays-Bas*, **88**, 1373 (1969).
82. S. W. Staley and N. J. Pearl, Unpublished results.
83. Cf. S. W. Staley and J. P. Erdman, *J. Amer. Chem. Soc.*, **92**, 3832 (1970).
84. (a) T. E. Hogen Esch and J. Smid, *J. Amer. Chem. Soc.*, **87**, 669 (1965); **88**, 307, 318 (1966). (b) J. B. Grutzner, J. M. Lawlor, and L. M. Jackman, *J. Amer. Chem. Soc.*, **92**, (1971), in press.

SUBJECT INDEX